THE VAULT

MARCUS HEARN

HARPER DESIGN
An Imprint of HarperCollinsPublishers

For Niamh, the girl of my dreams

Doctor Who is a BBC Wales production for BBC One

Daleks created by Terry Nation
Cybermen created by Kit Pedler and Gerry Davis
Silurians created by Malcolm Hulke
Autons and Sontarans created by Robert Holmes

HarperCollins books may be purchased for educational, business, or sales promotional use. For information please e-mail the Special Markets Department at SPsales@harpercollins.com

First published in 2013 by
Harper Design
An Imprint of HarperCollins Publishers
10 East 53rd Street
New York, NY 10022
Tel: (212) 207-7000
Fax: (212) 207-7654
www.harpercollinspublishers.com
harperdesign@harpercollins.com

Distributed throughout the world by
HarperCollins *Publishers*
10 East 53rd Street
New York, NY 10022

The Library of Congress Control Number can be obtained directly from the publisher

ISBN 978-0-06-228063-3

Commissioning editor: Albert DePetrillo
Managing editor: Joe Cottington
Project editor: Justin Richards
Design: Peri Godbold
New photography: Helen Solomon and Joe McIntyre
Production: Phil Spencer

Colour origination by Altaimage, London
Printed in the USA

Third printing, 2013

ABOUT THE AUTHOR

Since beginning his career at Marvel Comics in 1993, Marcus Hearn has written for *The Times*, *The Guardian* and *The Independent*, as well as contributing booklet notes, audio commentaries and documentaries to nearly 100 DVDs. His numerous books include authorised biographies of filmmakers George Lucas and Gerry Anderson, *The Avengers: A Celebration* and *Eight Days a Week*, the story of The Beatles' final world tour. He is an associate research fellow at Leicester de Montfort's Cinema and Television History Research Centre, and is the official historian of Hammer Film Productions.

PICTURE CREDITS

The author and publisher gratefully acknowledge the permission granted to reproduce the copyright material in this book. Every effort has been made to trace copyright holders and to obtain permission for the use of copyright material. The publisher apologises for any errors or omissions in the below list and would be grateful if notified of any corrections that should be incorporated in future reprints or editions of this book.

The specially commissioned prop, costume and toy photography in this book is copyright Flashpoint Media. All other material is copyright BBC, with the exception of the images on the following pages. Our thanks to their copyright holders:

Chris Achilleos (page 92), Jeremy Bentham (Innes Lloyd, 38), Richard Bignell (47), The Braxiatek Collection (257, 262, 266), David Bryher (Murray Gold, 288), Sonny Caldinez (60), Charles Carne (54, 55), Andrew Cartmel (217), Big Finish (239), The Big Issue Company (295), British Sky Broadcasting (228), Charles Buchan's Publications (*Kinematograph Weekly*, 17), The Chrysler Group (102), Grahame Flynn (2, 105, 188, 205, 221, 225), Derek Handley (21), The Hayward Gallery (283), Marcus Hearn (*30 Years in the TARDIS*, 226), Christopher Hill (77), David J Howe, Mark Stammers and Stephen James Walker (*The Frame*, 215), Waris Hussein (21), IPC Media (*Eagle*, 172), Mike Kelt (199), KLF Communications (214), Gary Leigh (*DWB*, 225), Jessica Martin (217), Mirrorpix (Verity Lambert, 20; Carole Ann Ford, 23; Stanley Franklin, 31), Joel Nation (Terry Nation, 96), News Group Newspapers (*The Sun*, 201), Northern & Shell (*Saturday*, 249), Ian O'Brien (79, 135), Oriole Records (25), Pinnacle Records (137), Chris Pocock and John Walker (193, 206, 214), *Radio Times* (17, 20, 46, 52, 54, 72, 76, 80, 111, 176, 183, 200, 207, 229, 235, 245, 266, 269, 278, 301), Record Shack Records (191), Rex Features (*This Sporting Life*, 15), Steve Roberts (*Resistance is Useless*, 226), Jan-Vincent Rudzki (36), Dez Skinn (140), Studiocanal (*Dr. Who and the Daleks*, 30; *Daleks' Invasion Earth 2150 A.D.*, 37; 167), Time Out Group (167), Wayland Publishers (148) and Colin Young (53, 158).

Doctor Who is a BBC Wales production for BBC One.

Executive producers: Brian Minchin, Steven Moffat, Faith Penhale, Caroline Skinner
Drama account manager: Edward Russell
Brand strategy manager: Matt Nicholls
Picture campaign managers: Francine Holdgate, Alexandra Thompson
Script editors: Richard Cookson, Derek Ritchie
Creative executive: Stephanie Milner

With special thanks to Mark Gatiss

Endpapers: Studio floorplan for *An Unearthly Child* © BBC, courtesy Waris Hussein.
Frontispiece: One of the prop umbrellas used by Sylvester McCoy as the Seventh Doctor.
Opposite page: The medical kit designed by Tony Oxley for *The Ark in Space* (1975).

CONTENTS

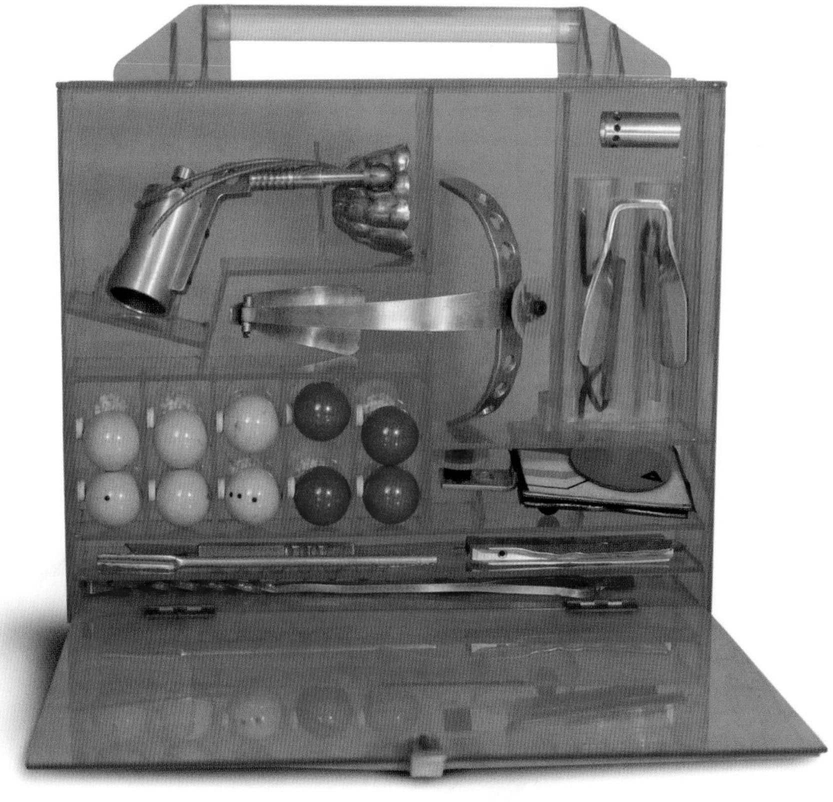

FOREWORD

Doctor Who takes place in a completely different world from ours, and the biggest difference isn't the spaceships or the time travel or the monsters, or the fact that the universe is kept safe by a loon with comedy hair, it's just this: the Doctor's adventures happen in a world where there isn't a television show called *Doctor Who*. Which sometimes makes it a bit difficult to write.

It's not the scenes where people from our world gasp in astonishment at their first sight of a Dalek or a Cyberman, when of course everybody knows what they are (if only from all the action figures lying around the living room), because I can just about cope with that. No, it's the TARDIS. It's the TARDIS disguised as a police telephone box. A police telephone box which would have been forgotten by everyone, except for the fact that the TARDIS looks like one. There is no clearer demonstration of *Doctor Who*'s extraordinary cultural impact than the fact the Doctor's time machine is cleverly disguised as an artefact now only recognised as a time machine.

And you sit there, typing away at a script, and someone is saying, "But what's this strange wooden box?", and the Doctor is busy explaining that his spaceship can cloak itself, but all of your mind is screaming, "It's the TARDIS, for God's sake, *what are you watching on Saturday night??*"

That's the special thing about *Doctor Who*. Most stories are just that – a momentary diversion from the real world. But some stories actually change the world around them. And in extreme cases, some TV shows escape from the box, hang in the air and enter the DNA of an entire nation.

I remember, a few years ago, just before the show made its triumphant return, I was talking to my very young son about my favourite TV series. Now let me be clear. On strict instructions from my wife (who'd had that stomach-lurch moment of realising she'd married a *Doctor Who* fan a few years previously) I'd never told Joshua anything about my lifelong obsession. Never shown him an episode. Never bought him a *Doctor Who* toy, or shown him a Dalek. He'd noticed the videos and DVDs on the shelves and maybe caught moments of old repeats, but really he knew nothing. Trouble was, he'd just told me his favourite show was *Sponge Bob* (excellent, if you haven't seen it) and he wanted to know mine.

Now, yes, I could have told him *The West Wing*, which is what I say when I want to sound clever, and is very nearly true. But one does not lie to one's son.

"*Doctor Who*," I said.

Joshua's eyes went to all the DVDs and videos.

"Have you ever seen it?" I asked him.

"A little bit. What's it about?"

Sue was out, so I seized my chance. I grabbed a DVD and showed him the cover. "That's Doctor Who", I said, pointing to Peter Davison, "and he fights the Daleks and other monsters."

Joshua frowned. "But he's not called Doctor Who – he's called the Doctor."

What?

What??

Sorry, but how could he know that without having watched it. Frankly, you could watch quite a lot of *Doctor Who* and still not know that. (Indeed, there are a number of episodes that confirm the exact opposite, but shh, we don't talk about those!)

But really, though, how did Joshua know that central, and slightly puzzling, fact about *Doctor Who*?

Many years later I put that question to Karen Gillan on the set of the show itself. "Oh, I'd never seen *Doctor Who* before I was in it," she answered, without shame. "But I knew all about it. Everyone knows about *Doctor Who* – children are *born* knowing about it."

Is that the terrible truth? Is *Doctor Who* now streaming in the air around us, like wi-fi?

Another conversation, from a few days ago, at a party. A man was booming away at me, and I'd just owned up to what I do for a living.

"Oh, I never watch that," he said. "Never have watched it. Dreadful."

"Okay," I said, wondering why people say this to me at parties, and how they think it will improve my evening.

"Good for the kids, though, I suppose," he continued.

"I suppose," I nodded, with my wine still in my glass and not all over his face.

"Did you write the one with the stone angels?"

"I did, yes."

"Which one was that?"

"You might be thinking of *Blink*, with David Tennant. And we did a two-parter with Matt Smith called *The Time of Angels* and we've just done another one called *The Angels Take Manhattan*."

"Never seen them!"

"Okay."

"Those Angels are scary though."

"Thanks."

Fear me, oh readers. For my Angels have climbed out of the screen and are now menacing my enemies without the aid of broadcast television. I like to think they took that man later that night, and he's now working as a farm labourer in 1502. But if not, and he's reading this, then hello again, and what an interesting choice of reading material.

Speaking of which, how does one tell the story of *Doctor Who*, this show of impossible contradictions? The cult TV show that is also a massive mainstream hit. The strictly British TV show, with an international audience of 77 million, most of whom don't live here. The show you don't even have to watch, to be watching.

As you flick through you'll see dates. But rest assured, this isn't anything as simple as a history. In proper TARDIS tradition it's a lot more timey-wimey than that. And that's because the story of this show is also the story of all television, and of the culture surrounding it – it's sprawling and complex and hilarious and impossible.

In short, it's a chronology of *Doctor Who* – but not necessarily in that order.

Steven Moffat
Executive producer, *Doctor Who*
July 2013

AUTHOR'S NOTE AND ACKNOWLEDGEMENTS

"A" "Dr Who + 100,000 B.C."
B "The Mutants"
C Inside the Spaceship.
D Marco Polo.
E Keys of Marinus.
F Dr Who + the Aztecs.
G Dr W + the Sensorites.
H Dr Who + the Reign of Ter
J Planet of Giants / Miniscule
K Dalek invasion of Earth
L Dr W + the Rescue (2)
M Dr W + the Romans.
N The Web Planet
P

Above: A 1965 note from the *Doctor Who* production office, listing titles for the first 13 stories.

Above right: In 1965 the Shepperton Studios Art Department sent this instruction to Shawcraft Models, the company building props for the film *Dr. Who and the Daleks*.

Every major anniversary of *Doctor Who* has been marked with a special publication, and the past weighed heavily on my considerations for this celebration of the 50th.

When I was a child in the 1970s and early '80s most retrospectives comprised little more than descriptions and photographs of old episodes, but I found them fascinating. As *Doctor Who*'s 50th year approached I wondered how to recreate the way I felt about those books and magazines. Nowadays, the internet makes such basic information readily available and DVD/Blu-ray allows similarly easy access to almost every surviving episode.

I was aware of a huge amount of material relating to *Doctor Who* that was either hidden away, or that had been virtually forgotten once its moment of usefulness or initial popularity had passed. There is no central archive of *Doctor Who*'s props, costumes, merchandise and other ephemera. After all these years it would be difficult to gather the most significant items in one place but, I realised, it would be possible to curate such a collection within the pages of a book.

And that's exactly what we've tried to do. *The Vault* is a journey through the greatest *Doctor Who* museum there never was – a place where the iconic sits side-by-side with the rare and the unseen.

Each chapter represents a year in the show's history, or longer for the periods when it was off the air. The chapter introductions set the scene, and the pictures (together with their captions) tell the linear story of *Doctor Who*. The essays themselves, however, weave in and out of its major innovations and themes. The inspiration for each one is a key event in the relevant year – for example, the story of the Daleks is told in 1965, when their popularity reached its peak, and the analysis of time travel can be found in 2007, the year the narrative-jumping *Blink* was broadcast.

Despite *The Vault*'s 'timey-wimey' nature, all bar two of the chapters include box-outs featuring titles, credits, production and broadcast dates for all the canonical stories screened in that year – 'canonical' meaning full-length, live-action, television-broadcast episodes of *Doctor Who*. For 1963 to 1966, when only individual episode titles were given on screen, stories are identified by the overall titles intended by the relevant production teams. In addition, the stories produced between 1963 and 1989 are prefixed by their internal BBC production codes.

Production dates for main unit recording and photography appear beneath each title. For the 1963-1989 stories these dates generally refer to the start of filming and the end of studio recording. There were often lengthy gaps between the two, and within studio recording schedules. In 1996 the *Doctor Who* TV movie became the last story to use film in its production. The way *Doctor Who* was made changed radically from 2005 onwards, and this is reflected in the recording dates that appear under the titles for these stories. As before, there are often gaps within the dates quoted for these years.

The production dates given throughout this book are not a definitive list of all work conducted (for example they don't include model shoots and pick-ups etc), but more a guide to the order in which stories were completed.

Writers, directors, producers and (for the 1963-1989 years) story/script editors are then listed. Writer pseudonyms are only explained where the true identity of authors can be definitively established.

Each story's listing ends with the dates or date of first UK broadcast. Within the main text, dates are generally given only if the relevant story was screened at a point subsequent to that chapter's year (and therefore not already covered in one of the episode guide box-outs).

Turning my imaginary museum of *Doctor Who* into published reality was only possible with the skill and dedication of designer Peri Godbold. She made more of my requests than I ever thought possible. Paul Taylor worked alongside her, digitally polishing numerous images.

Thanks also to Adrian Rigelsford for his unswerving loyalty both to me and this project. The photography of Helen Solomon and Joe McIntyre brought the artefacts to life, and Derek Handley's picture research yielded many of the other rarities. Jonathan Rigby and Derek Ritchie made helpful comments on the manuscript, and Jo Ware's research assistance saved me valuable time. Richard Bignell and Andrew Pixley answered my trickiest questions, while at *Doctor Who Magazine* Tom Spilsbury and Peter Ware provided phone numbers and general encouragement. Stephanie Milner at BBC Worldwide co-ordinated the picture research for the post-2005 episodes and Andrew Walker at the Doctor Who Experience in Cardiff kindly allowed us to photograph the exhibits. The staff of the BBC Written Archives Centre helped me uncover much of the information and some of the relics too.

My colleagues at BBC Books enabled all this to happen. My thanks to Albert DePetrillo, Joe Cottington and especially Justin Richards for their faith, patience and input.

Caroline Skinner, the former co-executive producer of *Doctor Who*, advised us in the early stages, and brand managers Matt Nicholls and Edward Russell oversaw the project for the BBC.

This book would not have been possible without the *Doctor Who* contributors and collectors who kindly allowed their props, costumes and other treasures to be photographed. Special thanks to: Andrew Beech, Lee Binding, John Bloomfield, Piers Britton, Nick Bullen, Steve Cambden, David Carlisle, Toby Chamberlain, Pat Godfrey, Mick Hall, Grahame Flynn, Christopher Hill, June Hudson, Mat Irvine, Joel Nation, Ian O'Brien, Chris Pocock, David Pratt, Jan-Vincent Rudzki, Gary Russell, Dez Skinn, Ken Trew, Mike Tucker, Alexandra Tynan, John Walker, Peter Ware and Colin Young.

Interviewees past and present are quoted in the text. My thanks to: Andrew Beech, Jeremy Bentham, Lee Binding, Andrew Cartmel, Phil Collinson, Terrance Dicks, Phil Ford, Graeme Harper, the late David Maloney, Richard Marson, Alan McKenzie, Steven Moffat, Gareth Roberts, Gary Russell, the late Ian Scoones, Dez Skinn, the late Elisabeth Sladen, Tom Spilsbury, Donald Tosh and Mike Tucker.

Elsewhere, I have quoted from interviews or drawn information from articles by the following: Jason Arnopp, Austen Atkinson, Alan Barnes, Jeremy Bentham, Nick Briggs, Benjamin Cook, Peter Griffiths, Jon Heckford, Mark Lawson, Gary Leigh, Ian McLachlan, Philip Newman, Andrew Pixley, Michael Stead and Mark Wyman. My thanks to them all.

Aside from *Doctor Who Magazine* and its various specials, pre-eminent among the published works I consulted were *The Doctor Who Production Guide: Volume Two – Reference Journal* by Keith A Armstrong, David Brunt and Andrew Pixley (Nine Travellers Publishing/Global Productions, 1997) and *Howe's Transcendental Toybox* by David J Howe and Arnold T Blumberg (Telos Publishing, 2000). My thanks to them too. Other sources are referenced in the text.

Most helpful of all has been Sharon, who took charge of the house and our schedule during the long months I spent exploring *The Vault*. Our children, Leo and Niamh, are not old enough to appreciate *Doctor Who*, but I'm working on it.

Finally, my thanks to Steven Moffat for answering my questions and writing such an engaging foreword. As he points out, this is not a conventional reference book. Its journey through 50 years of *Doctor Who* does, however, start at the beginning...

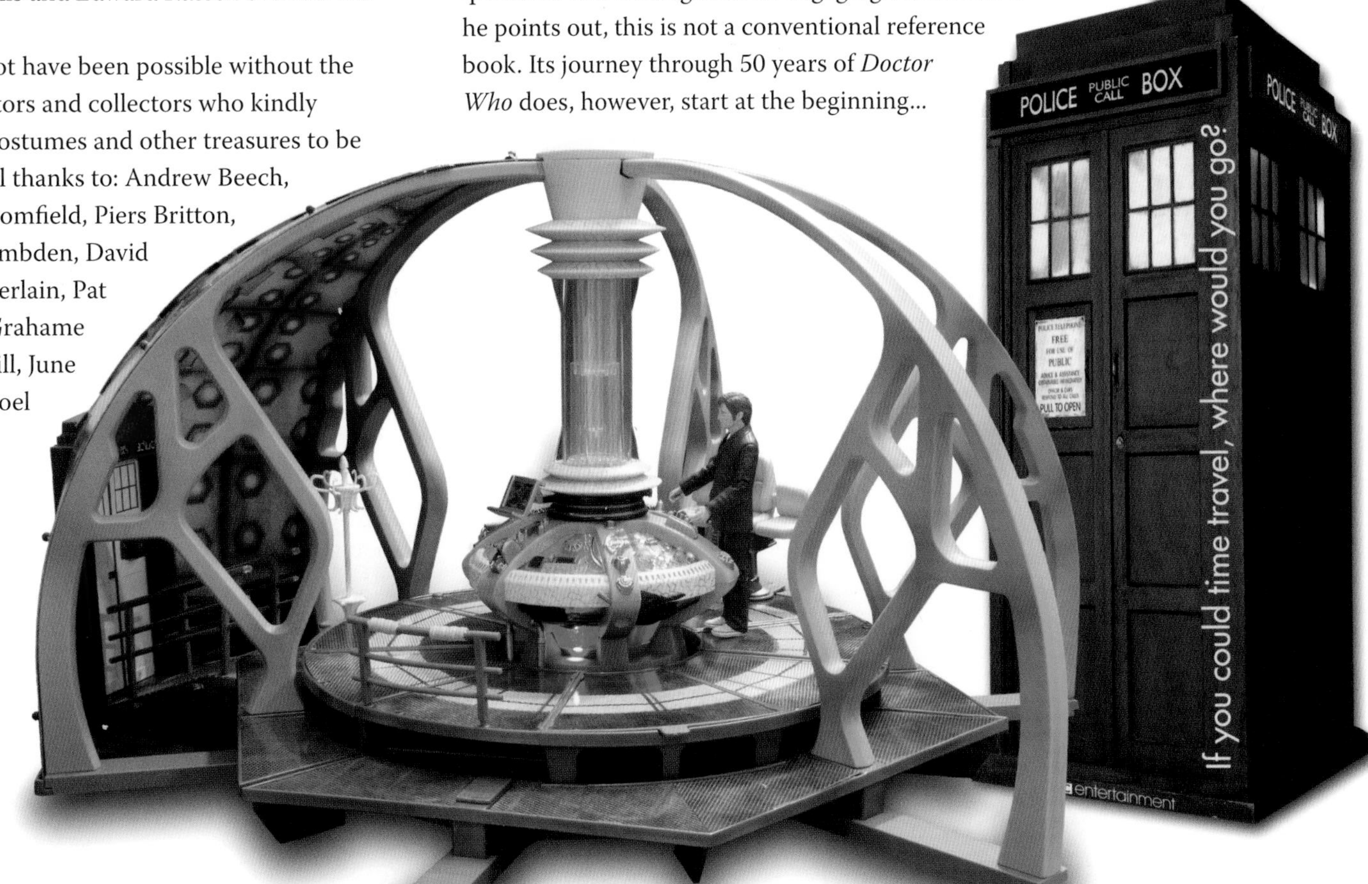

Below: An elongated TARDIS ticket to the 2005 press launch by Asian cable channel StarHub.

Below left: Character Options' TARDIS playset and Tenth Doctor action figure, released in 2006.

1963

The world's longest-running science fiction series had a troubled inception.

During its development in summer 1963 it was intended that *Doctor Who* should begin with a script by CE Webber called *The Giants*, in which the mysterious Doctor and his travelling companions were shrunk to miniscule size. The story was postponed, largely because it was deemed too technically demanding; *Doctor Who* was a production that would effectively be recorded as a live performance at the BBC's antiquated Lime Grove Studios.

Key elements from Webber's opening episode were incorporated into *An Unearthly Child*, the first instalment of Anthony Coburn's *100,000 BC*. *An Unearthly Child* remains one of the outstanding episodes of *Doctor Who*, but it's clear that the subsequent three episodes spent among the cave men and women of the Palaeolithic era must have become the series' opening adventure by default.

There were further problems when the original recording of *An Unearthly Child* was found lacking by Sydney Newman, the show's co-creator. The technical shortcomings were too obvious, the Doctor was too abrasive and his granddaughter, Susan, was too strange. Newman admonished producer Verity Lambert and director Waris Hussein, but gave them the chance to record the episode again.

The polished result of months of preparation was finally revealed when the new version of *An Unearthly Child* was screened just after 5.15 on Saturday 23 November. The timing was unfortunate, with much of the country still shocked and bewildered by the assassination of President Kennedy the previous day.

For all its faults, *100,000 BC* did at least fulfil Newman's brief – that the new series should feature semi-educational trips into the past. Newman considered the next story, the futuristic nightmare *The Mutants*, a flagrant contravention of his wish to avoid the worst excesses of pulp science fiction. He initially dismissed the Daleks, the story's alien antagonists, as "bug-eyed monsters". He was furious that Lambert and her story editor David Whitaker had steered the programme away from its original brief so early in its run.

History would soon prove Lambert's instincts correct. By the end of 1963 *Doctor Who* had only been on the air for six weeks, but it had already outgrown its creators.

THE FIREMAKERS

Above: The first items ever published to promote *Doctor Who* were postcards sent to viewers who wrote to the BBC asking for autographs of the cast. These examples show William Hartnell as the Doctor (above left), William Russell as Ian Chesterton (above) and Carole Ann Ford as Susan (left). The portraits depict the bedraggled characters from *The Firemaker*, the fourth episode of the first story, *100,000 BC*.

Below: The creation of *Doctor Who* was a collaborative effort, but three men were chiefly responsible for devising its original format. From left to right: Sydney Newman, Donald Wilson and CE 'Bunny' Webber.

Doctor Who's remarkable longevity is the result of a format that has evolved through necessity. That the format allowed such flexibility is at least partly due to the fact that it had no single creator. Within its establishing committee, however, a number of key voices emerged. Their opposing sensibilities regarding traditional and modern drama helped to lay the foundation for a programme that continues to thrive.

In March 1963 Donald Baverstock, the Chief of Programmes for BBC1, identified a gap in the Saturday evening schedule between sports programme *Grandstand* and teenagers' favourite *Juke Box Jury*. He decided that gap should be filled by an adventure serial with an appeal that would bridge both audiences. Baverstock asked Sydney Newman, the Corporation's Head of Drama Group (Television), to oversee its creation. Newman delegated the show's earliest development to Donald Wilson, the head of the Script Department.

The process began on 26 March when Wilson chaired a meeting with staff scriptwriters John Braybon, Alice Frick and Cecil Edwin 'Bunny' Webber. The previous year Braybon, Frick and another writer, Donald Bull, had variously compiled several feasibility studies into science fiction for Wilson. The spirit of these focus groups continued with explorations of themes inspired by literary science fiction. During the meeting Wilson proposed a time machine, "not only for going backward and forward in time, but into space, and into all kinds of matter."

Left: A subsequent set of postcards included these posed publicity portraits of Jacqueline Hill (left) William Russell (centre) and Carole Ann Ford (right). Unfortunately on these cards Carole Ann had to correct the mis-spelling of her name.

Bottom left: William Hartnell pauses by the freshly painted TARDIS prop during a camera rehearsal for the original recording of *An Unearthly Child* on 27 September.

Bottom centre: Lime Grove Studio D is dressed as the junkyard at 76 Totter's Lane for the recording of *An Unearthly Child*.

Below: Following an initial meeting chaired by Donald Wilson on 26 March 1963 the development of the new series proceeded apace. This report, sent by CE Webber to Wilson three days later, built elements of that discussion into a proposal for a series called *The Troubleshooters*. Parts of this proposal are identifiable in the ultimate format for *Doctor Who*.

Three days later, CE Webber suggested that "however we set up our serial we must come around to at least one scientist as the basic character." He recommended that "we create ad hoc villains for each story, as needed. It is the Western set-up in this respect: constant heroes, and a fresh villain every time."

By early May, the proposed series was referred to as *Doctor Who*, reportedly following a lunchtime meeting where caretaker producer/director Rex Tucker had written the title on a serviette. The regular cast of four characters was fleshed out by Webber. He devised male and female schoolteachers and, at Newman's insistence that there should be "a kid to get into trouble, make mistakes", added what he described as "A with-it girl of 15, reaching the end of her secondary school career, eager for life, lower-than-middle class." Heading the cast was the mysterious Doctor, described by Webber as "A frail old man lost in space and time. They give him this name because they don't know who he is... He has a 'machine' which enables them to travel together through time, through space, and through matter." Webber added that Doctor Who "remains always something of a mystery, and is seen by us rather through the eyes of the other three..."

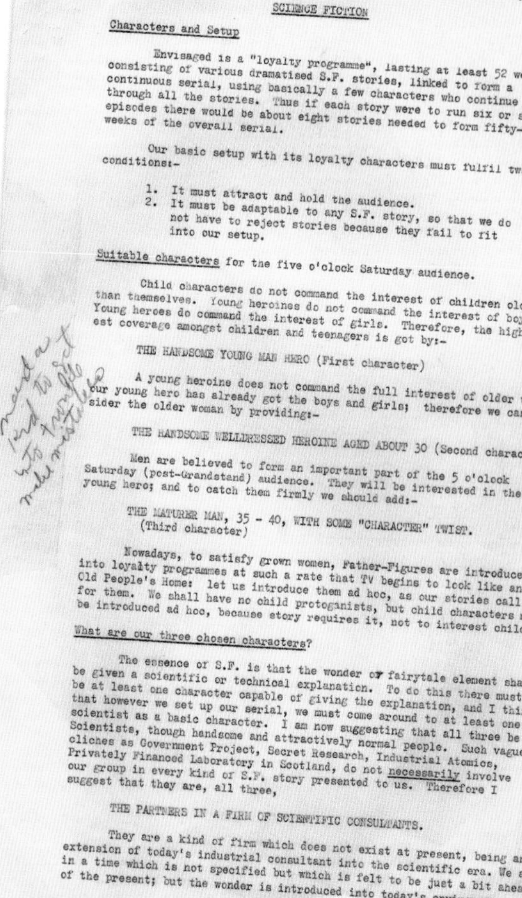

TO: Donald Wilson

FROM: C.E. Webber

29th March 1963

SCIENCE FICTION

Characters and Setup

Envisaged is a "loyalty programme", lasting at least 52 weeks, consisting of various dramatised S.F. stories, linked to form a continuous serial, using basically a few characters who continue through all the stories. Thus if each story were to run six or seven episodes there would be about eight stories needed to form fifty-two weeks of the overall serial.

Our basic setup with its loyalty characters must fulfil two conditions:-

1. It must attract and hold the audience.
2. It must be adaptable to any S.F. story, so that we do not have to reject stories because they fail to fit into our setup.

Suitable characters for the five o'clock Saturday audience.

Child characters do not command the interest of children older than themselves. Young heroines do not command the interest of boys. Young heroes do command the interest of girls. Therefore, the highest coverage amongst children and teenagers is got by:-

THE HANDSOME YOUNG MAN HERO (First character)

A young heroine does not command the full interest of older women; our young hero has already got the boys and girls; therefore we can consider the older woman by providing:-

THE HANDSOME WELLDRESSED HEROINE AGED ABOUT 30 (Second character)

Men are believed to form an important part of the 5 o'clock Saturday (post-Grandstand) audience. They will be interested in the young hero; and to catch them firmly we should add:-

THE MATURER MAN, 35 - 40, WITH SOME "CHARACTER" TWIST. (Third character)

Nowadays, to satisfy grown women, Father-Figures are introduced into loyalty programmes at such a rate that TV begins to look like an Old People's Home: let us introduce them ad hoc, as our stories call for them. We shall have no child protoganists, but child characters may be introduced ad hoc, because story requires it, not to interest children.

What are our three chosen characters?

The essence of S.F. is that the wonder or fairytale element shall be given a scientific or technical explanation. To do this there must be at least one character capable of giving the explanation, and I think that however we set up our serial, we must come around to at least one scientist as a basic character. I am now suggesting that all three be Scientists, though handsome and attractively normal people. Such vague cliches as Government Project, Secret Research, Industrial Atomics, Privately Financed Laboratory in Scotland, do not necessarily involve our group in every kind of S.F. story presented to us. Therefore I suggest that they are, all three,

THE PARTNERS IN A FIRM OF SCIENTIFIC CONSULTANTS.

They are a kind of firm which does not exist at present, being an extension of today's industrial consultant into the scientific era. We are in a time which is not specified but which is felt to be just a bit ahead of the present; but the wonder is introduced into today's environment.

2.

CLIFF — 27, red-brick university type, the teacher of applied science at Sue's school. Physically perfect, a gymnast, dexterous with his hands.

MISS MCGOVERN — 23, a history mistress at the same school. Middle class. Timid but capable of sudden courage. Admires Cliff, resulting in under-currents of antagonism between her and Sue.

These are the characters we know and sympathise with, the ordinary people to whom extraordinary things happen. The fourth basic character remains always something of a mystery ....

DR. WHO — A name given to him by his three earthly friends because neither he nor they know who he is. Dr. Who is about 650 years old. Frail looking but wirey and tough like an old turkey - is amply demorstrated whenever he is forced to run from danger. His watery blue eyes are continually looking around in bewilderment and occasionally a look of utter malevolence clouds his face as he suspects his earthly friends of being part of some conspiracy. He seems not to remember where he comes from but he has flashes of garbled memory which indicate that he was involved in a galactic war and still fears pursuit by some undefined enemy. Because he is somewhat pathetic his three friends continually try to help him find his way "home", but they are never sure of his motives.

THE SHIP

Dr. Who has a "ship" which enables them to travel together through space, through time, and through matter. When first seen, this ship has the appearance of a police box standing in the street, but anyone entering it finds himself inside an extensive electronic contrivance. Though it looks impressive, it is an old beat-up model which Dr. Who stole when he escaped from his own galaxy in the year 5733; it is uncertain in performance; moreover, Dr. Who isn't quite sure how to work it, so they have to learn by trial and error.

FIRST STORY

"The Giants"

Four epi... turbulent adventure in which prop... size are dramatized

Leaving the ... school where they work at the end of ... master, Cliff, and the history ... in the fog. She asks them ... man (Dr. Who) who is lost.

Top left: The Doctor and his 'ship' are described in this three-page outline prepared by Wilson, Webber and Newman on 16 May.

Top: William Hartnell and William Russell rehearse a scene from the first story in the TARDIS control room designed by Peter Brachacki. The set was largely constructed by specialist prop builders Shawcraft Models (Uxbridge) Ltd at a relatively astronomical cost of almost £4,000.

Above: A view of the junkyard set as it appeared in the opening sequence of *An Unearthly Child*.

Left: The closest thing to a *Doctor Who* toy available in 1963 was this 'police hut', a die-cast model originally produced by Dinky between 1936 and 1941.

The May document, entitled *Dr Who: General Notes on Background and Approach*, continued with Wilson and Webber's ideas about the narrative backbone of the series: "Dr Who's 'machine' fulfils many of the functions of science fiction gimmicks. But we are not writing science fiction." The ship's appearance should therefore not appear futuristic or far-fetched. "If we scotch this by positing something humdrum, say, passing through some common object in the street such as a night-watchman's shelter to arrive inside a marvellous contrivance of quivering electronics, then we simply have a version of the dear old Magic Door... The discovery of the old man and investigation of his machine would occupy most of the first episode."

The new programme would adhere to an episodic structure that was a staple of the era: a weekly punctuation of stories that was familiar to viewers and reassuring to producers seeking to ration their resources. Each instalment would

run to 25 minutes, enabling foreign broadcasters to add commercials while keeping each episode within a half-hour slot.

The May document stated that each new episode should "begin by repeating the closing sequence or final climax of the preceding episode." Sydney Newman's handwritten notes elaborated on this with the recommendation: "each episode to end with a very strong cliffhanger – curtain."

On 16 May Wilson, Webber and Newman were the credited authors of a three-page document that outlined a highly unusual format that wasn't shackled to any particular genre: "The series is neither fantasy nor space travel nor science fiction... Our central characters because of their 'ship' may find themselves on the shores of Britain when Caesar and his legionnaires landed in 44 BC; may find themselves in their own school laboratories but reduced to the size of a pinhead; or on Mars; or Venus etc etc."

In keeping with the increasingly idiosyncratic nature of the series, the 'ship' would initially have "the appearance of a police box standing in the street, but anyone entering it finds himself in an extensive electronic contrivance. Though it looks impressive it is an old beat-up model which Dr Who stole when he escaped from his own galaxy in the year 5733; it is uncertain in performance; moreover, Dr Who isn't quite sure how to work it, so they have to learn by trial and error."

With this groundwork in place, the series developed quickly under Newman's guidance. He invited Verity Lambert, who had been a production assistant at his previous post at ABC Television, to become the show's producer. David Whitaker joined as story editor and staff writer James Anthony Coburn incorporated Webber's ideas into the opening four-part story. In the process he gave the Doctor's ship the name TARDIS – Time And Relative Dimension In Space.

Left: William Hartnell as 'Dad' Johnson in Lindsay Anderson's film *This Sporting Life*, which was released in January 1963. Alongside this picture is one of the original publicity portraits of Hartnell as the Doctor, taken by Douglas Playle in September that year. Hartnell had been working on *Doctor Who* for several months before Verity Lambert told him he was cast on the strength of his performance as 'Dad'.

Bottom left: The travellers flee to the refuge of the TARDIS in *The Firemaker*.

Below: The regular cast met for the first time on 20 September 1963 when they took part in a photo session shot against specially mocked up sets at the BBC's Television Centre. It is not only the scenery that differentiates this picture from both versions of *An Unearthly Child* – William Russell appears without his overcoat and Jacqueline Hill is wearing a completely different costume. This was the first *Doctor Who* picture to appear in the *Radio Times*, the week before the new series began. The same picture represented *An Unearthly Child* in the *Radio Times* special published to mark *Doctor Who*'s tenth anniversary.

THE BRITISH BROADCASTING CORPORATION
HEAD OFFICE: BROADCASTING HOUSE, LONDON, W.1
TELEVISION CENTRE: WOOD LANE, LONDON, W.12
TELEGRAMS & CABLES: BROADCASTS, LONDON, TELEX * INTERNATIONAL TELEX 22182
TELEPHONE: SHEPHERDS BUSH 8000

Dear David,

I meant you to have this on monday morning, but I have found out one thing about the cave man that you might pass on to any learned anthropologists you know, - and I am sure you number many amongst your closest friends - it is this. They must have been very much smaller than ourselves. This fact I deduce, not from a close study of their implements, nor by using my Scobonomometer in Hachendorff's Test of the Plutonium content of their left elbows......but by knowing how bloody difficult it is to get into their skins.

And lastly, I rather think that wordwise, this one might be a little too long. I'm a lousy timer. See what you think.

Son of the son of the son of the son of the son of the ad infinitum,
firemaker,

Tony

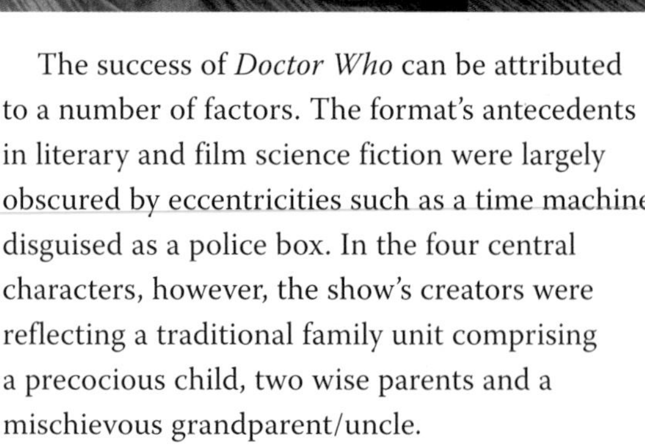

Above: Make-up artist Elizabeth Blattner attends to William Hartnell during filming for *The Firemaker* at Ealing Studios on 9 October.

Right: A letter from *100,000 BC* writer Anthony Coburn to story editor David Whitaker.

Far right: Hartnell and Carole Ann Ford meet the press at the Langham, opposite the BBC's Broadcasting House, on 21 November.

Below: Barbara is trapped by an unseen horror at the end of *The Dead Planet*, the first episode of *The Mutants*.

Director Waris Hussein was appointed and at the end of July the regular cast were issued with their contracts. The teenage Susan, now the Doctor's granddaughter, was played by Carole Ann Ford. History teacher Barbara Wright was played by Jacqueline Hill and science teacher Ian Chesterton was played by William Russell. The show's title role was taken by veteran character actor William Hartnell.

Filmed scenes for the first serial, *100,000 BC*, began shooting on 19 September. The bulk of the story would be recorded on videotape, and the recording of material for *An Unearthly Child* took place on 27 September. When Sydney Newman found the resulting episode to be lacking he ordered a remount. *An Unearthly Child* was rerecorded on 18 October, relegating the original version to the status of an untransmitted pilot. It was the rerecorded material, combined with the existing film footage, that comprised the first episode of *Doctor Who*, as broadcast on Saturday 23 November. The episode was watched by 4.4 million viewers, a modest figure undoubtedly compromised by a power blackout. Because of this power outage, the episode was repeated ahead of the next instalment the following week. This time it attracted 6.6 million viewers.

The success of *Doctor Who* can be attributed to a number of factors. The format's antecedents in literary and film science fiction were largely obscured by eccentricities such as a time machine disguised as a police box. In the four central characters, however, the show's creators were reflecting a traditional family unit comprising a precocious child, two wise parents and a mischievous grandparent/uncle.

This impression was reinforced at script stage by Coburn, who was apparently the person who decided to make Susan the Doctor's granddaughter. Coburn had justified a premise that depicted an old man jealously guarding a teenage schoolgirl, but his decision was unpopular with Newman and compromised one of the format's major assets – namely its ambiguity. The

KINEMATOGRAPH WEEKLY: OCT[...] 1963

22

TELEVISION

by TONY GR[...]

ITV can expect a jolt when the BBC launches its 'Dr. Who'

THE ITV companies can expect their first major jolt from the BBC Drama Group when Sidney Newman launches "Dr. Who," a 52-week family series, on November 23 at 5.25 p.m.

"Dr. Who" is a somewhat mysterious type of programme consisting in part of fantasy and realism. But Newman is backing it as a big rating success, and in fact initiated its format.

The story is about a strange doctor, played by William Hartnell, who creates a machine which goes backwards and forwards in time as well as going outside of time.

Accompanying him on various trips is a 15-year-old girl, Carole Ann Ford, and two young teachers, William Russell and Jacqueline Hill.

But while the premise of "Dr. Who" is fantastic, the treatment of various places and periods will be tackled factually and realistically.

The arrival at the time of the French Revolution will be as historically and naturalistically accurate as the landing on a new planet or a look into the future of the world in 100 years time.

In other words while "Dr. Who" will be informative and broadly educational, it will always be full of entertainment gimmicks and the type of showmanship that is part of the Newman flair.

And the BBC drama chief will create the world of "Dr. Who" in the confines of the Television Centre. There will be hardly any exterior shooting, but plenty of work for the two set designers of the series, Barry Newbury and Brachaki.

Full resources

Newman has always deprecated the failure of producers to use to the full the resources of a live television studio, and at ABC he was able to achieve astonishing results with the aid of such brilliant designers as Tim O'Brien and Voytek.

"Dr Who" will have at least two permanent directors, Chris Barry and Waris Hussein, and its producer is Miss Verity Lambert, who received her early training at ABC where she worked as PA to Dennis Vance and Ted Kotcheff.

Miss Lambert then went off to the States, directing shows for David Susskind, until her return to ABC early this year. In June she joined the serials department of the drama group under Donald Wilson, who has worked with Newman in setting up the project.

Producer Lambert told me this week: We think that "Dr. Who" will be something different in weekend family entertainment. We have some good writers who are exprienced in working on high-class series, for this show must please adults as well as children if it is to be successful. "Dr. Who" is a strange mysterious weird old man, and William Hartnell is giving a marvellous performance in the title role. None of the episodes will be self contained, but will be grouped together into four- or even eight-part serials. Only the four characters, Dr. Who, the young girl and the two teachers will be constant. The length of the serials will depend on the stories and locations and these will be varied in time and space.

With a perfect time spot of 5.25 p.m. on Saturday and with the full resources of the BBC backing Newman's pet project, one can prophesy with some confidence that with "Dr. Who" the BBC Drama Group should be making its first major ratings breakthrough against ITV.

And about time, too!

\+ + +

ASSOCIATED BRITISH PATHE has sold nearly 4,500 television programmes to overseas stations during the last twelve months, making it the company's most successful year so far.

Chris Towle, who heads the department under Macgregor Scott, told me this week there is a greater interest and demand for British programmes than ever before among the Commonwealth countries.

One of the most important deals signed recently was for 39 episodes of "The Avengers" to be sold to the Australian Broadcasting Commission. And further deals for the series are now being finalised with Canada, New Zealand and smaller Commonwealth stations including Malta, Gibraltar, Nigeria and Rhodesia.

ABC's variety show, "Big Night Out," which is not seen in London, has also been sold to Australia's ATV commercial channel. It comprises a complete series of 13 programmes.

Both programmes are taped and have to be converted to a 16-mm. negative, for sale outside this country.

But prices for all programmes whether originally film or converted film are about the same, Towle told me this week.

Spoof film

RADIO TIMES *November 21, 1963*

7

Your Weekend **Saturday**

DR. WHO

In this series of adventures in space and time the title-role will be played by William Hartnell

5.15

Dr. Who? That is just the point. Nobody knows precisely who he is, this mysterious exile from another world and a distant future whose adventures begin today. But this much is known: he has a ship in which he can travel through space and time—although, owing to a defect in its instruments he can never be sure where and when his 'landings' may take place. And he has a grand-daughter Susan, a strange amalgam of teenage normality and uncanny intelligence.

Playing the Doctor is the well-known film actor, **William Hartnell,** who has not appeared before on BBC-tv.

Each adventure in the series will cover several weekly episodes, and the first is by the Australian author **Anthony Coburn.** It begins by telling how the Doctor finds himself visiting the Britain of today: Susan (played by **Carole Ann Ford**) has become a pupil at an ordinary British school, where her incredible breadth of knowledge has whetted the curiosity of two of her teachers. These are the history teacher Barbara Wright (**Jacqueline Hill**), and the science master Ian Chesterton (**William Russell**), and their curiosity leads them to become inextricably involved in the Doctor's strange travels.

Because of the imperfections in the ship's navigation aids, the four travellers are liable in subsequent stories to find themselves absolutely anywhere in time—past, present, or future. They may visit a distant galaxy where civilisation has been devastated by the blast of a neutron bomb or they may find themselves journeying to far Cathay in the caravan of Marco Polo. The whole cosmos in fact is their oyster.

FOR DELIVERY VIA INTERNAL MAIL

CONFIRMATION OF MESSAGE DICTATED TO PHONOGRAMS, BROADCASTING HOUSE, LONDON

DATE 27.11.63

SERVICE CHARGEABLE TELEVISION

STAFF NO. (PERSONAL TELEGRAMS)

DELIVER TO:

DONALD WILSON RM 5075 TC EXT 3750

PRECEDENCE ORD LT

TO SYDNEY NEWMAN WARWICK HOTEL 65 WEST FIFTYFOURTH STREET NEW YORK CITY 19 USA

COPY TO

DOCTOR WHO OFF TO A GREAT START EVERYBODY HERE DELIGHTED REGARDS = DONALD+

B21 1735 SD

Top left: This article appeared in the edition of trade magazine *Kinematograph Weekly* dated 24 October 1963. It predicted that the series would be recorded at Television Centre and "exhibit the type of showmanship that is part of the Newman flair."

Above: "The whole cosmos in fact is their oyster" promised the *Radio Times* in this tantalising preview.

Left: A copy of the telegram sent by Donald Wilson to Sydney Newman on 27 November.

Below: The original, unseen version of the city and landscape from *The Dead Planet*. The model was created by independent prop-builders Shawcraft but rejected by designer Raymond Cusick.

fact that so little has ever been revealed about the Doctor and his TARDIS has allowed writers and audiences to project their own, evolving interpretations onto the series. This ambiguity was notably fostered in the short gap between the original and the ultimate recordings of *An Unearthly Child*. In the pilot episode Susan tells the incredulous Barbara that "I was born in the 49th century." In the broadcast version this was changed to the less specific "I was born in another time, another world..."

The central mystery that has intrigued viewers for decades is embodied in the title *Doctor Who*. It is a name that still conjures the spellbinding combination of science and mystery devised by Donald Wilson, Bunny Webber and Sydney Newman in the summer of 1963.

SEASON ONE

Producer: Verity Lambert
Associate producer: Mervyn Pinfield
Story editor: David Whitaker

A ***100,000 BC*** (aka ***An Unearthly Child***)
31 August – 8 November
written by Anthony Coburn (and CE Webber, uncredited)
directed by Waris Hussein

An Unearthly Child	23 November
An Unearthly Child	30 November (repeat)
The Cave of Skulls	30 November
The Forest of Fear	7 December
The Firemaker	14 December

B ***The Mutants*** (aka ***The Daleks***)
28 October – 10 January 1964
written by Terry Nation
directed by Christopher Barry and Richard Martin

The Dead Planet	21 December
The Survivors	28 December

1964

Doctor Who's first complete year of transmission saw the fulfilment of its creators' aim to present adventures set forwards, backwards and 'sideways' in time. *The Mutants*, the seven-part serial that introduced the Daleks, came to an end on 1 February, by which time *Doctor Who* was regularly watched by around ten million viewers every week.

Budgetary constraints partly prompted *Inside the Spaceship*, a claustrophobic two-parter set within the sentient TARDIS. The ship's doors finally opened to reveal the beginning of an epic adventure with *Marco Polo*. Along the way the Doctor gradually assumed a more benevolent, paternal role within the family of regular characters. *Marco Polo* and *The Aztecs*, both written by John Lucarotti, would become some of the best-loved purely historical stories of the entire series.

Many of 1964's science fiction adventures paled in comparison. *The Keys of Marinus*' quest scenario was a conspicuous example of the programme's reach exceeding its grasp. *The Sensorites* was a sedate follow-up, but presented its alien telepaths as more than just monsters. Along the way there was some intriguing character development for the otherwise neglected Susan.

The Reign of Terror took the Doctor and his companions to Paris during the French Revolution. William Hartnell was delighted, later citing the serial as his favourite story. The final episode brought *Doctor Who*'s first broadcast season to a close with the Doctor telling Ian that "Our destiny is in the stars, so let's go and search for it."

Recording continued, however, with *Planet of Giants* – a belated realisation of CE Webber's miniscule idea from 1963. Incorporating elements of a political thriller, Louis Marks' script played more like an early instalment of Sydney Newman's previous creation *The Avengers*, and was tightened from four episodes to three after recording was completed.

The series' first block of recording came to an end with an eagerly anticipated confrontation between the Doctor and his already notorious enemies in *The Dalek Invasion of Earth*. Terry Nation's typically ambitious rematch saw extensive location filming for the first time and completed recording in October. More than 12 million viewers would see the Doctor bid farewell to his granddaughter when the final episode was broadcast on Boxing Day.

The Daleks had secured the future of *Doctor Who*, and a second recording block was already underway.

DAY OF RECKONING

FEBRUARY 22–28

BBC tv Sound

Radio Times

SIXPENCE

DR. WHO

The four travellers in time and space return to Earth for a new adventure beginning on Saturday in Television

SEE PAGE 7

HUGH AND I

Laugh—with Terry Scott and Hugh Lloyd as they resume their interrupted series in Television on Saturday

PAGE 9

BENNY HILL

Laugh—with the many Bennys in his new show on Sunday afternoon in the Light

PAGE 15

ERIC SYKES

Laugh—with him and Hattie Jacques as they begin a series in Television on Tuesday

PAGE 27

IN THE LIGHT AND ON TELEVISION

World Heavyweight Championship

SONNY LISTON

v.

CASSIUS CLAY

SEE PAGE 29

IN THE LIGHT

The British, British Empire, and European Championship

HENRY COOPER

v.

BRIAN LONDON

SEE PAGE 21

Above: Although *Doctor Who* was due to receive a *Radio Times* cover for the week of its original broadcast, a lack of faith in the programme from BBC management meant that the 23-29 November 1963 edition led with a picture of radio comedian Kenneth Horne instead. Donald Wilson protested to the editor of the magazine, who finally gave the programme a front cover for the edition dated 22-28 February 1964. The photograph from *Marco Polo* proved controversial – William Russell's agent complained that it featured none of the regular cast except William Hartnell. Wilson shared his concerns, and passed them on to the editor.

Above right: Producer Verity Lambert pictured at Lime Grove Studios in December 1963. She is reviewing sound effects tapes prepared for *Doctor Who* by the BBC's Radiophonic Workshop.

As befits a programme about an exile, the earliest *Doctor Who* was made by outsiders. The show's original production team included a number of mavericks who created brilliant television but spent much of their first year together fighting the BBC establishment and, in some cases, each other.

Sydney Newman was an outspoken Canadian who arrived at the Corporation from commercial television in January 1963. Verity Lambert may not have been Newman's first choice for the producer of *Doctor Who* in June – he only turned to her after the more experienced Don Taylor and Richard Bates declined – but he later cited her appointment as the best decision he ever made regarding the series. She was certainly a brave choice; Lambert had joined the BBC as the Drama Department's only female producer. At just 27 she was also the youngest.

"I was very much aware that I was a complete oddity in the Drama Department," Lambert remembered. "I had to convince people who were running departments and giving us facilities, and a lot of them were very unconvinced for quite a long time. The Design Department in particular was absolutely horrendous and behaved abominably. It was completely unjustifiable."

Lambert found a more willing collaborator in Waris Hussein, the Indian-born director of *100,000 BC*. Hussein was just 24 when he started work on the programme and, like Lambert, was around 20 years younger than many of his colleagues at the BBC. He was acutely aware that this wasn't the only thing that set him apart. "As an Asian, I was a phenomenon," he says. "Nobody discussed it, almost out of self-consciousness, but that set off my own insecurities... It was a tough call for me. It was a secret ordeal."

Resistance to the series even extended to Australian staff writer Anthony Coburn, who wrote *100,000 BC*. According to Hussein, Coburn was disdainful of the new programme and disappeared once he'd delivered his scripts. "Verity was faced with a writer who didn't want anything to do with cavemen, and a director who didn't want anything to do with cavemen. David Whitaker, the script editor, was very caring and very much involved with what was going on, but I don't think he reckoned those first episodes very much either."

Newman fought with Lambert over the abstract title sequence and theme tune. A disagreement between Nation and Whitaker ended in a fist fight, and director Richard Martin lost his temper with Lambert. "One time, I was sweating to get the programme done, and she stopped the recording to object to a hat! I broke my little finger thumping the desk, telling her to get lost!"

Left: Sheet music for Ron Grainer's theme tune was published in 1964. The front cover seemed to credit Terry Nation as the series' sole writer.

Below: The tunic worn by Mark Eden in the title role of *Marco Polo*. The garment features customisations made for another *Doctor Who* appearance the following year, when it was worn by Julian Glover in *The Crusade*.

Inset: Jacqueline Hill, Carole Ann Ford, William Hartnell and Mark Eden at the camera rehearsal for the *Marco Polo* episode *Mighty Kublai Khan*.

Bottom left: Three of the most important contributors to *Doctor Who*'s early success. From left to right – director Waris Hussein, story editor David Whitaker and writer Terry Nation.

Right: Verity Lambert shares a tea break with William Hartnell and William Russell on the set of *Inside the Spaceship*.

Centre: Ixta (Ian Cullen) challenges Ian in *The Day of Darkness*, the fourth episode of *The Aztecs*. The fight was staged at Ealing Film Studios and benefited from more complex editing than was possible in the programme's videotaped sequences.

Below: The headdress worn by Cullen and designed by Daphne Dare.

From 1963 to 1964 Lambert also had to contend with hostility from the BBC management. *Doctor Who*'s freewheeling format demanded a cinematic sweep, but this was invariably curtailed by a budget of £2,300 per episode and similarly deficient studio facilities.

In 1964 Hussein returned to *Doctor Who* to direct the fondly remembered *Marco Polo*. He struggled to depict the TARDIS crew's epic journey across 13th century Cathay in "appalling" circumstances. "The best equipment was at Television Centre, but we were given the oldest studio at Lime Grove. It was worthy of being knocked down when we were working in it, and the camera equipment was horrendous."

The Aztecs was another highlight of the first series, perhaps surprisingly given that its director, John Crockett, didn't even own a television. ("Wouldn't have one in the house, old boy," he told fight arranger Derek

Right: Carole Ann Ford is menaced by one of the alien Voord (Peter Stenson) during a photo call for *The Keys of Marinus* on 10 April.

Left: Richard Jennings' original artwork for the frontispiece of *The Dalek Book*, which was published by Souvenir Press on 30 September. It was credited to David Whitaker and Terry Nation, and 'based on the Dalek Chronicles discovered and translated by Terry Nation'.

Centre: The first edition of *Doctor Who in an exciting adventure with the Daleks* was published on 12 November. The novel was written by Whitaker and based on Nation's original Dalek story.

Ware.) Its second episode, *The Warriors of Death*, marked the first time *Doctor Who* had been recorded at Television Centre. Newman's ambition was to move *Doctor Who* to Television Centre permanently. Meanwhile Lambert was fighting relegation to Lime Grove Studio G, an even more awkward space than the familiar but detested Studio D.

Lambert was further frustrated by the lack of commitment from Donald Baverstock to an extended run of episodes. In 1963 *Titbits* magazine hailed *Doctor Who*'s publicised 52-week run as making television history, but the reality in 1964 was rather different – Baverstock was only committing to 13 episodes at a time, placing enormous pressure on Lambert and Whitaker.

While seemingly fearless about what she put on screen, Lambert was scrupulously careful behind the scenes; *Inside the Spaceship* required next to no additional sets and helped her to amortise the considerable cost of building the TARDIS control room. The two-parter also rounded off the first 13 episodes of *Doctor Who*, creating a neat package to fit a quarter-year schedule should the series not be renewed.

It wasn't until 13 February 1964 that Baverstock finally agreed to the regular cast members being engaged for the full 52 weeks of their contracts. The catch-fire success of the Daleks undoubtedly played a part in his decision, and also gave Newman some leverage in negotiating better studio facilities.

Much of *The Reign of Terror* was made at Lime Grove Studio G (midway through, the overwhelmed director Henric Hirsch collapsed) but *Planet of Giants* was produced at Television Centre. *Doctor Who*'s first recording block ended with *The Dalek Invasion of Earth* at Riverside Studios.

Right: These sweet cigarettes were launched by Cadet in 1964 and remained in production until the end of the decade. Each box contained ten sweet cigarettes and one of 50 numbered picture cards by artist Richard Jennings. The cards depicted the Voord, as well as the golden-domed Emperor Dalek that Jennings had first illustrated in *The Dalek Book*.

Below: One of the must-have toys of Christmas 1964 – the original battery-operated Dalek produced by Louis Marx & Company.

A second recording block was now on the horizon. Personal disputes had been set aside and, for the time being at least, *Doctor Who* earned the respect of BBC management. In the background throughout all of this, however, there was a simmering resentment that would not be resolved.

Carole Ann Ford was a jobbing actress when Waris Hussein first spotted her on a monitor at Television Centre. By her own admission, playing Susan elevated her to the level of a pop star. Such was her newfound fame that she shared the screen with real pop stars such as Adam Faith and George Harrison when she appeared on teenagers' favourite *Juke Box Jury*. But such publicity only increased the risk of typecasting, and she didn't feel that the scripts she received were adequate compensation.

Susan was the 'unearthly child' of the opening episode, but since then Ford had been dismayed that the character's early intrigue and promise were largely forgotten. *Marco Polo*, her favourite story, gave Susan some memorable scenes with Ping-Cho (played by Zienia Merton), and in

Left: On 28 November Carole Ann Ford signs copies of *The Dalek Book* at the Gamages department store in London. She is inside the lower half of a 'Dalek Suit', a luxury item of merchandise produced by toy manufacturer Scorpion.

Below: Newcastle group The Go-Go's performed 'I'm Gonna Spend My Christmas With a Dalek' for this 7" record released in December. Producer Johnny Worth sensibly adopted the pseudonym 'Les Vandyke' for the writing credit.

VERITY.

19th NOVEMBER, 1964.

Terry Nation rang.

He says an song-writer called JOHNNY WORTH has been in touch with him, andwishes to bring out a Dalek song.

Johnny Worth is goingto write this song tonight and is bringing a group down from Newcastle next Wednesday. He would like to arrange a photo call next Wednesday, (23rdNov) if possible with some Daleks (Dalek holding a guitar etc).

(Incidentally, he has written more hit songs than anybody else, and had done many of Adam Faith's songs).

He would like to talk with you direct, and will ring you sometime tomorrow.

His no is: VANdyke 7987.
HASTINGS 2088.

(I have given Margaret a copy of this note).

P.T.O.

The Sensorites she received telepathic messages from the alien creatures. As the first series progressed, however, there no longer seemed to be anything particularly remarkable or mysterious about the Doctor's granddaughter.

Ford would have left sooner, but was contractually bound to the same 52-week run as her co-stars. William Hartnell wrote her a letter, trying to persuade her to stay, but she resolved to leave as soon as she could. She went straight from *Doctor Who* to a lead role in a pantomime opposite Diana Dors, but later voiced the opinion that the programme had damaged her long-term career.

David Whitaker's contract was also expiring, and his last duty as story editor was to rewrite Susan's final scene in *Flashpoint*, the closing episode of *The Dalek Invasion of Earth*. Susan has fallen in love with resistance fighter David Campbell (Peter Fraser), so the Doctor leaves them to build a new life in the ruins of 22nd century London.

It was an abrupt but poignant farewell to a character who had once seemed intrinsic to the programme. *Doctor Who* would continue without Susan, her departure the first indication that the show was now bigger than its cast.

SEASON ONE (continued)

Producer: Verity Lambert
Associate producer: Mervyn Pinfield
Story editor: David Whitaker

B *The Mutants* (continued)

The Escape	4 January
The Ambush	11 January
The Expedition	18 January
The Ordeal	25 January
The Rescue	1 February

C *Inside the Spaceship* (aka *The Edge of Destruction*) (7 – 24 January)

written by David Whitaker
directed by Richard Martin and Frank Cox

The Edge of Destruction	8 February
The Brink of Disaster	15 February

D *Marco Polo* (13 January – 13 March)

written by John Lucarotti
directed by Waris Hussein and John Crockett

The Roof of the World	22 February
The Singing Sands	29 February
Five Hundred Eyes	7 March
The Wall of Lies	14 March
Rider from Shang-Tu	21 March
Mighty Kublai Khan	28 March
Assassin at Peking	4 April

E *The Keys of Marinus* (March – 24 April)

written by Terry Nation
directed by John Gorrie

The Sea of Death	11 April
The Velvet Web	18 April
The Screaming Jungle	25 April
The Snows of Terror	2 May
Sentence of Death	9 May
The Keys of Marinus	16 May

F *The Aztecs* (13 April – 22 May)

written by John Lucarotti
directed by John Crockett

The Temple of Evil	23 May
The Warriors of Death	30 May
The Bride of Sacrifice	6 June
The Day of Darkness	13 June

G *The Sensorites* (May – 10 July)

written by Peter R Newman
directed by Mervyn Pinfield and Frank Cox

Strangers in Space	20 June
The Unwilling Warriors	27 June
Hidden Danger	11 July
A Race Against Death	18 July
Kidnap	25 July
A Desperate Venture	1 August

H *The Reign of Terror*

(15 June – 14 August)
written by Dennis Spooner
directed by Henric Hirsch (and John Gorrie, uncredited)

A Land of Fear	8 August
Guests of Madame Guillotine	15 August
A Change of Identity	22 August
The Tyrant of France	29 August
A Bargain of Necessity	5 September
Prisoners of Concergerie	12 September

SEASON TWO

Producer: Verity Lambert
Associate producer: Mervyn Pinfield
Story editor: Dennis Spooner
(unless otherwise stated)

I *Planet of Giants* (30 July – 11 September)

written by Louis Marks
directed by Mervyn Pinfield and Douglas Camfield
Story editor: David Whitaker

Planet of Giants	31 October
Dangerous Journey	7 November
Crisis	14 November

K *The Dalek Invasion of Earth*

(23 August – 23 October)
written by Terry Nation
directed by Richard Martin
Story editor: David Whitaker

World's End	21 November
The Daleks	28 November
Day of Reckoning	5 December
The End of Tomorrow	12 December
The Waking Ally	19 December
Flashpoint	26 December

L *The Rescue* (16 November – 11 December)

written by David Whitaker
directed by Christopher Barry

The Powerful Enemy	2 January
Desperate Measures	9 January

M *The Romans*

(17 November – 15 January 1965)
written by Dennis Spooner
directed by Christopher Barry

The Slave Traders	16 January
All Roads Lead to Rome	23 January
Conspiracy	30 January
Inferno	6 February

1965

An extraordinary year began with the broadcast of *The Rescue*, the first story in *Doctor Who*'s second recording block. David Whitaker's two-part mystery was designed to introduce new travelling companion Vicki (Maureen O'Brien), an orphaned teenager from the 25th century.

The Romans, a four-part historical by new story editor Dennis Spooner, was played as an exuberant farce. The eerie longueurs of *The Web Planet* seemed sedate in comparison. The conflict between the Menoptra and Zarbi was possibly the most over-ambitious scenario in the programme's entire history, and this expensive serial proved unpopular with critics. Normal service was resumed with the literate historical drama *The Crusade*.

The first episode of *The Space Museum* was compelling viewing, with a series of 'sideways' trips through time that culminated in the travellers confronting future versions of themselves as exhibits in glass cases. In *The Chase* the Daleks hunted the TARDIS in their own time machine, ultimately providing the Doctor with the means to return Ian and Barbara home. Their place would be taken by Steven Taylor (Peter Purves), an astronaut seeking refuge from a fiery skirmish between the Daleks and their spherical adversaries the Mechonoids.

Verity Lambert left the series after *The Time Meddler*, an innovative mix of historical adventure and science fiction. Peter Butterworth played the meddling monk of the title, a member of the Doctor's own (as yet unnamed) race.

A new producer, John Wiles, and a new story editor, Donald Tosh, were installed for the production of *Galaxy 4*, a morality tale enlivened by the Amazonian Drahvins, the ammonia-breathing Rills and their diminutive robots, which Vicki nicknames 'Chumblies'. The second recording block ended with *Mission to the Unknown*, a single-episode Dalek adventure that didn't feature any of the regular cast at all.

The third block started with *The Myth Makers*, a trip to 12th century Troy that saw Vicki replaced by the handmaiden Katarina (Adrienne Hill). Wiles and Tosh then braced themselves to tackle an unwelcome inheritance. *The Daleks' Master Plan* was conceived as a six-part story but extended to an unwieldy 12 at the behest of BBC management. The episode broadcast on 25 December saw the Doctor break the fourth wall and wish "a happy Christmas to all of you at home!"

It was a fitting end to a year of unprecedented experimentation.

DALEK NATION

VERITY. 14th December, 1964.

Please will you ring Perry Guinness urgently.

The Daily Mail Boys and Girls Exhibition want to borrow a Dalek from Monday, 28th December - Saturday, 9th January, 1965.

They are also making enquiries about a tape of the Dalek voices and, also, the Dalek operators.

B.H.Ext.2738.
68

(Perry Guinness's Sec. thought it would probably be best if we gave her the names of the operators, and she passed them on to the Daily Mail so they could contact them direct.

They are: Robert Jewell
Kevin Manser
Peter Murphy
Gerald Taylor
Nick Evans and Ken Tyllsen.

Above: In December 1964 the BBC loaned two of the Daleks from *The Dalek Invasion of Earth* to the *Daily Mail*'s annual Boys and Girls' Exhibition in Olympia, London. This Dalek was stationed on the route taken by visitors riding a converted goods transporter around the site.

Top right: One of two designs of *The Dalek Writing Pad* published by Newton Mills in 1965.

With responsibility for the 175 staff of the Series, Serials and Plays departments Sydney Newman was unable to maintain close links with all his productions. So it was that his first sight of the Daleks came when he saw the second serial, *The Mutants*, at home on television. Donald Wilson had angrily condemned the story at script stage, but Newman was absolutely livid. He summoned Verity Lambert to his office on Monday morning and accused her of betraying his trust by including the dreaded 'bug-eyed monsters' in *Doctor Who*. "But she very carefully explained that they weren't really BEMs," he told the *Daily Sketch* in 1969. "They were really humans who had lost their limbs, very sympathetic bits of ironmongery."

Newman's commitments to gritty social realism and a semi-educational tone were anathema to the creator of the Daleks, Terry Nation. "We were still in the fag end of the kitchen sink school, and I was fed up with people who had real problems," said Nation in 1965. "I wanted to produce entertainment with no moral or sociological values. There are very fine writers who will move the theatre forward, but I'm not one of them."

Newman had nevertheless underestimated the subtleties and subtexts in Terry Nation's script, initially overlooking the compelling back story and motivation that set the Daleks way apart from the

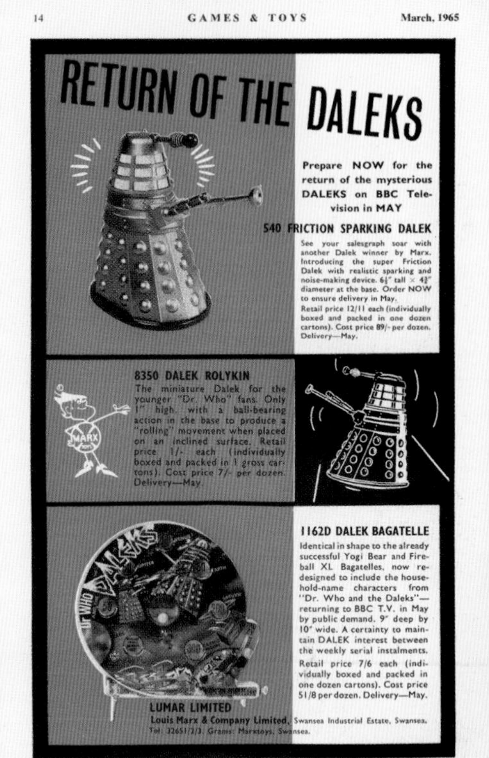

14 GAMES & TOYS March, 1965

RETURN OF THE DALEKS

Prepare NOW for the return of the mysterious DALEKS on BBC Television in MAY

540 FRICTION SPARKING DALEK

See your salesgraph soar with another Dalek winner by Marx. Introducing the super Friction Dalek with realistic sparking and noise-making device. 6¼" tall × 4¾" diameter at the base. Order NOW to ensure delivery in May.

Retail price 12/11 each (individually boxed and packed in one dozen cartons). Cost price 89/- per dozen. Delivery—May.

8350 DALEK ROLYKIN

The miniature Dalek for the younger "Dr. Who" fans. Only 1" high, with a ball-bearing action in the base to produce a "rolling" movement when placed on an inclined surface. Retail price 1/- each (individually boxed and packed in 1 gross cartons). Cost price 7/- per dozen. Delivery—May.

1162D DALEK BAGATELLE

Identical in shape to the already successful Yogi Bear and Fireball XL Bagatelles, now re-designed to include the household-name characters from "Dr. Who and the Daleks"—returning to BBC T.V. in May by public demand. 9" deep by 10" wide. A certainty to maintain DALEK interest between the weekly serial instalments.

Retail price 7/6 each (individually boxed and packed in one dozen cartons). Cost price 51/8 per dozen. Delivery—May.

LUMAR LIMITED

Louis Marx & Company Limited, Swansea Industrial Estate, Swansea. Tel: 32651/2/3. Grams: Marxtoys, Swansea.

B movie villains they superficially resembled. The Daleks were not robots at all, but pathetic radiation victims encased inside machines that enabled them to fulfil their vendetta against any life form that was different from their own. Analogies to the Nazis' ethnic cleansing were underlined by the creatures' chilling catchphrase 'exterminate'.

Newman had also failed to anticipate the Daleks' enormous appeal to the viewing public. The credit for this appeal must be shared between Nation and staff designer Raymond Cusick, whose utterly inhuman vision made an impact on viewers before the true horror of Nation's ideas was revealed. "I worked out the original drawing in three hours, did a couple of sketches in the office on Friday afternoon and took them home to finish on Sunday afternoon," said Cusick in 1965. "I realised they would have to have a human operator, so the basic design was shaped round a man sitting on a tricycle. I wanted midgets inside at first, because I thought it would be good to make the Daleks smaller than humans. I was making mechanical actors really."

The Daleks would alter the course of *Doctor Who* towards more science fiction/fantasy-based

Top: The cover and the first page of a sales brochure issued by BBC Enterprises in 1965. The remaining pages reproduced a selection of press commentary about the programme and offered foreign broadcasters 'more than 50 half-hour programmes of the weird adventures of Dr Who'.

Top right: This advert from Louis Marx and Co appeared in the March 1965 issue of trade magazine *Games & Toys*. The company's new line of Dalek merchandise was timed to coincide with the screening of *The Chase*, the third Dalek adventure, in May.

Above: A publicity still from *The Death of Time*, the second episode of *The Chase*, featuring the Doctor, Vicki (Maureen O'Brien), Rynian (Hywel Bennett), Ian, Barbara and Malsan (Ian Thompson).

Right: Terry Nation and his wife Kate demonstrate a selection of Louis Marx Daleks in August.

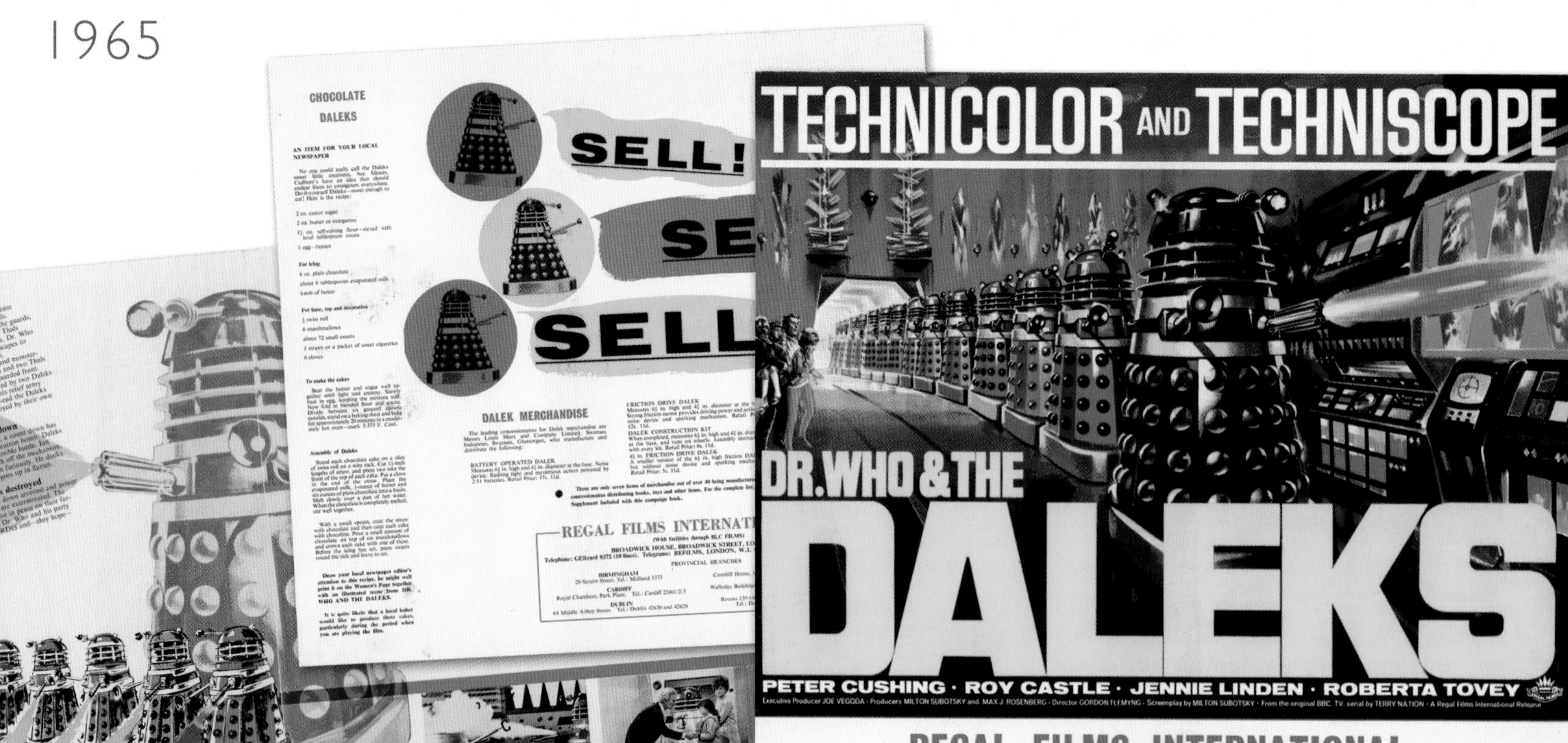

Above: *Dr. Who and the Daleks* opened in London on 25 June. This lavish campaign book positioned it in the vanguard of the Dalek merchandising campaign. Verity Lambert later dismissed the film as "shoddy" but the novelties of widescreen and colour helped make it a box-office hit.

Below: Raymond Cusick shows his daughter one of the Louis Marx Daleks inspired by his original design.

Centre: Verity Lambert and an obliging Mechonoid publicise *The Chase* at Ealing Film Studios on 14 April.

themes, and they would transform Terry Nation's life. The spoils from the first Dalek serial – including a generous cut of the ensuing licensing deals – contributed to the cost of an Elizabethan mansion in Kent. In summer 1964 Nation purchased Lynsted Park for £15,000, "most of it in hard cash". He celebrated his 34th birthday in the house, hosting a lavish party which numbered the cast of *Doctor Who* among the guests.

The second serial, *The Dalek Invasion of Earth*, was even more popular than the first. The Beatles appeared in a pre-recorded sequence in the third story, *The Chase*, and the term 'Dalekmania' was coined in obvious emulation of the group's hysterical reception in the early 1960s.

By 1965 the Daleks were a television phenomenon that threatened to overshadow every other element of *Doctor Who*. The spin-off feature films released in 1965 and 1966 are commonly referred to as 'the Dalek films', and the Doctor isn't even mentioned in the trailer for the second. In 1965 toy shops were flooded with licensed Dalek toys and other merchandise. Images of the Doctor and his companions barely featured on any of it.

Lambert tried to reduce the programme's dependence on the ubiquitous foes by hyping new monsters such as the ant-like Zarbi. In January 1965 their creator, Bill Strutton, told the *Daily Sketch* about the inspiration for his forthcoming serial *The Web Planet*. "I had to come up with something different from the robot-style Daleks," he admitted.

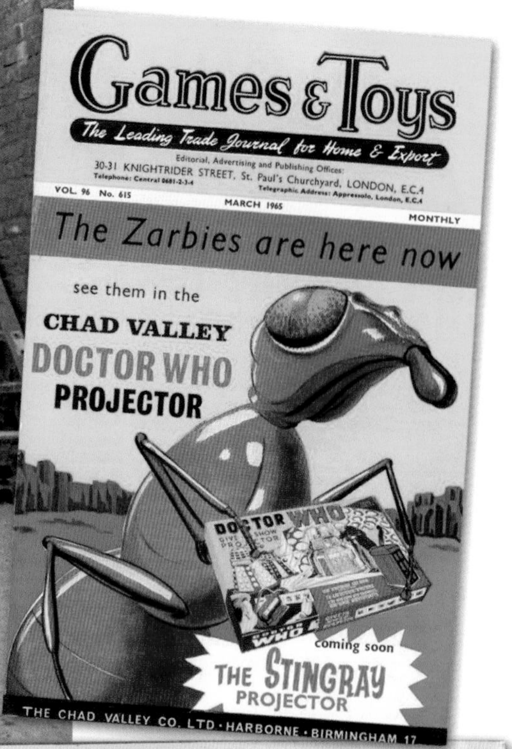

Nation tried to outdo himself with the Mechonoids, spherical robots who tackled the Daleks at the climax of *The Chase*, but again, these failed to catch on. The beehive-like Chumblies were actually operated by small actors, as per Raymond Cusick's original ambition for the Daleks, and appeared in William Emms' story *Galaxy 4*. They proved engaging enough for four episodes, but lacked the ruthlessness that children in particular seemed to admire in the Daleks.

In 1963 Nation had been one of David Whitaker's most industrious and reliable writers (he wrote all seven episodes of the first Dalek story in just four weeks) but by 1965 the success of his creations had started to compromise his output. Maintaining the business side of his Dalek empire was getting in the way of writing scripts,

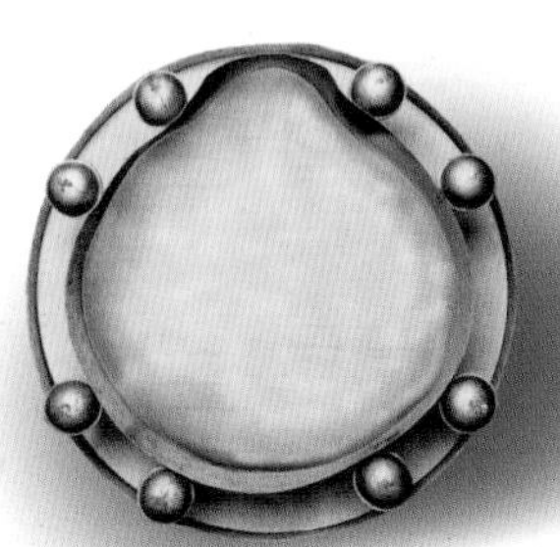

Top: The Uxbridge workshop of Shawcraft Models, the prop-makers who frequently supplied *Doctor Who*. The two staff are surrounded by creature costumes from *The Rescue*, *The Web Planet* and *The Dalek Invasion of Earth*.

Top Right: The Zarbi make the front cover of the March 1965 issue of *Games & Toys* magazine.

Above: A strip of slides from the Chad Valley Give-a-Show projector advertised on the front of *Games & Toys*.

Bottom left: *Doctor Who* was clearly making a cultural impact – this satirical cartoon from the *Daily Mirror*'s Stanley Franklin uses Daleks and Zarbi to reflect social tensions in British society.

Below: Views of a Chumbley in various modes of operation. These pictures were taken as reference for prospective toy manufacturers.

WIN A REAL DALEK

TV CENTURY 21

ADVENTURE IN THE 21s

No. 28 UNIVERSE EDITION

EVERY WEDNESDAY

DR. WHO and the DA

Peter Cushing and Roberta

PEACEFUL THALS AMBUSHED!

"FOLLOWING attempts to make peace with the Daleks, the Thals have been viciously ambushed and a fierce battle is still being fought."

This was the first frightening news of the Daleks to hit the Universe some eighteen months ago on B.B.C. Television.

Now the exciting film, "Dr. Who and the Daleks" is on the cinema screen in glorious Technicolour. You can see it at your local cinema soon.

SOLTURIAN NEWS AGENCY FLASH:—

Prince Jareth has been called in by Mirva to investigate another prophesy of Lurr. It is still not certain whether Lurr is continuing his opposition to the Daleks, or whether this new look into the future is concerned with other events.

(Full story page 20)

DALEKS! DALEKS! TO BE WON! 3 REAL FILM DALEKS TO BE WON! PLUS 450 OTHER DALEK PRIZES Pages 18 – 19

CORGI MODEL CLUB NEWS — EVERY WEEK PAGE 19

ELECTRIC TRAIN SETS TO BE WON! see page 8

TV CENTURY 21

7D

ADVENTURE IN THE 21st CENTURY

No. 36 UNIVERSE EDITION

EVERY WEDNESDAY

DATELINE: September 25, 2065

DALEKS SUFFER HEAVY LOSSES!

Atomic Rust Plagues Skaro

HEAVY losses are reported from Skaro where the super robot-like Daleks are fighting a losing battle against a cloud of deadly radio-active rust.

The beta-gamma contaminated rust consumes metal at the rate of a cubic inch a second and destroys a Dalek in little under half a minute.

Although the Black Dalek, who is in charge of the Dalek City while the Emperor is away, will reveal no information on the number of Daleks missing, the Space News Agency station S.N.A. 5 estimates that up to half the home army has been destroyed.

Unless effective action is taken, the Daleks will be destroyed.

Full news agency report on page 20.

STOP PRESS

HEAVY METEOR SHOWERS

The Space Meteorological Office issued the following warning to all spaceships travelling between Mars and Jupiter at 07.15 hours today: Heavy meteor showers are expected in space areas Mars–Jupiter. Space station will be closed to all traffic except for emergencies. Spaceships are warned to seek shelter at any of the large asteroids.

Dalek Tracking station PZ8 just before being destroyed by the rust cloud.

Picture by Regal Films International

DEATH MURDERED IN LOS ANGELES

LOS ANGELES Homicide is investigating the sudden death of fortune teller Irvin Death.

Mr. Death was killed when his crystal ball exploded.

Captain Amos Burke is in charge of the investigation and an arrest is expected very shortly.

(Full story on pages 2 and 3).

CORGI MODEL CLUB NEWS — EVERY WEEK PAGE 19

Above: *TV Century 21* was launched by producer Gerry Anderson in January 1965. By this time Anderson's programmes for ITV were shot entirely on film, in colour, and he wanted the comic representations of those shows to be similarly lavish. Anderson was no fan of the more modestly budgeted *Doctor Who*, and a childish strip based on the programme was already running in *TV Comic* anyway. The Daleks, however, appealed to his love of all things mechanical so he sought Terry Nation's permission to include them. A one-page strip, simply titled *The Daleks*, ran from the first issue until early 1967.

Top right: The Doctor examines the Taranium Core, surrounded by Bret Vyon (Nicholas Courtney), Katarina (Adrienne Hill) and Steven Taylor (Peter Purves) in this publicity still from *Devil's Planet*, the third episode of *The Daleks' Master Plan*.

Right: Part of a range of Dalek greetings cards produced by Newton Mills. Opening each card revealed a message that was only partially visible through the cut-outs on the cover.

Far right: These Daleks, manufactured by Cherilea Toys in 1965, came with interchangeable sections available in different colours.

and story editor Donald Tosh recalls that although the 1965/66 story *The Daleks' Master Plan* was credited to Nation and Dennis Spooner he made up the shortfall in Nation's scripts by writing much of the serial himself.

In Nation's mind the Daleks had outgrown *Doctor Who*, and in 1967 he took them out of the series altogether in an effort to sell them as the stars of their own US TV show. He was ultimately unsuccessful, and on New Year's Day 1972 they made a triumphant return to the programme that still hadn't found an adversary to supplant them. Future appearances would be carefully rationed ratings-grabbers, and there could be gaps of more than four years between Dalek serials.

In 1974 *Doctor Who* producer Barry Letts complained that Nation's latest Dalek proposal recycled concepts and plot devices that were familiar from previous adventures (in fact Nation had got away with this once already in 1973's *Planet of the Daleks*). Nation responded with the remarkable *Genesis of the Daleks*, the 1975 story that introduced the crippled psychopath Davros, the genetic engineer who would become the Daleks' figurehead for all their subsequent appearances in the show's original run.

The Daleks' evolution continued with other writers. Ben Aaronovitch's script for the

1988 story *Remembrance of the Daleks* finally eliminated the creatures' most conspicuous weakness. In the cliffhanger to Part One the Doctor (Sylvester McCoy) scales a flight of stairs to evade a Dalek, only for it to rise up and relentlessly pursue him through mid-air.

Nation's last script for the programme was *Destiny of the Daleks* in 1979, but he continued to review and approve other writers' Daleks scripts up to the *Doctor Who* TV movie broadcast in 1996, and Dalek-inspired merchandise until his death in 1997.

Since 2005 the Daleks have appeared in every series, now redesigned and augmented by computer-generated imagery but still eminently faithful to Nation and Cusick's original conception. Russell T Davies, the head writer and executive producer who oversaw *Doctor Who*'s comeback, helped to re-establish the Daleks' once fearsome reputation by casting them in the Time War that provided the largely unseen backdrop to the new series. Relatively little has been revealed about those cataclysmic events, but it has been established that the Daleks were the protagonists in a conflict that massacred both them and the Doctor's own race, the Time Lords. The Daleks, however, live on.

Nation's scripts may not have found favour with every production team, but the Doctor's never-ending battle with the Daleks is still one of the central tenets of the programme. And long after the toy shop bonanza of 1965 Daleks remain the most consistently popular items of licensed merchandise. For these reasons, if for no other, Terry Nation was surely the most influential of all *Doctor Who*'s writers.

Above left: This wooden jigsaw was part of a series produced by Thomas Hope & Sankey Hudson for the department store Woolworths in 1965. In this example the Doctor and the Daleks appeared as self-contained pieces that could be presented on stands. The artist, William Howarth, also painted the cover of the first *Doctor Who Annual* (above), published by World Distributors in September.

SEASON TWO
(continued)
Producer: Verity Lambert
Story editor: Dennis Spooner
(unless otherwise stated)

N ***The Web Planet*** 4 January – 26 February
written by Bill Strutton
directed by Richard Martin

The Web Planet	13 February
The Zarbi	20 February
Escape to Danger	27 February
Crater of Needles	6 March
Invasion	13 March
The Centre	20 March

P ***The Crusade*** 16 February – 26 March
written by David Whitaker
directed by Douglas Camfield

The Lion	27 March
The Knight of Jaffa	3 April
The Wheel of Fortune	10 April
The Warlords	17 April

Q ***The Space Museum***
11 March – 23 April
written by Glyn Jones
directed by Mervyn Pinfield

The Space Museum	24 April
The Dimensions of Time	1 May
The Search	8 May
The Final Phase	15 May

R ***The Chase*** 9 April – 4 June
written by Terry Nation
directed by Richard Martin

The Executioners	22 May
The Death of Time	29 May
Flight Through Eternity	5 June
Journey into Terror	12 June
The Death of Doctor Who	19 June
The Planet of Decision	26 June

S ***The Time Meddler*** 10 May – 2 July
written by Dennis Spooner
directed by Douglas Camfield

The Watcher	3 July
The Meddling Monk	10 July
A Battle of Wits	17 July
Checkmate	24 July

T ***Galaxy 4*** 22 June – 6 August
written by William Emms
directed by Derek Martinus (and Mervyn Pinfield, uncredited)
Story editor: Donald Tosh

Four Hundred Dawns	11 September
Trap of Steel	18 September
Air Lock	25 September
The Exploding Planet	2 October

T/A ***Mission to the Unknown***
25 June – 6 August
written by Terry Nation
directed by Derek Martinus
Story editor: Donald Tosh

Mission to the Unknown	9 October

SEASON THREE
Producer: John Wiles
Story editor: Donald Tosh

U ***The Myth Makers*** 27 August – 8 October
written by Donald Cotton
directed by Michael Leeston-Smith

Temple of Secrets	16 October
Small Prophet, Quick Return	23 October
Death of a Spy	30 October
Horse of Destruction	6 November

V ***The Daleks' Master Plan***
27 September – 14 January 1966
written by Terry Nation and Dennis Spooner
directed by Douglas Camfield

The Nightmare Begins	13 November
Day of Armageddon	20 November
Devil's Planet	27 November
The Traitors	4 December
Counter Plot	11 December
Coronas of the Sun	18 December
The Feast of Steven	25 December

1966

The premature demise of Katarina and the dynamic Sara Kingdom (Jean Marsh) in *The Daleks' Master Plan* left a vacancy in the TARDIS at the beginning of *The Massacre of St Bartholomew's Eve*. This four-part historical is the year's overlooked gem and boasts an accomplished performance from Hartnell as the Doctor's doppelgänger, the Abbot of Amboise. New travelling companion Dodo Chaplet (Jackie Lane) made her debut in the final episode.

Following the HG Wells-inspired *The Ark*, and the sinister surrealism of *The Celestial Toymaker*, Dodo and Steven were cast as a pair of travelling entertainers in *The Gunfighters*. This notorious adventure took the series to a new frontier in more ways than one, dropping the TARDIS crew into a Western where saloon ballads served as a commentary on the occasionally outlandish action.

The first conspicuous innovation from the latest producer Innes Lloyd and story editor Gerry Davis saw individual episode titles dropped. Steven was given a noble send-off, uniting the survivors of a dystopian power struggle at the end of *The Savages*. Dodo, however, would be written out in slapdash style midway through the next serial.

The newly built Post Office Tower was a totemic presence in *The War Machines*. The serial was an early glimpse of a more contemporary format, and while the curmudgeonly Doctor was an incongruous figure in Swinging London his new companions Ben (Michael Craze) and Polly (Anneke Wills) seemed refreshingly authentic.

Behind the scenes, more fundamental revisions were afoot. There had been plans to discard the faltering and irascible Hartnell during *The Celestial Toymaker*, but the actor would remain in the title role a while longer. He continued into the Cornish swashbuckler *The Smugglers*, returning for the fourth recording block in *The Tenth Planet*. The story introduces the disturbing Cybermen, but even this is overshadowed by the climactic sequence where the Doctor's body is miraculously 'renewed'.

Audiences had already seen another actor in the role of Doctor Who – Peter Cushing had played the character in feature films based on the first two Dalek stories – but it was the means of recasting the lead role on television that was the programme's latest stroke of genius. The process would become known as 'regeneration', and it would ensure *Doctor Who*'s longevity.

500 YEAR
DIARY

SCIENCE AND FICTION

From its very earliest days *Doctor Who* was envisaged as a science fiction series, but ever since the format moved away from Bunny Webber's idea about 'partners in a firm of scientific consultants' the programme has had a rather superficial relationship with science itself.

This approach was satirised as early as 1965 in the opening scene of the film *Dr. Who and the Daleks*. Susan (Roberta Tovey) is obscured behind a copy of *Physics for the Inquiring Mind* while the similarly intense Barbara (Jennie Linden) studies *The Science of Science*. The camera then pans across the room to where their grandfather, Doctor Who (Peter Cushing), is engrossed in a copy of the boys' adventure comic the *Eagle*. "Most exciting!" he exclaims.

It's not even clear whether our hero is a doctor of science. The First Doctor states numerous

Above: The public's appetite for Dalek merchandise declined in 1966, but there was still room for items such as this book, published by Souvenir Press in March.

Right: An original model of Mavic Chen's spaceship the *Spar*, designed by Raymond Cusick for *The Daleks' Master Plan*. This 12-part story came to an end in January 1966.

Below: The Doctor (William Hartnell) struggles towards the sanctuary of the TARDIS in *The Destruction of Time*, the final episode of *The Daleks' Master Plan*.

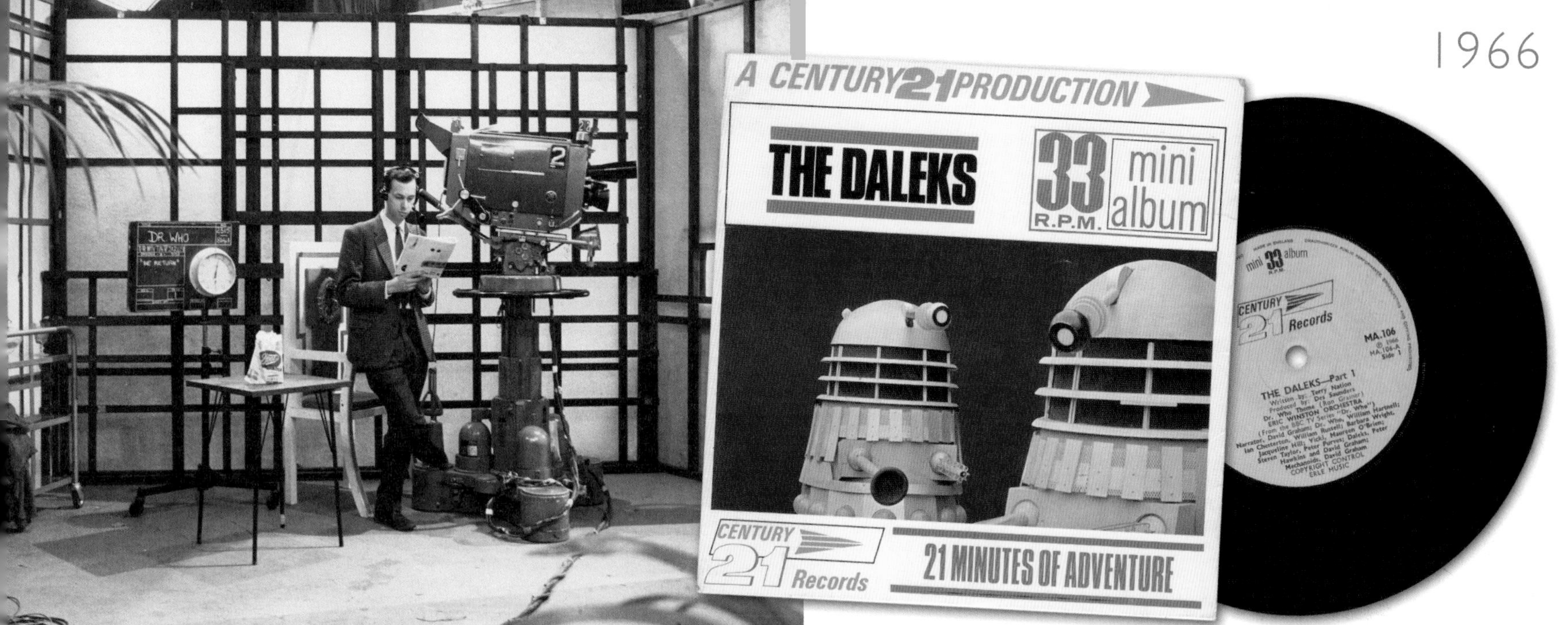

times that he's not a doctor of medicine, although he contradicts himself in his second and eleventh incarnations. He's more likely to proclaim vaguely that he is a doctor qualified in many things, and provides the most accurate description of his expertise in the second episode of the 1964 story *The Aztecs*. "They call me the Doctor," he tells Cameca. "I'm a scientist, an engineer, I'm a builder of things."

This line was almost certainly prompted by a memo from Sydney Newman to Verity Lambert. "May I encourage you to do something in future episodes of *Dr Who* to glamourise the title, occupation etc of an engineer," he wrote on 10 April 1964. "Engineers, of course, are people who repair cars, aeroplane engines, run atomic energy plants etc. Another way of putting it is an emphasis on the applications of science rather than on pure science by itself."

Newman had already advised Lambert to take out a subscription to *New Scientist*, and although she indulged him the magazine's contents had little impact on the show. "I didn't understand a word of it!" she later said.

When *Doctor Who* moved into colour producer Barry Letts revived the subscription to *New Scientist.* Letts' own scientific interests were evident in some of the scripts he co-wrote. *The Time Monster* (1972) sought no social relevance in a plot about the misuse of a device called TOMTIT (Transmission Of Matter Through Interstitial Time). But, more memorably, *The Green Death* (1973) pitted a militant group of self-sufficient ecologists against a secretive corporation that is callously dumping toxic waste. The company's domination by a mad computer firmly roots the story in the techno paranoia of the era, but the

Above left: Preparing to record *The Return*, the third episode of *The Ark*, at Riverside Studios on 4 March.

Above: As well as publishing a weekly comic featuring the Daleks, producer Gerry Anderson was the head of a record company that issued this adaptation of the final episode of *The Chase*, a Dalek story from the previous year.

Left: The feature film *Daleks' Invasion Earth 2150 A.D.* opened in London on 22 July, backed by a promotional campaign for the Sugar Puffs breakfast cereal that included three-and-a-half million special packets and a television commercial. This selection of Sugar Puffs advertising material was designed to be separated along the perforated lines.

Below: Peter Cushing as Doctor Who, inside the redesigned TARDIS control room from the second film.

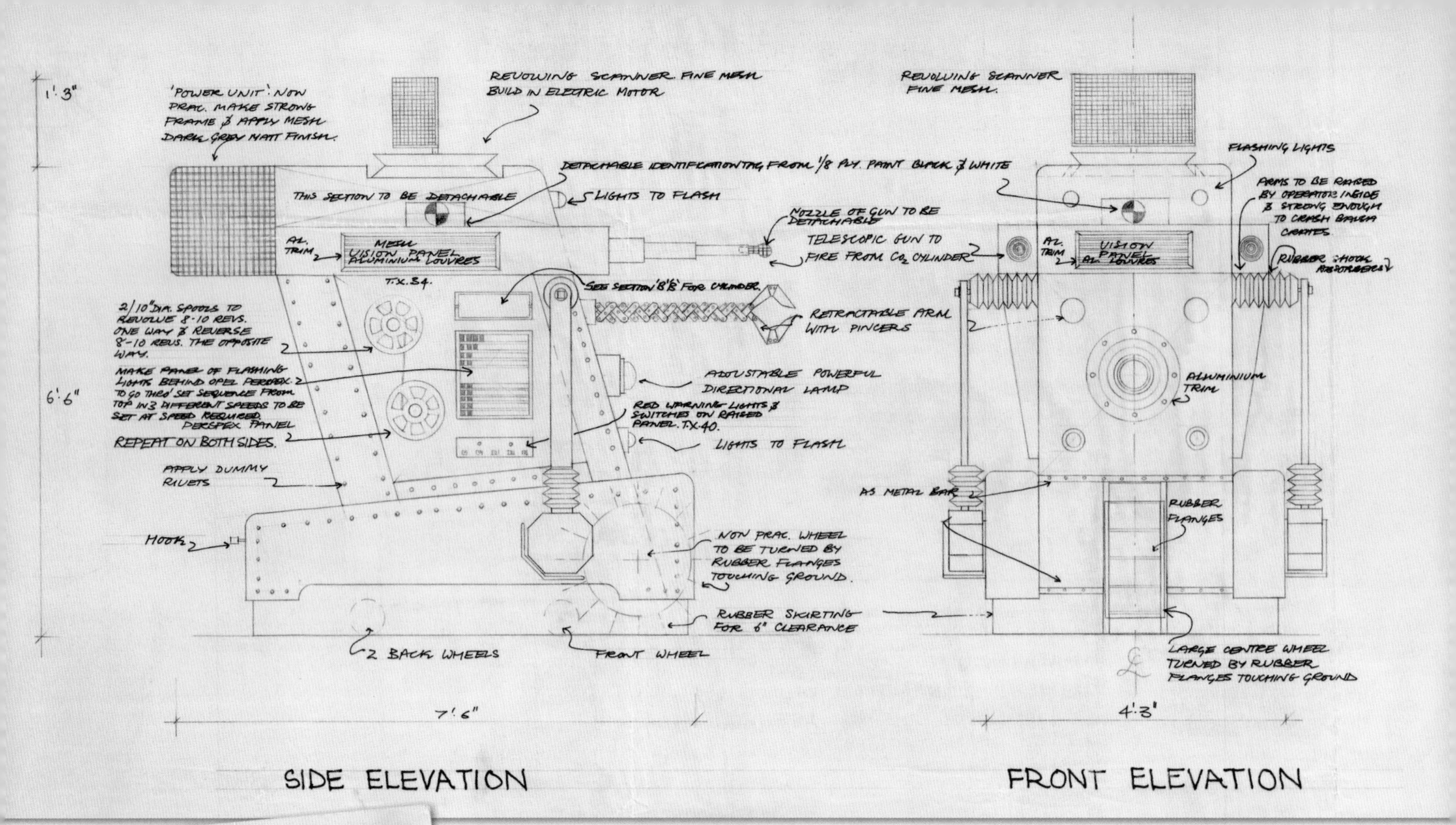

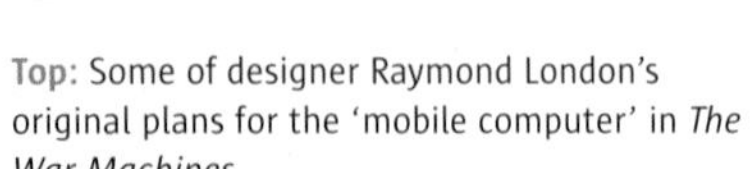

Top: Some of designer Raymond London's original plans for the 'mobile computer' in *The War Machines*.

Above: A finished prop takes to the streets for location filming in May.

Above right: Producer Innes Lloyd, story editor Gerry Davis and scientific advisor Kit Pedler.

Right: Steven and Dodo (Jackie Lane) find the way to the TARDIS blocked by three sinister ballerinas (Beryl Braham, Ann Harrison and Delia Lindon) in *The Dancing Floor*, the third episode of *The Celestial Toymaker*.

consequences of industrial negligence are just as disturbing today.

The programme's next flirtation with 'real' science was prompted by script editor Christopher H Bidmead. Viewers of stories broadcast in 1980 and 1981 were challenged by such complex subjects as tachyonic particles (*The Leisure Hive*), a Charged Vacuum Emboitment (*Full Circle*) and entropy (*Logopolis*).

For the most consistently successful employment of scientific principles we have to go back to 1966, when producer Innes Lloyd decided to ground the programme by appointing a scientific advisor. Lloyd's story editor Gerry Davis

Left and below: Both sides of an elaborate brochure issued by Stanmark Productions in summer 1966 to promote their *Doctor Who* radio series. Peter Cushing played Doctor Who in the pilot, *Journey Into Time* by Malcolm Hulke, but no further episodes were recorded.

met a number of candidates, including physician Alex Comfort (later to write *The Joy of Sex*) and astronomer Patrick Moore. "I tried out a little fiction with them to see if they were flexible," said Davis in 1988. "If, instead of just saying, 'No, it can't happen,' they said 'Well, it doesn't happen this way but there's another way it could happen...' That was what I wanted; someone who would be adaptable so I could say to writers if they were stuck, 'Well, phone up the scientific advisor and he will help you.'"

The man they eventually hired was Christopher 'Kit' Pedler, a surgeon and research scientist who was the head of the Electron Microscopy Department at the University of London. "The research department he created was rather like a sci fi adventure itself, attempting to reproduce vision for the blind," recalled Davis in 1986. "It was hoped that a TV camera might convert images into the kind of electrical impulses the retina normally sends to the cortex of the brain – enabling the blind to see without eyes."

After a promising initial meeting Pedler was asked to devise a story for *Doctor Who*. London's tallest building, the Post Office Tower, was visible from both Pedler's and Davis' offices. It provided Pedler with the inspiration for a story about a sentient super-computer that attempted to enslave humanity using a hypnotic signal transmitted through telephones. *The War Machines* was ultimately scripted by Ian Stuart Black and screened in summer 1966. It was the first story to depict present-day London since 1963, and predicted an internet-like system where computers around the world could be linked together to form a 'network' of processing power and data storage.

Above: In May 1966 a picture from *Daleks' Invasion Earth 2150 A.D.* featured on the cover of religious magazine *Annunciation*.

Left: The second *Doctor Who Annual* was published in September, with cover artwork by Walter Howarth. The draft text was corrected by the programme's story editor Donald Tosh, who advised the publishers that no one should ever address the Doctor with the surname 'Who', and that he "never kills anything or anybody. His stick is never used as a sort of super ray gun."

Pedler made his first foray into scriptwriting with *The Tenth Planet*, the story that opened *Doctor Who*'s fourth recording block in August 1966. *The Tenth Planet* saw a beleaguered group at the South Pole defending themselves from a mysterious new threat – the Cybermen.

The script was co-written by Davis. "He was afraid that medicine would become a matter of machines; that you'd hook people up to computers and they would be the doctors and nurses… Kit originally came up with that idea for *Doctor Who*, then I said, 'Well, suppose people get totally cybernetic, where they lose their souls. What would happen…?' and that's how the Cybermen were born between the two of us."

This methodical collaboration and commitment to realism was a far cry from the instinctual approach that had left Terry Nation unable to tell inquisitive journalists how he dreamt up the Daleks. Yet in the process Pedler and Davis had created the first of the Doctor's adversaries to seriously rival the Daleks in popularity.

Pedler and Davis' next story, *The Moonbase*, began production in January 1967 and saw the Cybermen try to conquer a future Earth by first crippling an isolated weather control station on the moon. Their third and final *Doctor Who* collaboration inverted their 'base under siege' template, this time using the science of archaeology as the means to excavate *The Tomb of the Cybermen*.

A series about a man who travels in a time machine disguised as a police box was probably

Top: *Doctor Who and the Invasion From Space* was published by the Manchester-based World Distributors. The book was similar in format to the company's *Doctor Who* annuals but featured just one story, by JL Morrissey.

Left: Costume designer Sandra Reid (left) and production assistant Edwina Verner dress the Cybermen during filming for *The Tenth Planet* at Ealing Studios in September.

Far left: Sandra Reid's original sketches for the Cybermen. The 'handle' attachments on either side of the head are a design feature that has remained consistent throughout their many reinventions.

Right: Patrick Troughton as the Second Doctor, with the Paris Beau hat he wore in his earliest episodes.

Far right: This proposal for the appearance and character of the Second Doctor was prepared by the production team in summer 1966. They describe a man haunted by his experiences in the "galactic war" that caused him to leave his home planet.

Bottom: *The Dalek Outer Space Book* was published by Souvenir Press and Panther Books in September. Featuring stories by Terry Nation, Brad Ashton and Russ Winterbotham, this was the last of the decade's annual-style publications devoted to the Daleks.

always going to be too fanciful a vehicle for Pedler and Davis' topical concerns. Pedler came up with the ideas for two further Cybermen stories in 1968 – *The Wheel in Space* and *The Invasion* – but Davis bowed out as the series' story editor in 1967.

Pedler and Davis would concoct numerous nightmare scenarios from contemporary science in their early 1970s series *Doomwatch*, but the Cybermen remained arguably the most successful application of this technique.

Appropriately, the Cybermen are frequently upgraded to reflect the latest technology. In 2006 *The Rise of the Cybermen*/*The Age of Steel* gave them a new origin story and a new appearance, but at their core they remain faithful to the horrifying medical concept that first intrigued their creators. As such, the Cybermen endure as a symbolic reminder of an era when science came before fiction.

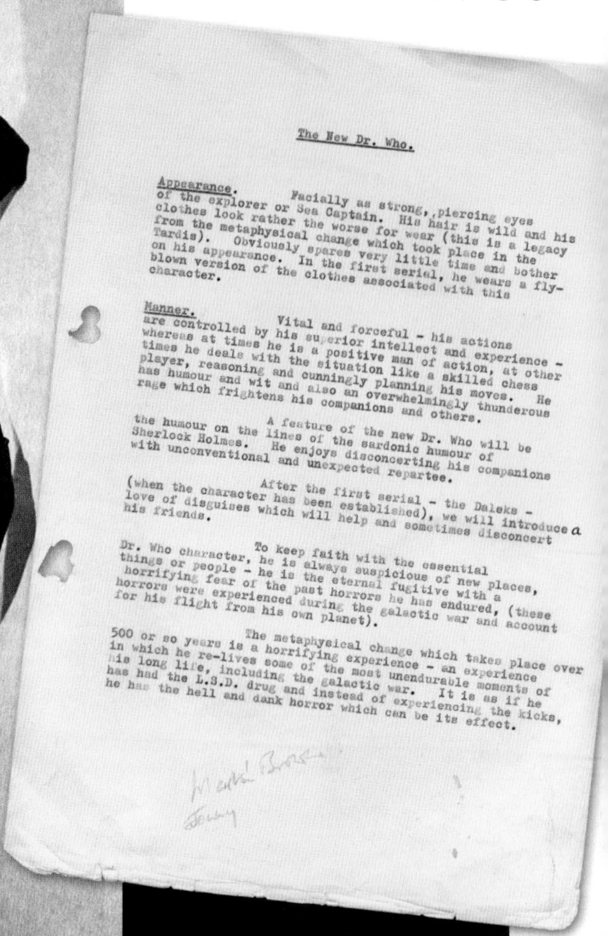

The New Dr. Who.

Appearance. Facially as strong, piercing eyes of the explorer or Sea Captain. His hair is wild and his clothes look rather the worse for wear (this is a legacy from the metaphysical change which took place in the Tardis). Obviously spares very little time and bother on his appearance. In the first serial, he wears a fly-blown version of the clothes associated with this character.

Manner. Vital and forceful - his actions are controlled by his superior intellect and experience - whereas at times he is a positive man of action, at other times he deals with the situation like a skilled chess player, reasoning and cunningly planning his moves. He has humour and wit and also an overwhelmingly thunderous rage which frightens his companions and others.

A feature of the new Dr. Who will be the humour on the lines of the sardonic humour of Sherlock Holmes. He enjoys disconcerting his companions with unconventional and unexpected repartee.

After the first serial - the Daleks - (when the character has been established), we will introduce a love of disguises which will help and sometimes disconcert his friends.

To keep faith with the essential Dr. Who character, he is always suspicious of new places, things or people - he is the eternal fugitive with a horrifying fear of the past horrors he has endured, (these horrors were experienced during the galactic war and account for his flight from his own planet).

The metaphysical change which takes place over 500 or so years is a horrifying experience - an experience in which he re-lives some of the most unendurable moments of his long life, including the galactic war. It is as if he has had the L.S.D. drug and instead of experiencing the kicks, he has the hell and dank horror which can be its effect.

SEASON THREE
(continued)
Producer: Innes Lloyd
Story editor: Gerry Davis
(unless otherwise stated)

V *The Daleks' Master Plan* (continued)
Producer: John Wiles
Story editor: Donald Tosh

Volcano	1 January
Golden Death	8 January
Escape Switch	15 January
The Abandoned Planet	22 January
The Destruction of Time	29 January

W *The Massacre of St Bartholomew's Eve*
3 January – 11 February
written by John Lucarotti and Donald Tosh
directed by Paddy Russell
Producer: John Wiles
Story editor: Donald Tosh (up until third episode)

War of God	5 February
The Sea Beggar	12 February
Priest of Death	19 February
Bell of Doom	26 February

X *The Ark* 24 January – 11 March
written by Paul Erickson and Lesley Scott
directed by Michael Imison
Producer: John Wiles

The Steel Sky	5 March
The Plague	12 March
The Return	19 March
The Bomb	26 March

Y *The Celestial Toymaker*
2 February – 8 April
written by Brian Hayles
directed by Bill Sellars

The Celestial Toyroom	2 April
The Hall of Dolls	9 April
The Dancing Floor	16 April
The Final Test	23 April

Z *The Gunfighters*
28 March – 6 May
written by Donald Cotton
directed by Rex Tucker

A Holiday for the Doctor	30 April
Don't Shoot the Pianist	7 May
Johnny Ringo	14 May
The OK Corral	21 May

AA *The Savages* 27 April – 3 June
written by Ian Stuart Black
directed by Christopher Barry

Episode 1	28 May
Episode 2	4 June
Episode 3	11 June
Episode 4	18 June

BB *The War Machines*
22 May – 1 July
written by Ian Stuart Black, based on an idea by Kit Pedler
directed by Michael Ferguson

Episode 1	25 June
Episode 2	2 July
Episode 3	9 July
Episode 4	16 July

SEASON FOUR
Producer: Innes Lloyd
Story editor: Gerry Davis

CC *The Smugglers* 19 June – 29 July
written by Brian Hayles
directed by Julia Smith

Episode 1	10 September
Episode 2	17 September
Episode 3	24 September
Episode 4	1 October

DD *The Tenth Planet* 30 August – 8 October
written by Kit Pedler and Gerry Davis
directed by Derek Martinus

Episode 1	8 October
Episode 2	15 October
Episode 3	11 October
Episode 4	29 October

EE *The Power of the Daleks*
26 September – 26 November
written by David Whitaker (and Dennis Spooner, uncredited)
directed by Christopher Barry

Episode One	5 November
Episode Two	12 November
Episode Three	19 November
Episode Four	26 November
Episode Five	3 December
Episode SIx	10 December

FF *The Highlanders*
11 November – 24 December
written by Elwyn Jones and Gerry Davis
directed by Hugh David

Episode 1	17 December
Episode 2	24 December
Episode 3	31 December

1967

Installing Patrick Troughton as William Hartnell's replacement was part of an overall revitalisation. Following *The Highlanders*, in which Jamie Macrimmon (Frazer Hines) made his debut, the historical stories were dropped.

The programme took a little while to adapt to Troughton's puckish characterisation. *The Underwater Menace* adopted a melodramatic B movie style and continued the Second Doctor's short-lived penchant for disguises. *The Moonbase* also played like pulp science fiction, but a credible futuristic scenario (courtesy of Kit Pedler, the programme's scientific advisor) and even more frightening Cybermen set the tone for what was to follow.

Over-ambitious special effects impaired *The Macra Terror*, but giant crabs gave way to the rather more horrifying Chameleons in *The Faceless Ones.* Ben and Polly elected to stay on present-day Earth at the end of that story, which segued seamlessly into *The Evil of the Daleks.* Terry Nation was developing the Daleks' own series for US television, and David Whitaker's profound narrative was designed to depict the Doctor's final confrontation with his greatest enemy. As the Daleks and their Emperor seemingly wiped themselves out in a civil war on Skaro, the orphaned Victoria (Deborah Watling) joined the Doctor and Jamie as their new travelling companion.

The fourth recording block came to an end with the third Cybermen story in less than a year. With the Daleks out of the show, *The Tomb of the Cybermen* was ample proof that the silver giants were the perfect new adversary for the diminutive Second Doctor. It also showed that this Doctor's occasionally skittish behaviour concealed a dark, manipulative curiosity.

This was the era of monsters and the most prolific use of the 'base under siege' template. The fifth recording block took the crew on location to Wales, doubling as the Tibetan home of the robotic Yeti in *The Abominable Snowmen. The Ice Warriors* introduced yet more aggressive giants in a story that exemplified prevailing anxieties about encroaching technology.

The final serial to be broadcast that year was *The Enemy of the World*. In this latest dystopian future the role of the monster was played by Patrick Troughton as the callous dictator Salamander.

The Doctor had appeared to gain a split personality, but the series was now pursuing a very singular direction.

PAST TENSE

Top: Regular cast Michael Craze, Patrick Troughton and Anneke Wills on location in Frensham Ponds for *The Highlanders*, which brought the run of historical stories to an end. In the final episode they were joined by Frazer Hines as Jamie Macrimmon.

Above: Michael Elwyn as Lieutenant Algernon Ffinch in *The Highlanders*.

In the final episode of *The Highlanders*, broadcast in January 1967, the TARDIS dematerialises from 18th century Culloden. This departure marked a turning point for the series. Prompted in part by the new Doctor, Patrick Troughton, production of purely historical stories came to an end.

Frequent trips into Earth's history had been part of *Doctor Who*'s original format, with Sydney Newman particularly keen that such stories should fulfil his brief that the series try to educate its audience. However, over the next three years scripts for the historical stories would prove to be a dramatic cul-de-sac, fraught with difficulties that few writers could surmount.

Furthermore, they had proved unpopular with the audience. On 21 August 1964 story editor David Whitaker wrote to viewer Pauline Ammesley, addressing her concerns about recent episodes. "I am afraid I cannot agree with you that children are not interested in the historical subjects of *Doctor Who*. We find that there is very little difference in their letters and in our audience surveys, and perhaps it is that the past subjects do have some bearing on lessons or essays that the children are having to do."

His letter continued with a surprising admission: "This is fortunate because we only regard the historical stories as necessary make-weights between the science fiction ones."

Whitaker's comment about surveys was a reference to Audience Research Reports, which presented opinions from panels of viewers in summaries that were only intended for internal circulation at the BBC. In some cases this was just as well. The November 1965 report on *Temple of Secrets*, the first episode of *The Myth Makers*, expressed disappointment with the Trojan War setting. "There were a number of criticisms to the effect that episodes in which Dr Who was mixed up in ancient history were decidedly less entertaining than his excursions in the future and adventures on other planets."

The June 1966 report on *The OK Corral*, the final episode of *The Gunfighters*, represented a nadir in terms of viewers' opinions and the Reaction Index of their opinions – a score of just 30 remains the lowest in the programme's history. "The OK Corral story in general and this final episode in particular came under the same sort of critical barrage from viewers in the sample," began the report, which described the production as a "tenth rate Western" before pointing out that "viewers on the whole seemed pretty disgusted with a story that was not in the science fiction genre they associate with *Doctor Who*, and which was not in itself, in their opinion, convincing or exciting."

This report, and others in a similar vein, must have played a part in the decision taken by Innes Lloyd and Gerry Davis to scrap the historical stories. They were, in any case, proving extremely difficult to get right.

The show and its guest cast were so intrinsically English that depicting foreign cultures was

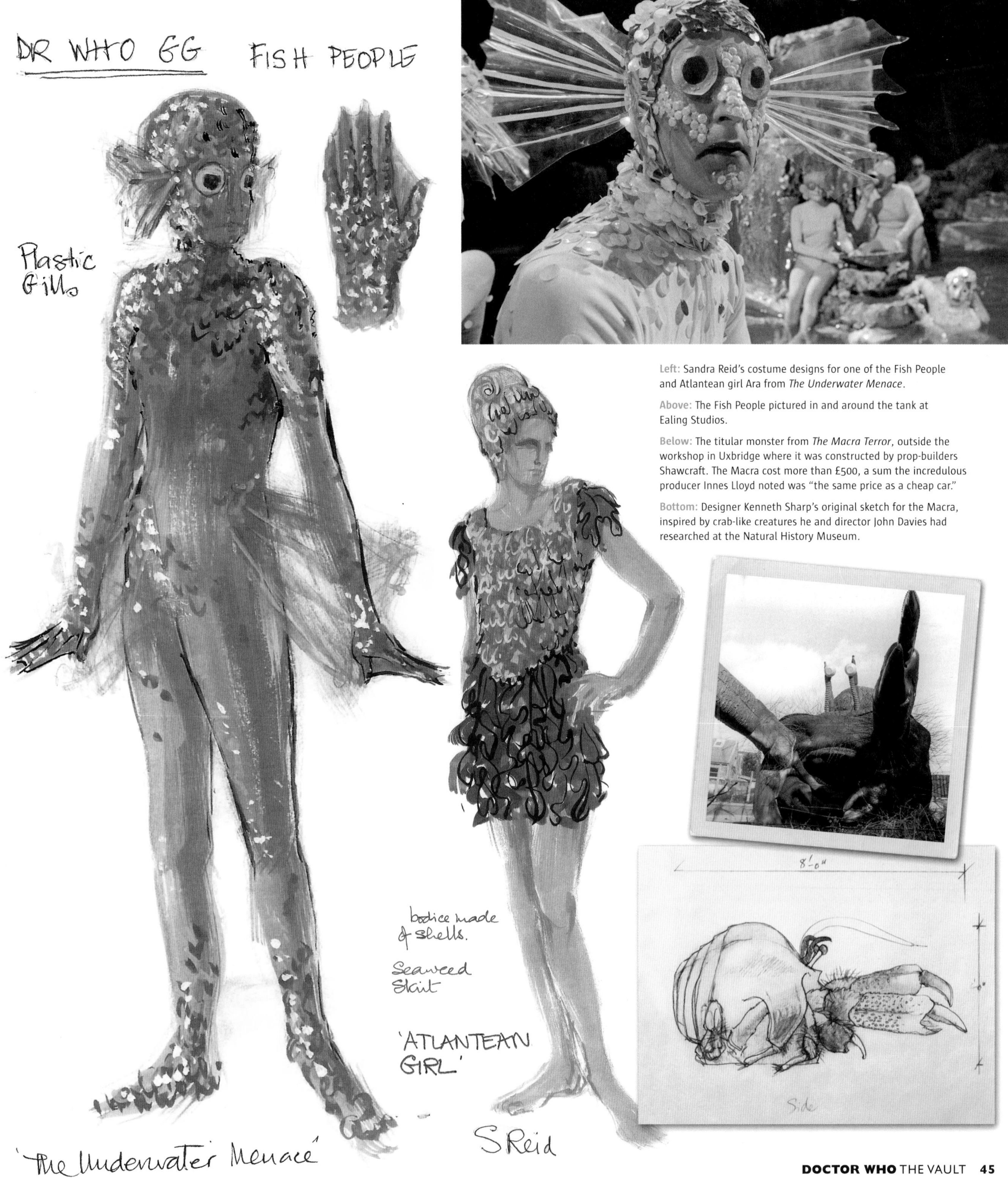

Left: Sandra Reid's costume designs for one of the Fish People and Atlantean girl Ara from *The Underwater Menace*.

Above: The Fish People pictured in and around the tank at Ealing Studios.

Below: The titular monster from *The Macra Terror*, outside the workshop in Uxbridge where it was constructed by prop-builders Shawcraft. The Macra cost more than £500, a sum the incredulous producer Innes Lloyd noted was "the same price as a cheap car."

Bottom: Designer Kenneth Sharp's original sketch for the Macra, inspired by crab-like creatures he and director John Davies had researched at the Natural History Museum.

Below: A Cyberman helmet, created for *The Moonbase* and reused in *The Tomb of the Cybermen*. This particular prop had a 'speaking' role and features a mouth flap that could be operated by the actor's chin. The circular arrangements of holes at either side of the helmet were added to aid the actor's breathing.

Below right: The 31 August 1967 edition of the *Radio Times* promoted *The Tomb of the Cybermen* with this striking front cover and an article that included an extract from the Doctor's diary. "My instincts tell me that terrible danger awaits us!" he wrote.

sometimes more challenging than creating alien planets. John Ringham's portrayal of Tlotoxl, the High Priest of Sacrifice in *The Aztecs*, owes an obvious debt to Shakespeare's Richard III but does little to compromise the programme's otherwise meticulous recreation of 15th century Mexico. Two years later, however, Laurence Payne's pronunciation of the line "Kinda like cheese in a mousetrap" makes him sound more like Noël Coward than Johnny Ringo in *The Gunfighters*.

Another problem was that so many of the most significant moments in history are characterised by the type of violence and bloodshed that would have been impossible to show at 5.15 on a Saturday evening. The camera averts its gaze from a human sacrifice in *The Aztecs*, but there was no escaping the implications of stories with titles such as *The Reign of Terror* and *The Massacre of St Bartholomew's Eve*. Divorced from fantasy settings, the acts of violence committed during the French Revolution and the persecution of the Huguenots must have seemed more disturbing than a blast from a Dalek.

Story editor Donald Tosh maintains that the scripts for such stories bore a special responsibility to viewers. "Historical accuracy was *hugely* important," he says. "You must not change the facts of history as far as they are reported because if you do you're in danger of misleading the next generation about what happened back then. That is unforgivable. The programme was still educational insofar as it was possible to establish historical accuracy."

Tosh's producer John Wiles was similarly fond of these trips into the past but conceded they were problematic. "The temptation with the historical ones was to write almost self-contained things,"

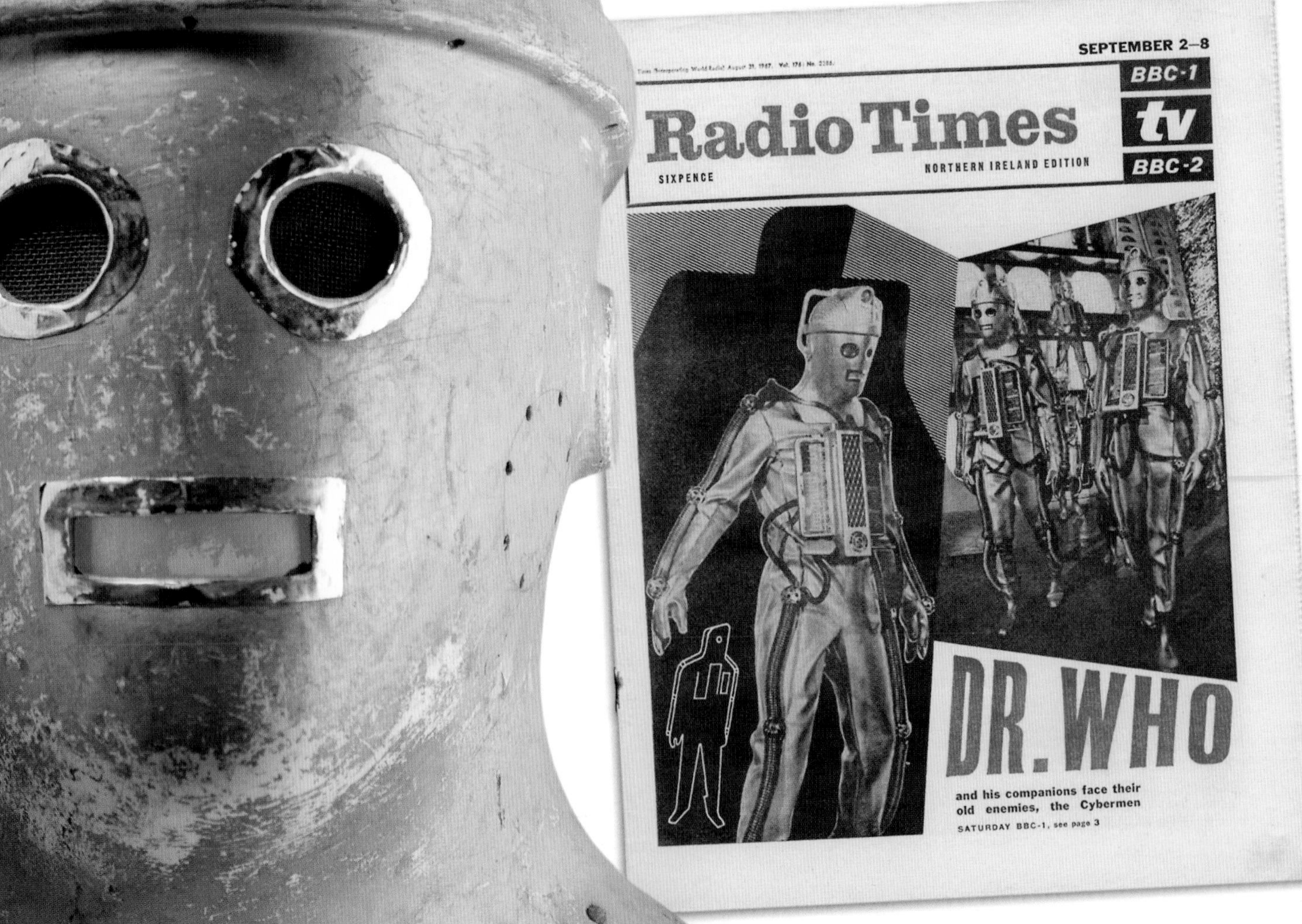

he said in 1985. "The Trojan War is a marvellous story. The difficulty was actually to fit Doctor Who into it."

The problems were recognised by Dennis Spooner, Tosh's predecessor as story editor, and he was the first to experiment with a solution by combining historical settings with science fiction scenarios. Spooner's *The Time Meddler* contained all the educational elements that viewers would have expected about the Viking and Norman invasions of England in 1066, but only as the backdrop to a more whimsical tale about a member of the Doctor's own race attempting to pervert the course of history for his own mischievous ends.

Things got noticeably more relaxed from here on. In *The Myth Makers* the Doctor breaks a once cardinal rule by influencing events, in this case suggesting the Greeks smuggle their soldiers into Troy inside a hollow wooden horse. Writer Donald Cotton was even more inventive with *The Gunfighters*, which staged the events leading up to the gunfight at the OK Corral as a comedy with musical interludes,

Above left: Unable to use an image of Patrick Troughton in their campaign for Sky Ray ice lollies, Wall's commissioned artist Patrick Williams to create a new likeness for the Doctor.

Above: A series of 36 *Doctor Who* picture cards could be collected with Sky Ray lollies and pasted into the above book.

Below: A counter display designed to hold order forms for *Dr Who's Space Adventure Book*.

Below left: A selection of the picture cards, illustrated by Patrick Williams.

Top left: The Emperor dominates the control room on Skaro in *The Evil of the Daleks*.

Top right: The cover of this year's *Doctor Who Annual* was by Walter Howarth and Ron Smethurst.

Above: Victoria Waterfield (Deborah Watling) and Ruth Maxtible (Brigit Forsyth) in *The Evil of the Daleks*, a story that combined scenes set in the present day and the 19th century with an epic science fiction plot.

Right: Sandra Reid's costume design for Ruth Maxtible, complete with fabric sample.

including a Greek chorus in the style of a saloon bar cabaret act.

The Smugglers, the First Doctor's final historical adventure, was set in 17th century Cornwall, but as story editor Gerry Davis later admitted this was more an effort to appropriate a Robert Louis Stevenson style of storytelling than to evoke a realistic picture of the era.

Culloden, the last pitched battle fought on British soil, was appropriately where *Doctor Who*'s historical adventures came to an end with *The Highlanders*. Dennis Spooner's template for pseudo-historical stories would prove more durable, and David Whitaker picked up the baton with *The Evil of the Daleks*. This 1967 story dovetailed such diverse locations as Gatwick Airport, Victorian England and the Daleks' home planet Skaro.

The 1960s ended with *The War Games*, which again combined the best of both worlds by applying a science fiction twist to its appropriation of soldiers from the First World War, the American Civil War and other conflicts.

Of the following decades, some of the series' most successful stories have followed the example set by *The Time Meddler* by placing incongruous characters and technology in historical settings. *The*

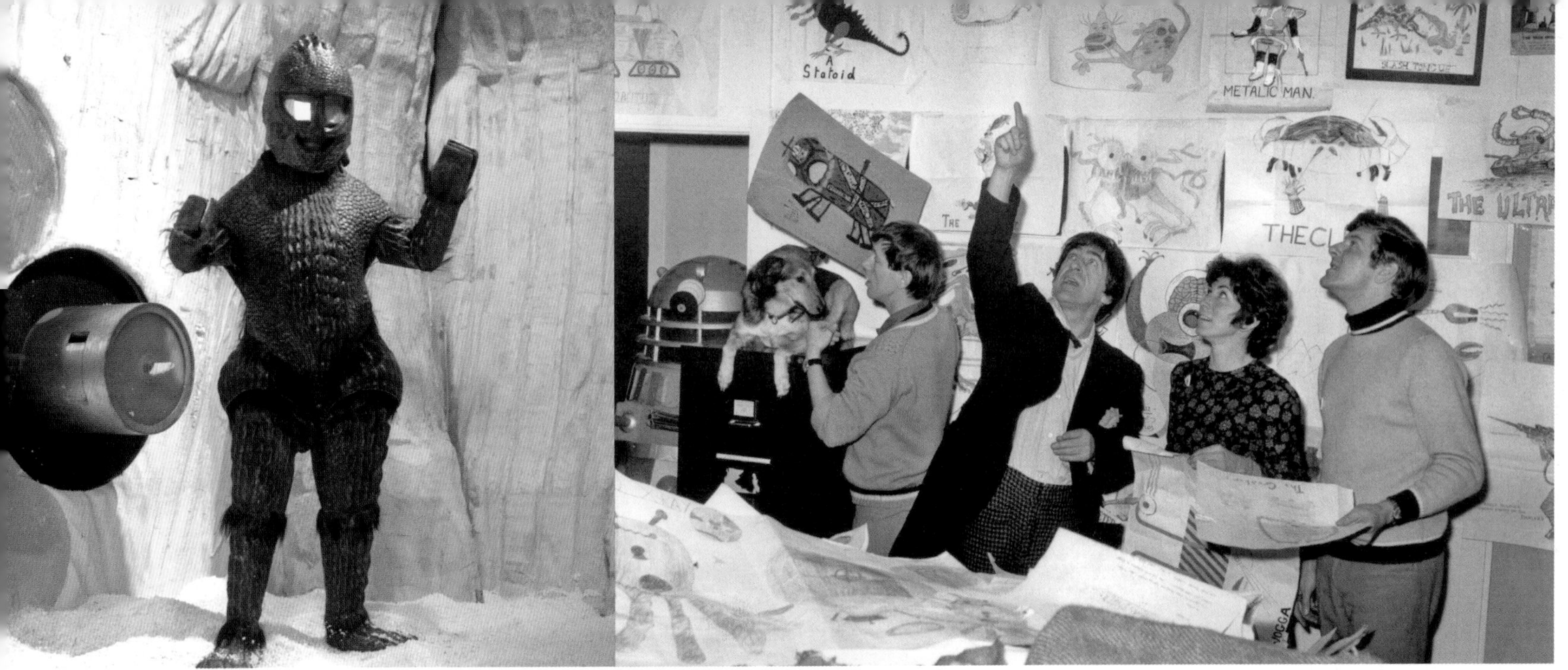

Time Warrior, *The Talons of Weng-Chiang* and *The Curse of Fenric* are just three of the most accomplished. When *Doctor Who* returned in 2005 a new type of celebrity-based pseudo-historical became a staple of the series, with the Doctor meeting Charles Dickens (*The Unquiet Dead*), Queen Victoria (*Tooth and Claw*), William Shakespeare (*The Shakespeare Code*), Agatha Christie (*The Unicorn and the Wasp*) and Vincent Van Gogh (*Vincent and the Doctor*). Both Winston Churchill and Adolf Hitler have also appeared.

After *The Highlanders* the programme's next purely historical story didn't occur until 1982. *Black Orchid* was a murder mystery set against a backdrop of 1925. It was met with indifference from viewers now conditioned to expect at least some element of science fiction in every story. To this day, 'straight' history has yet to repeat itself in *Doctor Who*.

Above left: Brian Hayles' script for *The Ice Warriors* described the creatures as cyborgs, but costume designer Martin Baugh realised them as armoured reptilians.

Above: *Blue Peter* received more than a quarter of a million entries to its 'Design a Monster' competition. In December Patrick Troughton joined John Noakes, Valerie Singleton and Peter Purves to help choose the winners.

Below left: A claw from an original Ice Warrior costume.

SEASON FOUR
(continued)
Producer: Innes Lloyd
Story editor: Gerry Davis
(unless otherwise stated)

FF *The Highlanders* (continued)

Episode 4	7 January

GG *The Underwater Menace*
12 December 1966 – 28 January
written by Geoffrey Orme
directed by Julia Smith

Episode 1	14 January
Episode 2	21 January
Episode 3	28 January
Episode 4	4 February

HH *The Moonbase*
17 January – 25 February
written by Kit Pedler
directed by Morris Barry

Episode 1	11 February
Episode 2	18 February
Episode 3	25 February
Episode 4	4 March

JJ *The Macra Terror*
15 February – 25 March
written by Ian Stuart Black
directed by John Davies

Episode 1	11 March
Episode 2	18 March
Episode 3	25 March
Episode 4	1 April

KK *The Faceless Ones* 10 March – 6 May
written by David Ellis and Malcolm Hulke
directed by Gerry Mill
Associate producer: Peter Bryant (up until third episode)

Episode 1	8 April
Episode 2	15 April
Episode 3	22 April
Episode 4	29 April
Episode 5	6 May
Episode 6	13 May

LL *The Evil of the Daleks*
20 April – 24 June
written by David Whitaker
directed by Derek Martinus (and Timothy Combe, Episode 7 Dalek fight film sequence)
Story editor: Peter Bryant (from Episode 4)

Episode 1	20 May
Episode 2	27 May
Episode 3	3 June
Episode 4	10 June
Episode 5	17 June
Episode 6	24 June
Episode 7	1 July

SEASON FIVE
Producer: Innes Lloyd
Story editor: Peter Bryant
(unless otherwise stated)

MM *The Tomb of the Cybermen*
12 June – 22 July
written by Kit Pedler and Gerry Davis
directed by Morris Barry
Producer: Peter Bryant
Story editor: Victor Pemberton

Episode 1	2 September
Episode 2	9 September
Episode 3	16 September
Episode 4	23 September

NN *The Abominable Snowmen*
23 August – 14 October
written by Mervyn Haisman and Henry Lincoln
directed by Gerald Blake

Episode One	30 September
Episode Two	7 October
Episode Three	14 October
Episode Four	21 October
Episode Five	28 October
Episode Six	4 November

OO *The Ice Warriors*
25 September – 25 November
written by Brian Hayles
directed by Derek Martinus

One	11 November
Two	18 November
Three	25 November
Four	2 December
Five	9 December
Six	16 December

PP *The Enemy of the World*
5 November – 6 January 1968
written by David Whitaker
directed by Barry Letts

Episode 1	23 December
Episode 2	30 December

1968

After a climactic confrontation between the Doctor and Salamander in *The Enemy of the World*, incoming producer Peter Bryant and story editor Derrick Sherwin pursued variations on the claustrophobic stories that had worked so well the previous year.

The Yeti returned in *The Web of Fear*, once more controlled by the Great Intelligence and now an incongruous presence in the London Underground. The introduction of the Doctor's great ally Colonel Lethbridge-Stewart (Nicholas Courtney) further distinguishes this fondly remembered story.

Fury from the Deep conjured more horror in confined spaces, this time inside an impeller shaft beneath a gas-drilling rig. The serial is noteworthy for its kitchen sink pretensions and disturbing scenes of possession, as well as our first sight of the Doctor's multi-purpose gadget, the Sonic Screwdriver.

With the traumatised Victoria deciding to bring her travels with the Doctor to an end, there was space in the TARDIS for 21st century astrophysicist Zoe Heriot (Wendy Padbury), who joined the crew at the end of *The Wheel in Space*. Though the 'base under siege' format was now exhausted, the Cybermen remained a formidable foe and there was a hidden highlight in the first use of the Doctor's pseudonym 'John Smith'.

Writers Mervyn Haisman and Henry Lincoln took their names off the next serial, *The Dominators*, following its abridgement by Derrick Sherwin. A further dispute, over the ownership of the robotic Quarks, later threatened this story's status as the opening adventure of the sixth broadcast season.

The fifth recording block closed on a happier note: *The Mind Robber* was the programme's latest, and arguably most successful, foray into surreal territory.

The Invasion was a sobering start to the sixth recording block, as Lethbridge-Stewart (now Brigadier) and UNIT (the United Nations Intelligence Taskforce) did battle with the Cybermen. This template for an Earthbound *Doctor Who* was the last serial to benefit from the prescient input of Kit Pedler.

The eerie sound design of Brian Hodgson and the BBC Radiophonic Workshop was a distraction from the underwhelming visuals in *The Krotons*. The script itself contained only hints that newcomer Robert Holmes would later be celebrated as one of *Doctor Who*'s greatest talents.

U.N.I.T

VIEW FINDERS

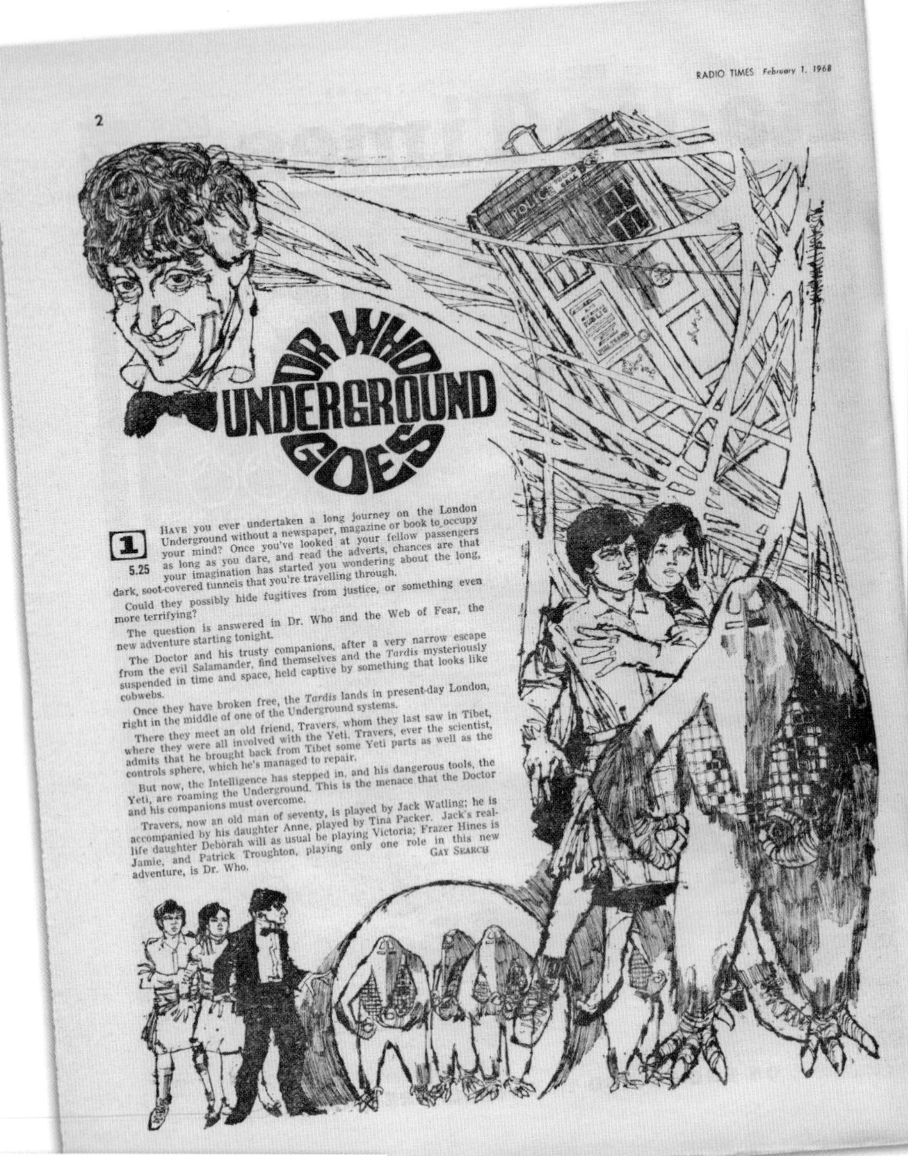

RADIO TIMES February 1, 1968

2

DR WHO GOES UNDERGROUND

1 5.25

Have you ever undertaken a long journey on the London Underground without a newspaper, magazine or book to occupy your mind? Once you've looked at your fellow passengers as long as you dare, and read the adverts, chances are that your imagination has started you wondering about the long, dark, soot-covered tunnels that you're travelling through.

Could they possibly hide fugitives from justice, or something even more terrifying?

The question is answered in Dr. Who and the Web of Fear, the new adventure starting tonight.

The Doctor and his trusty companions, after a very narrow escape from the evil Salamander, find themselves and the *Tardis* mysteriously suspended in time and space, held captive by something that looks like cobwebs.

Once they have broken free, the *Tardis* lands in present-day London, right in the middle of one of the Underground systems.

There they meet an old friend, Travers, whom they last saw in Tibet, where they were all involved with the Yeti. Travers, ever the scientist, admits that he brought back from Tibet some Yeti parts as well as the controls sphere, which he's managed to repair.

But now, the Intelligence has stepped in, and his dangerous tools, the Yeti, are roaming the Underground. This is the menace that the Doctor and his companions must overcome.

Travers, now an old man of seventy, is played by Jack Watling; he is accompanied by his daughter Anne, played by Tina Packer. Jack's real-life daughter Deborah will as usual be playing Victoria; Frazer Hines is Jamie, and Patrick Troughton, playing only one role in this new adventure, is Dr. Who.

Gay Search

Doctor Who's strict production turnaround has never allowed room for auteurs, but during the series' original run there were a handful of directors who left an indelible mark on the programme.

The role of the director is to interpret the script and bring the episode to the screen. The longest-serving was Christopher Barry, who co-directed the first Dalek serial and made 43 episodes in total between 1963 and 1979. David Maloney directed 45, including such landmarks as *The War Games* (1969), *Genesis of the Daleks* (1975) and *The Deadly Assassin* (1976). The most industrious of all, however, was Douglas Camfield, who directed 52 episodes of *Doctor Who* between 1964 and 1976.

Camfield created a signature style while stretching the show's meagre resources and dignifying sometimes indifferent material. He became the first *Doctor Who* director to achieve any kind of celebrity status, when his 1965 wedding to actress Sheila Dunn was reported by the *Daily Mail*. Camfield's colleague, designer Raymond Cusick, had even offered to lend him a few of his Daleks for the ceremony.

The stories of Camfield's working methods speak volumes about the way *Doctor Who* was made in its early years, and studying his episodes reveals how much could be achieved under difficult circumstances. *Day of Armageddon*, the second episode of *The Daleks' Master Plan*, was recorded at Studio 3 in Television Centre on

Opposite below: Douglas Camfield and his wife, actress Sheila Dunn.

Opposite left: In February the *Radio Times* previewed *The Web of Fear* with this article by Gay Search.

Opposite right: Glamorous visitors to the Schoolboys and Girls' Exhibition at Olympia, London, pose with an empty Yeti costume.

Right: A concept illustration of a Quark by costume designer Martin Baugh. It was hoped the robots from *The Dominators* would be as popular as the Daleks, but they never appeared in the series again.

Below: The Quarks were operated by children, and are pictured here during location filming at Gerrards Cross Quarry in April.

Bottom: The Servo Robot prop from Episode 1 of *The Wheel in Space* is now missing its legs.

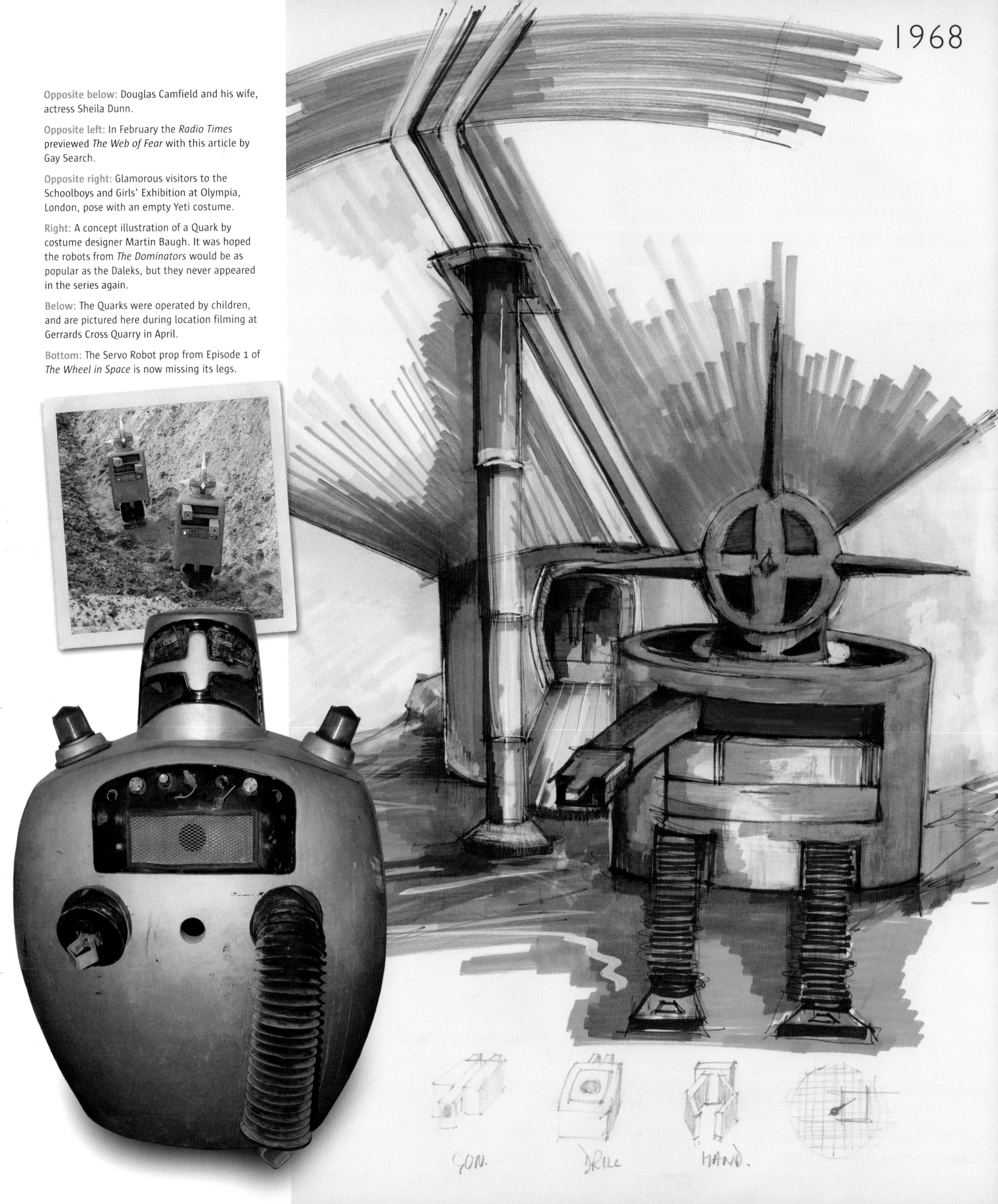

1968

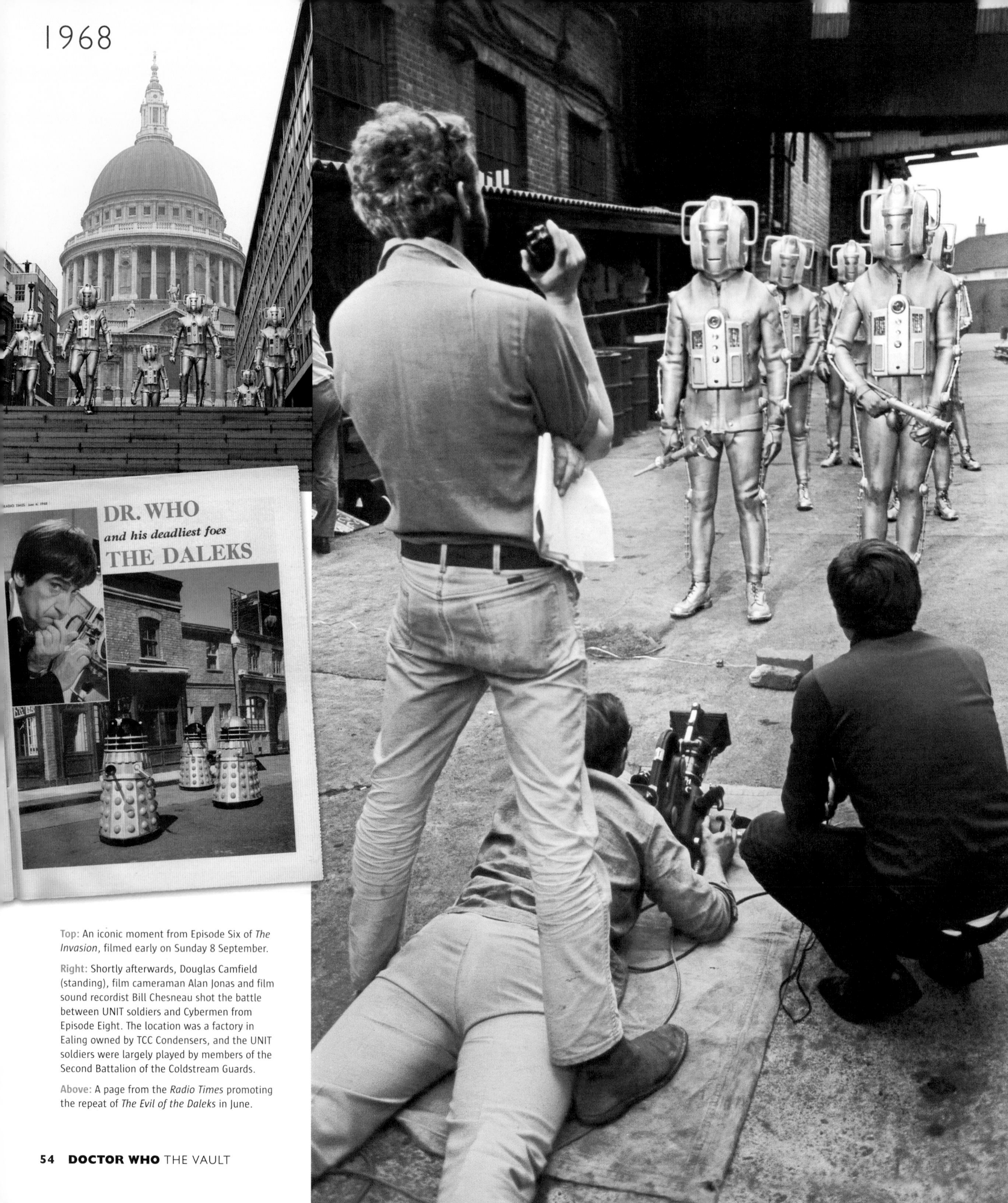

DR. WHO

and his deadliest foes

THE DALEKS

Top: An iconic moment from Episode Six of *The Invasion*, filmed early on Sunday 8 September.

Right: Shortly afterwards, Douglas Camfield (standing), film cameraman Alan Jonas and film sound recordist Bill Chesneau shot the battle between UNIT soldiers and Cybermen from Episode Eight. The location was a factory in Ealing owned by TCC Condensers, and the UNIT soldiers were largely played by members of the Second Battalion of the Coldstream Guards.

Above: A page from the *Radio Times* promoting the repeat of *The Evil of the Daleks* in June.

From: Douglas Camfield
Serials, Drama Group
BBC-tv

"DOCTOR WHO AND THE INVASION"

CHARACTER BREAKDOWN FOR CASTING

ATTACHED IS A BRIEF DESCRIPTION OF THE PARTS FOR WHICH I WILL WELCOME SERIOUS AND CONSIDERED SUGGESTIONS.

PLEASE SEND SUGGESTIONS TOGETHER WITH UP-TO-DATE PHOTOGRAPHS OR SPOTLIGHT NUMBERS FOR EASY ASSESSMENT. I WILL DO MY BEST TO RETURN ALL PHOTOGRAPHS.

BASIS OF STORY

Doctor Who and his travelling companions find themselves on earth about the year 1976 A.D. Tobias Vaughn, megalomanic electronics tycoon has allied himself with the Cybermen and is planning to take over the Earth as dictator.

The Doctor teams up once more with Army officer Lethbridge-Stuart from "THE WEB OF FEAR" story, who is now a Brigadier, and with the helf of new-girl, Isobel, and various military gentry takes on the terrible mission of saving the world from the threatened Cyber invasion.

29 October 1965. Camfield had just 75 minutes, from 8.30 to 9.45pm, to record the 25-minute episode. His camera script provided only a basic description of a scene set in the Dalek Conference Room: 'The delegates are beginning to gather. The Dalek Supreme is near his dais. A Dalek glides through the door and up to the Supreme.'

In practice, however, Camfield found time to realise Scene 39 in an impressively choreographed shot. The Dalek sweeps past various aliens, traversing almost the entire floor space before delivering its report. The Dalek Supreme then spins round to deliver his summary of the situation straight to the viewers. The studio's depth is fully exploited and the frame is filled with intriguing movement, although there are no cuts and the camera hardly shifts at all.

Camfield was similarly inventive in *Escape Switch*, the serial's tenth episode, where a shot of the shimmering sun dissolves to reveal a point of light reflected on the dome of a Dalek.

Such meticulous set-ups were only made possible through careful time management. Camfield had served in the army before joining the BBC and adopted a regimental attitude to recording sessions. Cusick remembered that making *The Daleks' Master Plan* felt at times like National Service. "I remember one afternoon in the studio he said, 'It's 22-and-a-half minutes past three, so we should be on shot 52.' He used to call me 'Major'. He said 'That would be your rank if this was a military operation.' He was Colonel Camfield."

This tongue-in-cheek approach endeared him to the crew, while his efficiency made him a popular choice with the production office. "Douglas was without a doubt the best at coping with our mini budgets," recalls the programme's then story editor Donald Tosh. "You would say to Dougie, 'I'm terribly sorry, but we've only got fourpence ha'penny in the till,' and he would say, 'That's OK, we'll have a go.' He planned it very well, despite the endless problems we had. In fact I've got no doubt that he would have made a very good producer of *Doctor Who*."

Camfield may not have been entirely serious about apportioning ranks to his colleagues, but his penchant for all things military made him a natural choice for the 1968 story *The Web of Fear*, which pitted Colonel Lethbridge-Stewart and his beleaguered soldiers against robotic Yeti. The sets designed by David Myerscough-Jones were so authentic that London Underground mistakenly accused the BBC of unauthorised shooting on their premises.

Later in 1968 Camfield returned to direct *The Invasion*, the story that promoted Lethbridge-Stewart to Brigadier and put him in command of the United Nations Intelligence Taskforce. An unprecedented amount of location filming saw Camfield in his element. He had proved himself an inventive director working with the multi-camera set-ups in television studios, but he preferred

Top left: Camfield briefs *Invasion* cast members Sally Faulkner, Frazer Hines, John Levene and Wendy Padbury.

Top: The first page of Camfield's casting notes for *The Invasion*.

Above: *The Web of Fear* designer David Myerscough-Jones created his convincing reconstructions of the London Underground at Ealing Film Studios.

Top: The cover and inside pages from this year's *Doctor Who Annual*. Walter Howarth's cover was the first to include one of the series' companions as well as the Doctor.

Above: Pages from *Attack of the Daleks* and *Pursued by the Trods*, *Doctor Who* strips written by Roger Noel Cook and illustrated by Patrick Williams for the 1968 *TV Comic Annual* (right).

taking a single camera on location. In his hands the results of these very different approaches could sometimes look indistinguishable.

Some of the most dynamic location material was suggested by Camfield's production assistant Chris D'Oyly-John. "I said to Dougie, 'The Cybermen should be placed against an instantly recognisable bastion of London.' He immediately bought it because it worked visually, and I suggested St Paul's. We filmed on a Sunday, with St Paul's in the background and a manhole cover in the foreground. The cover shot straight up in the air and out of it swarmed Cybermen."

These shots were quite unlike anything that had been seen in the programme before, with the camera fixed in low positions to film Cybermen rampaging across London. The final episode's battle sequence was similarly impressive, with an energy and pace rarely seen in low budget productions of the era.

Camfield wanted to "go out on a high", and after completing the 1976 story *The Seeds of Doom* he never directed another *Doctor Who*. His production assistant at that time was Graeme Harper. "Dougie's action sequences were great, but I think it's also worth pointing out that he loved actors. He knew that the artists in front of the camera were the people telling your story. And he knew how to excite those actors into doing what he dreamt of. In rehearsal he would tell actors what he

Right: Plans for the Krotons by costume designer Bobi Bartlett, submitted to producer Peter Bryant when Serial WW was still being described by its working title *The Space Trap*.

Inset: One of the finished costumes, pictured during location filming for *The Krotons* at a quarry in Malvern in November.

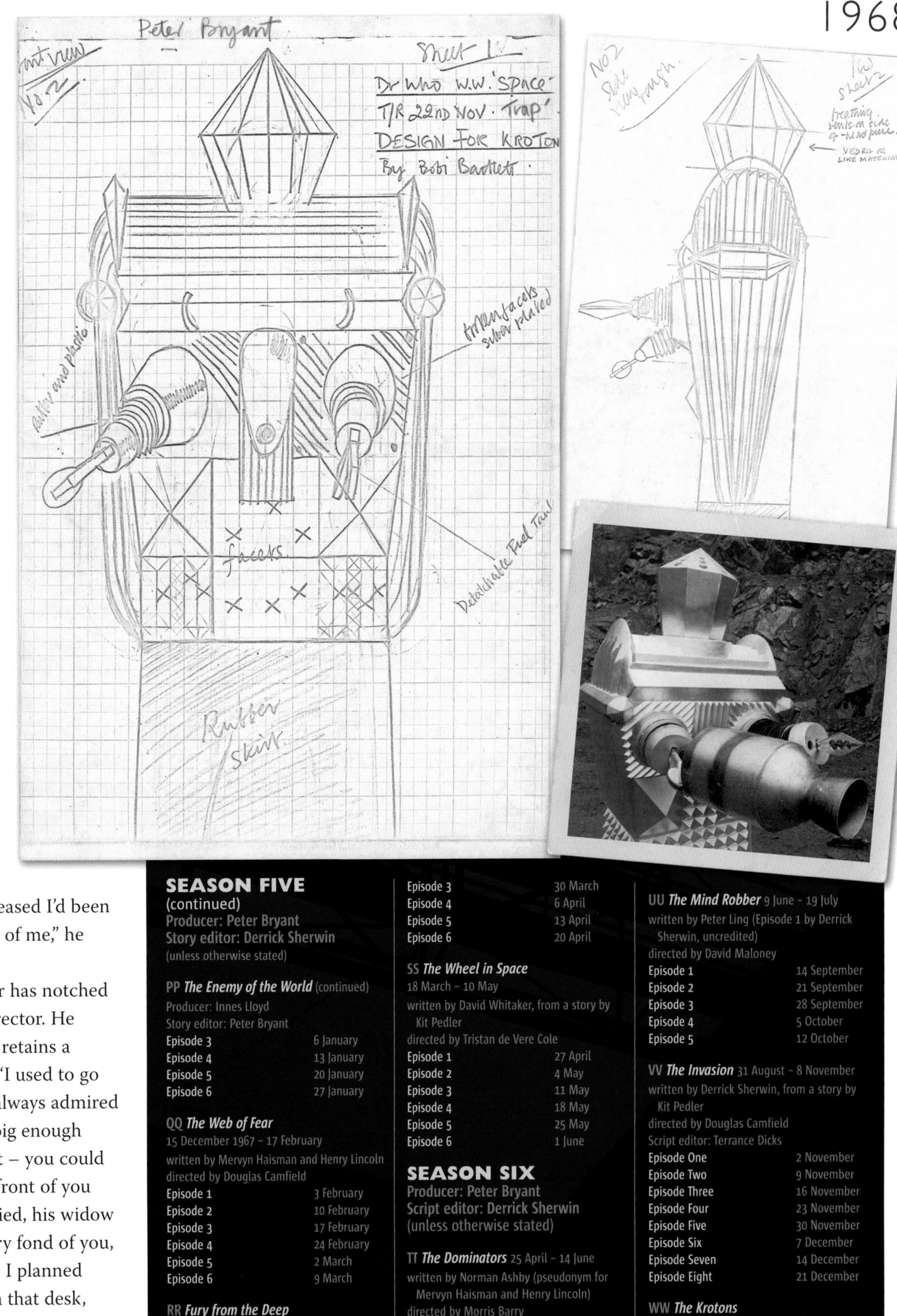

wanted to achieve, but he wouldn't tell them how to play it – he would allow them to play the characters the way they wanted, and if it wasn't quite right he would adjust that. He would never shoot anything until everyone was happy."

Dedicated rehearsal days and recording in multi-camera studios came to an end when *Doctor Who* finished its original run in 1989. The modern episodes are still made to a tight budget and a strict schedule but, in common with most television drama, the series is now exclusively recorded using the single-camera technique.

Camfield died of a heart condition in 1984, shortly after Harper graduated from the BBC directors' course. "When I became a director Dougie told me he was pleased I'd been given a break and that he was proud of me," he recalls.

Since *Doctor Who*'s return Harper has notched up more episodes than any other director. He cites Camfield as an inspiration and retains a treasured memento of his old boss. "I used to go to Dougie's house quite a lot, and I always admired his Victorian partner's desk. It was big enough for your studio plans and your script – you could unfold it all and have everything in front of you when you were planning. After he died, his widow Sheila rang me and said, 'He was very fond of you, and he wanted you to have his desk.' I planned most of my *Doctor Who* episodes on that desk, and I still use it today. When I'm working there, doing exactly what he used to do, I sometimes wonder how he would go about a particular scene. Because of that desk I can feel him in the room."

SEASON FIVE
(continued)
Producer: Peter Bryant
Story editor: Derrick Sherwin
(unless otherwise stated)

PP *The Enemy of the World* (continued)
Producer: Innes Lloyd
Story editor: Peter Bryant

Episode 3	6 January
Episode 4	13 January
Episode 5	20 January
Episode 6	27 January

QQ *The Web of Fear*
15 December 1967 – 17 February
written by Mervyn Haisman and Henry Lincoln
directed by Douglas Camfield

Episode 1	3 February
Episode 2	10 February
Episode 3	17 February
Episode 4	24 February
Episode 5	2 March
Episode 6	9 March

RR *Fury from the Deep*
4 February – 29 March
written by Victor Pemberton
directed by Hugh David

Episode 1	16 March
Episode 2	23 March
Episode 3	30 March
Episode 4	6 April
Episode 5	13 April
Episode 6	20 April

SS *The Wheel in Space*
18 March – 10 May
written by David Whitaker, from a story by Kit Pedler
directed by Tristan de Vere Cole

Episode 1	27 April
Episode 2	4 May
Episode 3	11 May
Episode 4	18 May
Episode 5	25 May
Episode 6	1 June

SEASON SIX
Producer: Peter Bryant
Script editor: Derrick Sherwin
(unless otherwise stated)

TT *The Dominators* 25 April – 14 June
written by Norman Ashby (pseudonym for Mervyn Haisman and Henry Lincoln)
directed by Morris Barry

Episode 1	10 August
Episode 2	17 August
Episode 3	24 August
Episode 4	31 August
Episode 5	7 September

UU *The Mind Robber* 9 June – 19 July
written by Peter Ling (Episode 1 by Derrick Sherwin, uncredited)
directed by David Maloney

Episode 1	14 September
Episode 2	21 September
Episode 3	28 September
Episode 4	5 October
Episode 5	12 October

VV *The Invasion* 31 August – 8 November
written by Derrick Sherwin, from a story by Kit Pedler
directed by Douglas Camfield
Script editor: Terrance Dicks

Episode One	2 November
Episode Two	9 November
Episode Three	16 November
Episode Four	23 November
Episode Five	30 November
Episode Six	7 December
Episode Seven	14 December
Episode Eight	21 December

WW *The Krotons*
10 November – 13 December
written by Robert Holmes
directed by David Maloney
Script editor: Terrance Dicks

Episode One	28 December

1969

Patrick Troughton had been dissatisfied with *Doctor Who*'s variable scripts and punishing schedule for some time. These feelings, and an instinctual fear of typecasting, prompted his decision to resign during his third year. His departure was announced in *The Times* and London's *Evening News* on 7 January.

The decision came as a disappointment to the production office. The BBC's Audience Research Department would report a favourable response to the current serial, *The Krotons*, and while viewing figures had not returned to the heights of 1964 and '65 the recent rotation of monsters had helped to rebuild the show's popularity.

That loyalty would falter in 1969. *The Seeds of Death* saw an attempt by the Ice Warriors to first contaminate and then repopulate the Earth via a hijacked teleport station on the moon. The six-part serial was broadcast early in the year, only after the uncredited Terrance Dicks rewrote the last four episodes.

Scripts were thin on the ground, and when another one fell through Robert Holmes enhanced his reputation with Dicks by coming up with *The Space Pirates* at short notice. One of the decade's less distinguished efforts, *The Space Pirates* nevertheless boasts some impressive model work and early glimpses of Holmes' skill for creating memorable characters with names such as Milo Clancey and Dom Issigri.

Derrick Sherwin was in the producer's chair when Season Six drew to a close. The abandonment of not one but two scripts left Dicks and Malcolm Hulke the unenviable task of filling ten episodes with their replacement story *The War Games*. The writers also assumed responsibility for returning the errant Doctor to his (still unnamed) home planet, as his fellow Time Lords prosecuted him for interfering in the affairs of other worlds.

The Doctor's impassioned defence included cameos from many of the evil creatures he had fought. As a consequence, the Time Lords' sentencing was relatively lenient – Jamie and Zoe were restored to their own times with their memories wiped, while the Doctor was subjected to a change of appearance and exile on 20th century Earth.

Jon Pertwee had already been contracted to take over as Troughton's replacement, but *The War Games* didn't conclude with a transformation from one actor to the next. The protesting Second Doctor simply receded into darkness as the black and white era of *Doctor Who* came to an end.

FELLOW TRAVELLERS

Above: Sonny Caldinez and Wendy Padbury during a break in recording *The Seeds of Death*. Caldinez was the series' longest-serving Ice Warrior, also playing Turoc in *The Ice Warriors*, Ssorg in *The Curse of Peladon* (1972) and Sskel in *The Monster of Peladon* (1974).

Top right: Frazer Hines signs autographs for fans.

The War Games, which came to an end in June 1969, drew a line under numerous aspects of *Doctor Who*. Not least of these was the time-travelling career of Jamie Macrimmon (Frazer Hines), whose lengthy occupancy of the TARDIS earns him the accolade of most prolific companion.

The Doctor's companions, or 'assistants' as they're sometimes called, have been an integral part of the programme since Sydney Newman wrote "Need a kid to get into trouble" in the margin of a March 1963 proposal document. The kid in question would become Susan (Carole Ann Ford), the Doctor's granddaughter, although once the show was underway it became clear that it was the Doctor, and not his companions, who would act as the chief protagonist.

Which is not to say that companions don't fill important supportive roles. Opinions vary over exactly what constitutes a companion. (Do you have to have travelled in the TARDIS? What's the minimum total of stories you need to have appeared in?) But there is a broad consensus that they number around 50. Most of them belong in at least one of a group of clearly defined categories.

Although Susan was the Doctor's granddaughter, her place in the Doctor's affections was soon supplanted by the orphaned Vicki (Maureen O'Brien). The *Radio Times* introduced the character to readers in December 1964, adopting the now familiar vernacular in the headline 'A New Companion for Dr Who'. Dodo (Jackie Lane), Victoria (Deborah Watling), Adric (Matthew Waterhouse), Nyssa (Sarah Sutton) and Ace (Sophie Aldred) were amongst subsequent companions who, to a lesser or greater degree, were also treated like surrogate grandchildren.

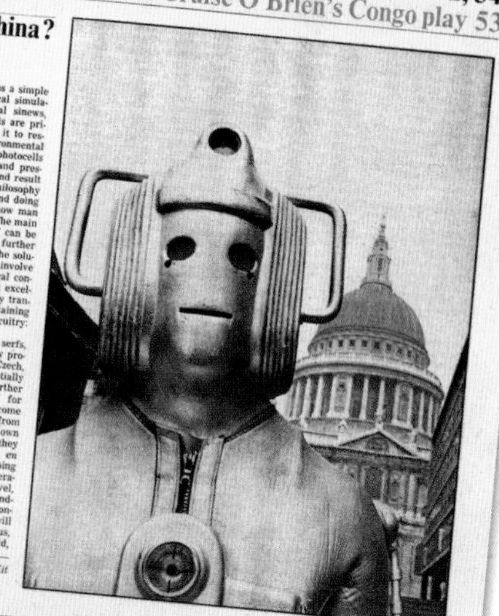

The Listener

Thursday 10 July 1969 Volume 82 No 2102 1s. 3d.

Robots, computers, people: Kit Pedler and David Paterson 33
Ian Gilmour: the case for private members' legislation 37
C.M.Woodhouse: conversations with the Greek Colonels 39
Alex Comfort on aging. Tom Driberg on Beaverbrook 42, 54
Christopher Ricks on Conor Cruise O'Brien's Congo play 53

Deus ex Machina? by Kit Pedler

A robot is commonly regarded as a simple machine—usually a morphological simulation of man—made from metal sinews, muscles and wires. Added to this are primitive sense organs which allow it to respond crudely to relevant environmental energy sources. Thus there are photocells for eyes, microphones for ears, and pressure transducers for touch. The end result of this rather charming design philosophy is a 'tin man' which clumps around doing nothing in particular except to show man how graceful he is in comparison. The main lines of development of 'tin men' can be fairly accurately predicted. Their further refinement is based essentially on the solution of technical problems and will involve no significant change in philosophical concept. Thus we may end up with an excellent functional homunculus, properly transistorised, microminiaturised, containing all the most advanced monolithic circuitry: a marvel of useless endeavour.

So let us forget about robots as serfs, which is the way they were originally proposed in Capek's *RUR* (*robotnik*, in Czech, means a serf). Such robots are essentially in the 'Golem' image and have no further interest except as ingenious dolls for grown-ups. They will certainly become more capable, and may even evolve from climbing stairs and seeking their own power requirements to a level where they are able successfully to cook *pigeon en cocotte*, or seek out the week's shopping requirements. They are of the first generation and can evolve only to a certain level, where they will still remain an understandable and wholly controllable machine, constituting no sort of a threat. They will remain self-evidently clumsy, ungracious, totally dependent, and above all stupid,

One of the Cybermen, designed by Kit Pedler for the BBC-1 'Dr Who' series

In the early years of the programme the Doctor acted as the senior figure at the head of the TARDIS family. Aside from Susan, his first travelling companions were inquisitive schoolteachers Ian Chesterton (William Russell) and Barbara Wright (Jacqueline Hill). They completed the family unit, playing the roles of protective parents who, through their respective specialisations in science and history, also helped to educate the audience during the TARDIS' trips into the future or the past.

Just as importantly, Ian and Barbara were the most readily identifiable members of the TARDIS crew. We were introduced to Susan and her ship through their eyes in the first episode, and it is through their experiences that we learned more about the enigmatic Doctor.

Ian fulfilled an additional role – that of the show's action hero. With an elderly character as the lead, any rough and tumble generally fell to the resourceful schoolteacher. When he and Barbara left in *The Chase*, this role was filled by space pilot Steven Taylor (Peter Purves) and subsequently by Able Seaman Ben Jackson (Michael Craze). As younger actors were cast as the Doctor the role of the male sidekick became largely redundant. Harry Sullivan (Ian Marter) could do little but get under the feet of the Fourth Doctor, and Mickey Smith (Noel Clarke) only

Top left: The model of the launch site and rocket from Episode Two of *The Seeds of Death*. The model sequences were shot at Ealing Studios on 13 December 1968.

Above: Kit Pedler, *Doctor Who*'s former scientific advisor, contributed the cover story to this issue of *The Listener* in July.

Centre: Slaar (Alan Bennion), the leader of the Ice Warriors in *The Seeds of Death*.

Below: One of the original claws from Slaar's costume.

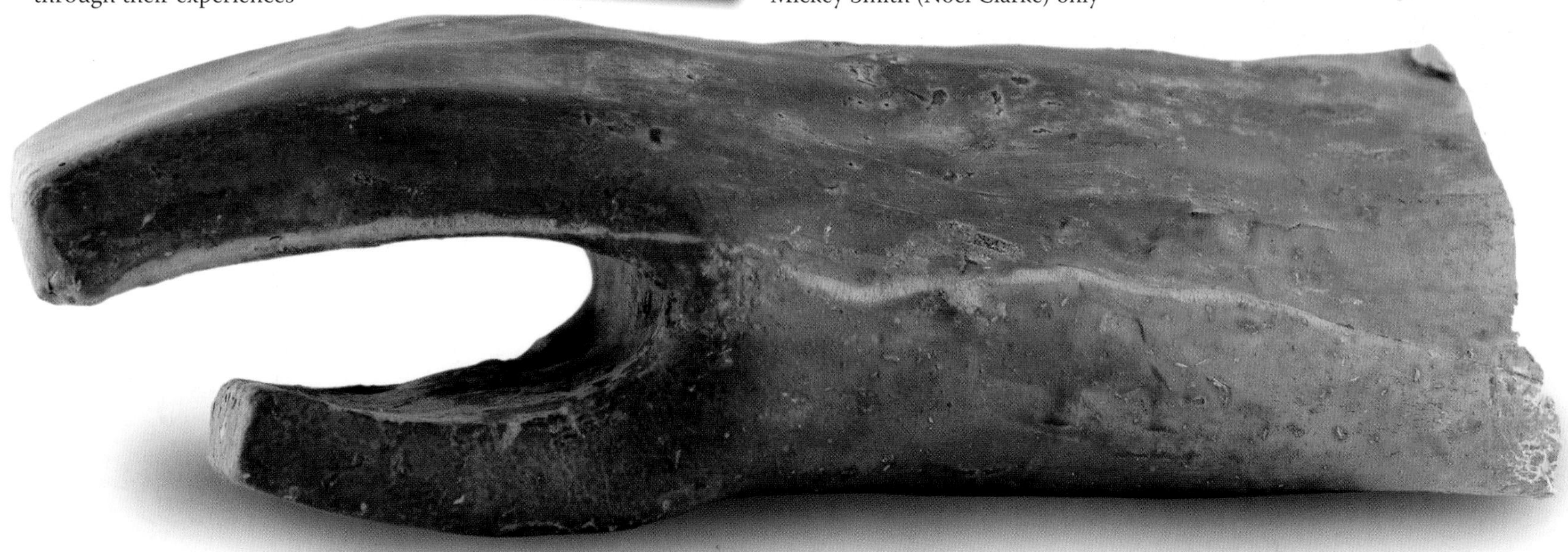

Below: Caven (Dudley Foster), the ruthless leader of *The Space Pirates*.

Bottom left: The original helmet worn by Foster in the story.

Bottom right: Costume designer Nick Bullen's sketch for Dom Issigri, the argonite miner kidnapped by Caven in *The Space Pirates*.

evolved into a more capable character as a result of his experiences with the Doctor.

The type of companions most commonly associated with the programme are vulnerable, screaming girls, but there have been very few characters who can be chiefly described in these terms. The first was dolly bird secretary Polly (Anneke Wills). Both she and Victoria became renowned for their piercing screams, some of which became an essential weapon against the seaweed monster in Victoria's final story *Fury from the Deep*.

Polly and Mel (Bonnie Langford), another of the show's screaming girls, belong to the category of companions who have lent continuity to the show across changes of lead actor. At the beginning of *The Power of the Daleks* Polly and Ben briefly took central roles while they, and the audience, decided whether the new Doctor was an interloper. Similarly Mel became a touchstone following the Sixth Doctor's abrupt regeneration into the Seventh in *Time and the Rani* (1987).

Perhaps the most enduring model is the feisty but occasionally vulnerable fellow explorer. The type of character exemplified by investigative journalist Sarah Jane Smith (Elisabeth Sladen) has reappeared in diverse guises – notably Leela (Louise Jameson), Romana (Mary Tamm and Lalla Ward), River Song (Alex Kingston) and Amy Pond (Karen Gillan).

None of these, however, would appear in as many episodes as Jamie Macrimmon. The

Top left: Jamie (Frazer Hines), the Doctor (Patrick Troughton) and Zoe (Wendy Padbury) arrive at Beacon Alpha 4 in Episode One of *The Space Pirates*.

Top right: Nick Bullen's *Space Pirates* designs included these sketches for Zoe and a Space Corps trooper.

Above: Major Ian Warne (Donald Gee) and General Hermack (Jack May) in *The Space Pirates*.

Left: One of Bullen's original sketches for Dom Issigri's daughter, Madeleine. The design was changed in consultation with actress Lisa Daniely, with the ultimate version far left.

Right: The cover of the *Doctor Who Annual* published in 1969, and a portrait of Zoe by David Brian from one of its strip stories, *Robot King*.

Below: The War Chief (Edward Brayshaw) in *The War Games*.

Bottom: Designer Roger Cheveley's sketch for the SIDRAT space-time machines used to abduct humans for *The War Games*.

Bottom right: Nick Bullen retrieved costumes of a Cyberman, an Ice Warrior and other adversaries for the Doctor's defence in Episode Ten of *The War Games*.

DR. WHO
ROBOT KING
WHEN DOCTOR WHO, JAMIE AND ZOE TRAVEL 5000 YEARS INTO THE FUTURE TO HOPE CITY, THEY FIND THE METROPOLIS DEVASTATED AND DESERTED–THAT IS DESERTED EXCEPT FOR ITS SINISTER GUARDIAN...

character's longevity can be partly explained by the fact that during the late 1960s new *Doctor Who* was on air for around 43 weeks of the year, but it's also true that Jamie crossed many of the most useful categories for a successful companion. His 18th century ignorance of futuristic technology made him a useful cipher for scriptwriters and a character that similarly baffled audience members could empathise with. On other occasions Jamie could play the roles of action hero, surrogate grandchild and fearless explorer with equal aplomb.

Frazer Hines' engaging portrayal contributed enormously to Jamie's success. Patrick Troughton was a notoriously publicity-shy Doctor, but in October 1968 Hines released the single 'Who's Dr Who?' and two months later joined co-star Wendy Padbury and some Cybermen for a signing at Selfridges. He even donated a pig called Whoey to Chessington Zoo.

Hines' long-term commitment to the soap opera *Emmerdale Farm* prevented Jamie from taking part in the tenth anniversary reunion

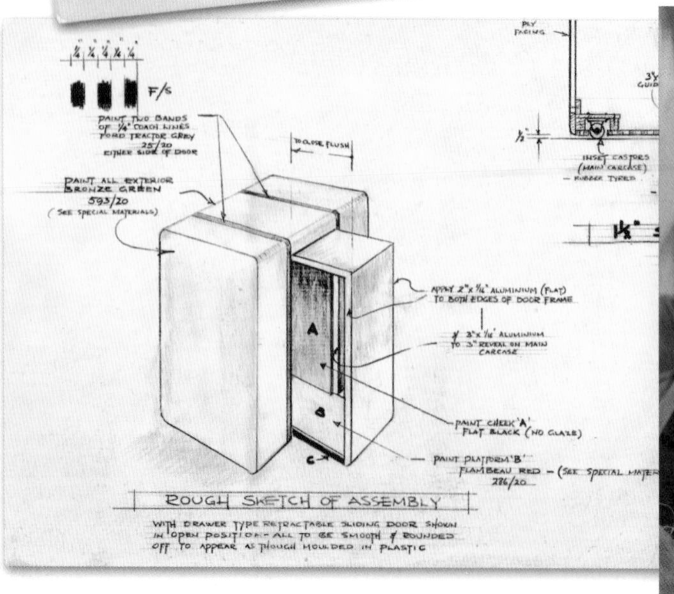

The Three Doctors (1972-73) but he made a cameo appearance in *The Five Doctors* (1983) and played an integral role in *The Two Doctors* (1985).

There has been a new accent on the programme's sidekicks since 2005, when *Doctor Who* resumed with *Rose*, a story – like the very first episode – named after the Doctor's companion. We are reacquanited with the Doctor and his TARDIS through shop girl Rose Tyler (Billie Piper), who initially fulfils the same dramatic function as Ian and Barbara. Rose's relationship with the Doctor soon develops into something more complex, however. In echoes of *Pygmalion,* the balance of power shifts between the two. In an inversion of the old formula, the Doctor ends up learning from his more emotionally literate companion.

Clara Oswald (Jenna Coleman) was introduced in the first of various personas in *Asylum of the Daleks* (2012) and represented another innovation for the show – the intrigue surrounding her true identity made her more of a protagonist than many of her predecessors. During 2013 the Doctor's obsession with discovering her true identity led to a different emphasis on the companion. Clara's role as a plot motor initiated a new era for *Doctor Who* companions, bringing a fresh – and yet essentially faithful – interpretation to a handwritten note made almost 50 years before.

Above: *TV Comic* bridged the gap between *The War Games* and *Spearhead from Space* with five stories depicting the exiled Doctor's adventures on Earth. The Time Lords finally catch up with him in *The Night Walkers*, written by Roger Noel Cook and illustrated by John Canning. On the last page, animated scarecrows drag the Doctor into the TARDIS so his change of appearance – the final part of his sentence – can take place.

SEASON SIX
(continued)
Producer: Peter Bryant
Script editor: Terrance Dicks
(unless otherwise stated)

WW *The Krotons* (continued)

Episode Two	4 January
Episode Three	11 January
Episode Four	18 January

XX *The Seeds of Death*
13 December 1968 – 7 February 1969
written by Brian Hayles
directed by Michael Ferguson

Episode One	25 January
Episode Two	1 February
Episode Three	8 February
Episode Four	15 February
Episode Five	22 February
Episode Six	1 March

YY *The Space Pirates* 7 February – 28 March
written by Robert Holmes
directed by Michael Hart
Script editor: Derrick Sherwin

Episode One	8 March
Episode Two	15 March
Episode Three	22 March
Episode Four	29 March
Episode Five	5 April
Episode Six	12 April

ZZ *The War Games*
23 March – 12 June
written by Terrance Dicks and Malcolm Hulke
directed by David Maloney
Producer: Derrick Sherwin

Episode One	19 April
Episode Two	26 April
Episode Three	3 May
Episode Four	10 May
Episode Five	17 May
Episode Six	24 May
Episode Seven	31 May
Episode Eight	7 June
Episode Nine	14 June
Episode Ten	21 June

1970

The seventh series of *Doctor Who* represented the most radical overhaul in the show's history.

The most obvious innovation was that the programme was now made in colour. For the scenes recorded on videotape, incoming producer Barry Letts embraced the technology of CSO (Colour Separation Overlay) – a forerunner of today's green-screen technique that allowed the imposition of new backgrounds behind actors and props.

Following the Doctor's exile to Earth by the Time Lords the programme was no longer advertised in the *Radio Times* as 'An adventure in space and time'. Jon Pertwee's Doctor travels to alien-related crime scenes in a vintage motor car called Bessie. Although still an anti-authoritarian figure, he was now attached to UNIT and accompanied on his investigations by the formidable Dr Liz Shaw (Caroline John).

The new series ran to just 25 episodes. The pressure on the lead actor was further alleviated by other regular characters that, aside from Liz Shaw, included UNIT's Brigadier Lethbridge-Stewart (Nicholas Courtney) and Sergeant Benton (John Levene), both last seen in *The Invasion*.

The opening story *Spearhead from Space* was shot entirely on 16mm film and pitted the Third Doctor against the Nestene Consciousness, a creature with the ability to create living plastic. The sequence where Auton mannequins burst through a department store window is justifiably regarded as iconic. The seven-part stories that followed possess an air of measured sophistication unique to this season. *Doctor Who and the Silurians* challenged the limitations of the show's earthbound setting with an alien threat whose presence actually predated that of mankind. *The Ambassadors of Death* was a triumph of style over content, its muddled story of benign visitors from Mars skilfully overshadowed by some impressive action sequences and a memorable score from composer Dudley Simpson.

The season peaked with the eerie *Inferno*, in which a reckless drilling operation threatens both this world and a fascistic parallel Earth. By the time it came to an end in June 1970 the positive reaction to the first two stories had already ensured that *Doctor Who* would be renewed for a further series. Liz Shaw would not be returning – there was only room for one know-it-all at UNIT HQ – but in every other respect the reinvented *Doctor Who* was ready for the 1970s.

SCENE SHIFTING

A photograph of Jon Pertwee made the front page of the *Daily Express* on Saturday 21 June 1969, the day the final episode of *The War Games* was broadcast. Reporter Martin Jackson described the actor as "A new, swinging Dr Who... no frock coat and baggy trousers – and above all, *sophisticated*. For the BBC has discovered that it is not so much the children who switch on to *Dr Who* – most of the eight million viewers are adults. So out go the child actors, the young space travellers. And in come two new partners – a middle-aged brigadier and slick-chick female scientist."

The programme's new format had been devised by Peter Bryant and Derrick Sherwin, but the responsibility for delivering their vision largely fell to Barry Letts, the producer who took over during Pertwee's second story, *Doctor Who and the Silurians*.

Letts began his career as an actor, with several notable roles in Ealing films before making his

Above: Jon Pertwee on location for *Spearhead from Space* at the TCC Condensers factory in Ealing, 19 September 1969.

Right: Barry Letts and Terrance Dicks in the *Doctor Who* production office, spring 1970.

Far right: Jon Pertwee's first photocall, on 17 June 1969. He told journalist Brian Dean that the next series would be "set on Earth in the 1980s. I won't be wearing the Victorian clothes the other Doctor Whos have used. I will be in a modern-day suit."

television debut in the BBC's 1950 production *The Gunpowder Plot*. The director was Rex Tucker and his fellow cast members included Patrick Troughton as Guy Fawkes.

Acting provided too unpredictable an income for a man with a family to support, so Letts became a scriptwriter as well. In 1966 he left acting behind to join the BBC staff as a director. One of his earliest assignments was the 1967 *Doctor Who* story *The Enemy of the World*.

In 1969 Shaun Sutton, the head of the BBC's Television Drama Group, asked him if he would consider producing *Doctor Who*. "At that time the viewing figures were so low that the line was in danger of going off the bottom of the graph," remembered Letts. Furthermore he had not spent long as a television director, and was reluctant to embark on yet another career so soon.

Appeased by the promise that he could still direct the occasional *Doctor Who* story, Letts decided to take the job. After the briefest of handovers with Bryant and Sherwin, his reign as *Doctor Who* producer officially began on 20 October 1969. While he settled into his new office he nervously reflected on the fact that his time on the producers/directors' course had included just one day on how to prepare a budget.

Top centre: Jon Pertwee and Caroline John unveil the Doctor's new car, Bessie, at a photocall on the Thames Embankment on 13 January 1970.

Top right: Caroline John models Liz Shaw's *Spearhead from Space* costume at Television Centre on 10 September 1969.

Right: Bessie in 2013, one of the exhibits at the Doctor Who Experience exhibition in Cardiff.

Right: Gerald Abouaf's original sketch and refined final head design for the Silurians, based on discussions with director Timothy Combe. The ears were inspired by the open gills of deep sea fish.

Below: Christine Rawlins, the costume designer for *Doctor Who and the Silurians*, supervises an early fitting. While the Visual Effects Department constructed the Silurians' heads, hands and feet, the rest of their bodies were made by the Costume Department.

Bottom left: A Silurian foot, one of the few items from the original costumes to have survived.

Bottom right: The finished Silurian mask was fitted with a 12-volt battery to illuminate the third eye, which was operated off camera by a visual effects assistant.

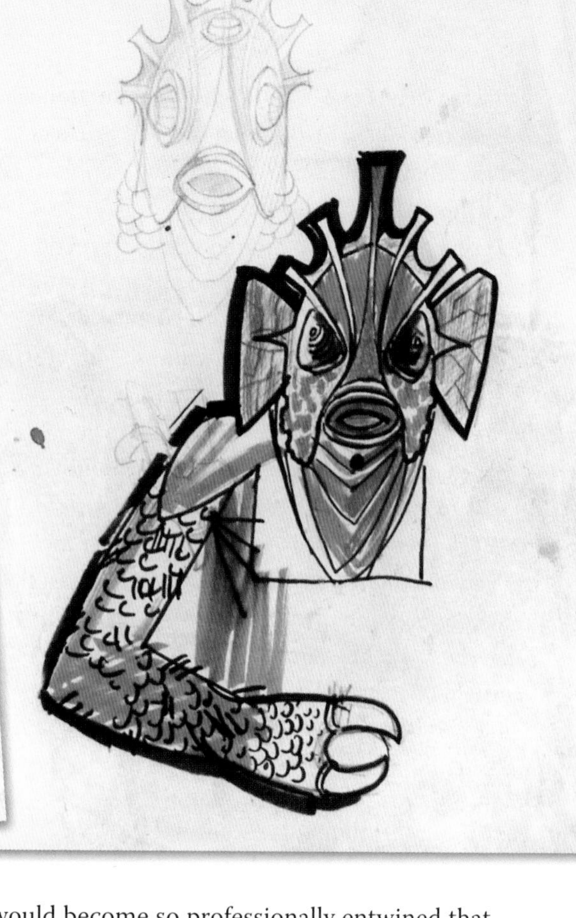

Spearhead from Space was already in the can when Letts arrived. The remaining three stories were victims of a budget squeeze and were consequently strung out to seven episodes apiece. *Doctor Who and the Silurians* was midway through recording when Letts assumed his duties. Malcolm Hulke, the story's author, was among those who commiserated with Letts over the fact he had inherited a format saddled with such geographical and narrative restrictions.

Letts was determined to make *Doctor Who* work, and he found a kindred spirit in the neighbouring office. Letts and script editor Terrance Dicks would become so professionally entwined that it's sometimes difficult to establish where the influence of one ends and the other begins. In his 2007 memoir *Who and Me*, Letts wrote: "We soon discovered that we looked at the script problem that we'd had dumped in our laps in very much the same way. But it was more than that. Simply put, each of us understood what the other was saying... Any reference I made to an idea, or a fact in any field, was instantly picked up. And vice versa. Once we started talking we couldn't stop. And we haven't stopped for the last 37 years."

Letts made his first change to a script while reviewing the final scene in *Doctor Who and the*

Silurians. He replaced the Doctor's anger at the Brigadier's murder of the creatures with a more compassionate response. "The Doctor, if he's to be true to himself, must be more humane than many a human," he later said.

During production of the next story, *The Ambassadors of Death*, another piece of the jigsaw fell into place when Letts adopted a suggestion by director Michael Ferguson that the programme should hire a stunt team called Havoc. They would become a mainstay of the show for the next two years.

When director Douglas Camfield was hospitalised during the studio recording of *Inferno*, Letts stepped in to handle the remaining episodes. Considering it all part of the job, he left Camfield's name on the credits.

It had been a difficult recording block, but Letts had already taken steps to make things easier. "Instead of doing one episode a week, we did two a fortnight," he recalled. "That may sound the same, but it isn't. The cast would rehearse two episodes concurrently for the whole of the first week and most of the second week, having two days' camera rehearsal in the studio, with a two-and-a-half hour recording on the second day."

This approach brought dramatic advantages. "The sets only had to be put up and struck once in two weeks instead of twice. The cast had time to learn their lines and let their characters grow. Everybody had time to think."

Letts also increased the efficiency of studio time by shooting out-of-sequence within stories,

Below left: Two finished Silurian masks, alongside the batteries used to illuminate their third eyes.

Bottom left (left to right): The head of the Tyrannosaurus Rex from *Doctor Who and the Silurians* was sculpted by Anna Braybrooke; Gerald Abouaf's initial sculpt for the Silurian mask; Abouaf removes a mask from the mould, which was taken from his finished sculpt.

Below: The Tyrannosaurus Rex costume was built by the Visual Effects Department. Bertram Caldicott, the department's store man, was drafted in to play the creature.

The visual effects team plus monster. James Ward is in front of the rocket, with Anna Braybrooke on his right, and Peter Day on his left

'Keeping a fatherly eye on what's going on' – Producer Barry Letts (left) with secretary Sandra Brenholz and script editor Terrance Dicks

the face appear irradiated when the special electronic colour overlay process was used,' Marion explained: 'On some parts of his face we used tissue and wet latex. Only the human eyes remained recognisable.'

It's all a bit uncomfortable for the actor, but Marion says she tried to leave it to the last minute before he was due to go on – 'so that he wouldn't have to suffer too long.' Removing it is less complicated: 'It all comes off very quickly, with a special solvent that we use.'

The costume supervisor, Christine Rawlins, works closely with the visual effects department. Designing costumes for the alien astronauts was a relatively straightforward job.

'Because the story was set late in the 70s, I designed a simplified version of the sort of spacesuit we know today,' she explained.

Silurian costumes required more imagination: Christine was responsible for the bodies, heavy scaled rubber skins. She did all she could to minimise discomfort. But despite her efforts, veteran monster Pat Gorman did suffer. 'You lose weight pretty fast in some of these costumes,' he said. 'They are pretty wet and sticky, but you do get used to it eventually ... All I wanted afterwards was a long soak in the bath!'

Pat has played a Yeti and a Cyberman as well as the Silurian. For the Cyberman he wore a diver's wet suit, painted silver, which stopped air getting in – or out. 'It was a bit claustrophobic because the glass-fibre head had to be screwed on from the back, so if there was no one around to help me out, I was stuck in there!' Undaunted by all these parts, Pat has another monstrous part in *Dr Who* lined up.

The rest of the monsters, and other peculiarities required for *Dr Who*, are the creations of the Visual Effects team of designers. Rockets, computers, explosions, Silurian heads, space helmets – all these and other oddities pour out of their workshop at the back of Television Centre. Tools of their trade range from pots of glue to buckets of dry ice. Says boss Jack Kine: 'Our motto is "anything that can be imagined can be made" – it has to be! This place may look like a schoolboy's dream but it can be very hairy and hysterical.'

James Ward designed the Silurians' heads, and it was the only girl in the team, Anna Braybrooke, who worked on the Silurian monster itself. Peter Day was responsible for the rocket in the

Don Houghton ... he wrote *Inferno*, the story beginning this week

RADIO TIMES DATED 7 MAY 1970

PAGE 53

DR WHO'S WHO'S WHO

Saturday
5.15 BBC1 Colour

Dr Who wouldn't be so remarkable without the enemies, and the enemies wouldn't be so spectacular without the people who stitch up and dab on the horror. On your left, a handful of Silurians, the Silurian Monster, 40 feet high in his stocking soles, a couple of deceptive astronauts, and a Cyberman, with Dr Who – intrepid and gallant – floating in adversity. On your right, the make-up and costume team at work on the astronauts who featured in the last series, 'The Ambassadors of Death.'

Stripped of their anonymous space-age uniform and their helmets, the alien astronauts are not a pretty sight, thanks to Marion Richards, make-up supervisor on *Dr Who*. Creating the alien within the space suit meant building up the actor's face 'with very fine latex rubber, with blue make-up foundation underneath, to make

Dr Who make-up supervisor Marion Richards (centre) at work on an alien astronaut with Christine Rawlins, costume supervisor.

Radio Times
Favourite No.4
Jon Pertwee-Dr Who

RADIO TIMES DATED 11 JUNE 1970

Top: The 7 May edition of the *Radio Times* went behind the scenes on *Doctor Who* with Christine Rawlins, make-up artist Marion Richards, *Inferno* writer Don Houghton and other members of the production team.

Above: The following month Jon Pertwee was the *Radio Times*' colour 'Favourite No.4'.

Right: One of *The Ambassadors of Death* recreates a menacing pose for the *Radio Times*.

clustering disconnected scenes with particular set or make-up requirements. These were the beginnings of the working practices that are still employed by the programme today.

Letts and Dicks could now implement the most consistent approach to the series since the days of Verity Lambert and David Whitaker. With Letts also contributing as a writer and director it was perhaps inevitable that some of his personal philosophies would rub off.

"Talking of moral passion might sound a bit pompous, but being aware of it also makes for good storytelling," he said. "When people used to come up with a story, or Terrance and I thought of a story, and we couldn't quite see where we were going with it we would say, 'Let's go back to the basics and ask ourselves, what is the story about? What point is the story making?' If it's just an adventure chase-about then it's very difficult to

Above: The endpapers and front cover of the *Doctor Who Annual* published in September. The cover artwork was by Ron Smethurst.

Right: Berry Wiggins & Co – an oil refinery and bitumen manufacturer in Werburgh, Kent – was the filming location for *Inferno* in March and April. Director Douglas Camfield can be seen far right.

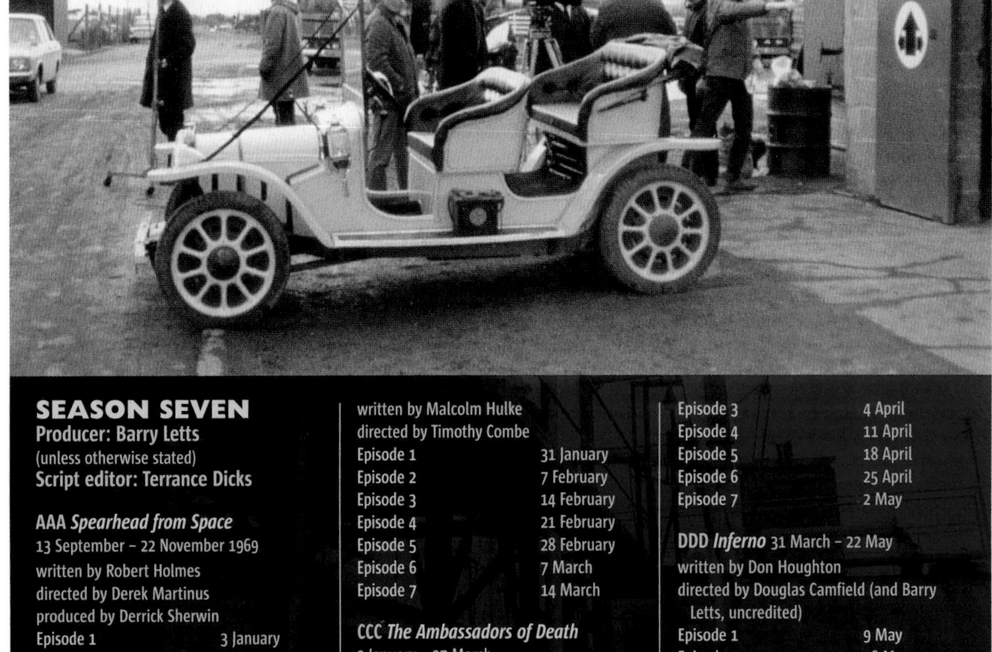

make a good story because all you're doing is just inventing new incidents. On the other hand, if you go back to brass tacks and say to yourself, 'The point of this story is, for instance, just because a chap has green skin doesn't mean he should be treated as an inferior' then immediately things start to fall into place, so that if an incident arises within the plot you can ask, 'Is this leading the story in that direction?' It is an enormous help in the structuring of stories to have a point or a theme to the whole thing."

Letts implemented this thematic approach, and other elements of his grand design for *Doctor Who*, over the summer of 1970. He created a solid foundation for the future and left a lasting impression on friends and colleagues. "It's no exaggeration to say that his arrival saved my job," says Terrance Dicks, "and probably saved the show as well."

SEASON SEVEN
Producer: Barry Letts
(unless otherwise stated)
Script editor: Terrance Dicks

AAA *Spearhead from Space*
13 September – 22 November 1969
written by Robert Holmes
directed by Derek Martinus
produced by Derrick Sherwin

Episode 1	3 January
Episode 2	10 January
Episode 3	17 January
Episode 4	24 January

BBB *Doctor Who and the Silurians*
12 November 1969 – 26 January 1970
written by Malcolm Hulke
directed by Timothy Combe

Episode 1	31 January
Episode 2	7 February
Episode 3	14 February
Episode 4	21 February
Episode 5	28 February
Episode 6	7 March
Episode 7	14 March

CCC *The Ambassadors of Death*
3 January – 27 March
written by David Whitaker (with Trevor Ray and Malcolm Hulke, uncredited)
directed by Michael Ferguson

Episode 1	21 March
Episode 2	28 March
Episode 3	4 April
Episode 4	11 April
Episode 5	18 April
Episode 6	25 April
Episode 7	2 May

DDD *Inferno* 31 March – 22 May
written by Don Houghton
directed by Douglas Camfield (and Barry Letts, uncredited)

Episode 1	9 May
Episode 2	16 May
Episode 3	23 May
Episode 4	30 May
Episode 5	6 June
Episode 6	13 June
Episode 7	20 June

1971

The *Radio Times* listing for Episode One of *Terror of the Autons* promised the arrival of the Doctor's "most deadly enemy". The Master, played by Roger Delgado, was devised by Barry Letts and Terrance Dicks as a devious visiting Time Lord who would help to rationalise the now frequent invasions of southern England.

In response to this threat, the Doctor could now call upon even more members of UNIT. New assistant Jo Grant (Katy Manning) brought a vulnerable femininity which better suited Pertwee's paternal portrayal of the Doctor. Joining the Brigadier and Sergeant Benton was Captain Mike Yates (Richard Franklin), introduced as a potential love interest for Jo.

Terror of the Autons proved controversial with its depiction of sinister plastic policemen alongside murderous facsimiles of household objects such as an armchair, artificial daffodils and a telephone flex. The programme's new realism took it to a prison for *The Mind of Evil* and a nuclear power station in *The Claws of Axos*. Letts and Dicks had reluctantly inherited the show's current format from their predecessors, and were already thinking of ways to fully reinstate the Doctor's TARDIS. For *Colony in Space* the Time Lords intervened, sending the Doctor on a mission to the planet Uxarieus. Ironically, it would prove less eventful than most of the show's earthbound adventures.

The season would be dominated by Roger Delgado, whose understated portrayal of the Master is the benchmark for all subsequent interpretations of the role. Letts and Dicks were so enamoured of their new creation that he was written into every story this year. While never less than a serious threat, the Master's efforts to torment the Doctor were soon shown to be a battle of wits between two old sparring partners.

The show's prevailing tenets and family atmosphere are encapsulated in the closing moments of *The Daemons*. The local church is in ruins, any apparently supernatural activity has been dismissed with nominally scientific explanations and the Master is finally captured by UNIT. The Brigadier and Yates head to the pub for a pint, leaving the Doctor and Jo in a celebratory dance around the village maypole.

As Season Eight came to an end the debonair Jon Pertwee was clearly leading one of television's favourite repertory companies.

MASTER PLAN

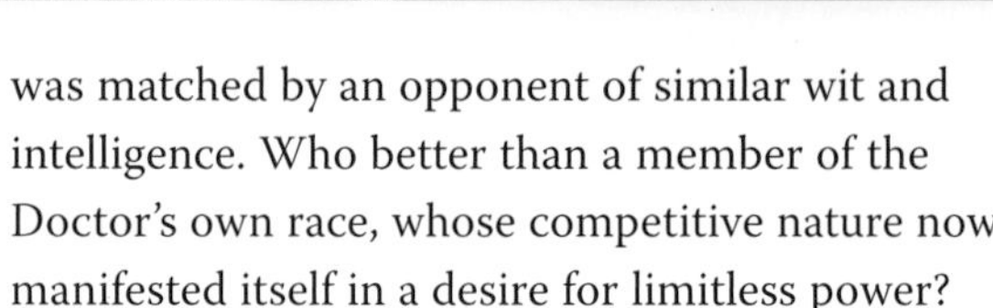

NORTH 2-8 January One Shilling/5p
BBC Radio Leeds

Cliff's new series
Saturday BBC1, see p 7

You and your doctor
Tuesday, Wednesday, Friday BBC1, see p 3

Words and Music
BBC Radio Leeds
See pp 5, 68

Radio Times

Gielgud and Richardson
Together in Hassan
Wednesday BBC2, see p 13

Killing statistics
The facts behind smoking
Tuesday Radio 4, see p 19

Come Dancing again
Monday BBC1

GREAT NEW DOCTOR WHO ADVENTURE! SATURDAY BBC1 COLOUR

DOCTOR WHO in The Terror of The Autons

Above: The *Radio Times* announces the arrival of the Master. "To me, *Doctor Who* was like a strip story on television," says the magazine's art editor David Driver. He would return to this motif in future issues.

Top right: Katy Manning and Jon Pertwee pictured during the location shoot for *Terror of the Autons* at the Roberts Brothers Circus on 18 September 1970.

The first episode of *Terror of the Autons*, the opening story in Season Eight, begins with a materialising TARDIS – an unfamiliar sound since the Doctor's exile to Earth. Its occupant calmly introduces himself to his first victim: "I am usually referred to as the Master... universally."

The Master was devised by Terrance Dicks and Barry Letts in summer 1970, prompted by Dicks' observation that it was about time the Doctor was matched by an opponent of similar wit and intelligence. Who better than a member of the Doctor's own race, whose competitive nature now manifested itself in a desire for limitless power?

It's easy to see why Dicks and Letts often referred to the Master as Moriarty to the Doctor's Holmes. In his capacity as UNIT's scientific advisor the Third Doctor now resembled an extra-terrestrial version of Conan Doyle's consulting detective, with Jo cast as his scatterbrained Watson and the Brigadier as the obstinate Lestrade. The Master was an intergalactic 'Napoleon of crime', but for several years would torment the Doctor by largely confining his activities to Earth.

To play the Master, Barry Letts cast Roger Delgado, an old friend from his acting days. "He had the enormous capacity for villainy – and charm – that the part of the Master demanded."

As befitted his scholarly title, the Master was a brilliant scientist, as well as being an expert in hypnotism and disguise. Delgado imbued the character with an air of understated menace at odds with the moustache-twirling melodrama of Dudley Simpson's four-note theme. "Roger understood that evil is best shown not by histrionics, but by being calm," remembered Pertwee. "He had these very controlled movements as the Master. He knew that just the act of standing still could be sinister."

Pertwee was surprised to see the Master dressed in a modern suit – very much the style

he had wanted for his Doctor before he was overruled by Letts' predecessors. The Master's most identifiable look was inspired by cinematic conceptions of urbane villainy – his Nehru jacket and black leather gloves closely resembled the uniform of Bond characters Dr No and Ernst Blofeld.

Top left: In October 1970 location filming for *The Mind of Evil* took place at RAF Manston. This picture shows the Bloodhound missile that doubled for the Thunderbolt in the story.

Above: Costume designer Ken Trew's original sketch for the Master. At this early stage Trew was referring to *Terror of the Autons* by its working title, *The Spray of Death*.

Left: In 1971 Kellogg's Sugar Smacks offered 'The timeless energy of Dr. Who' at breakfast time, and one of six different badges in every packet.

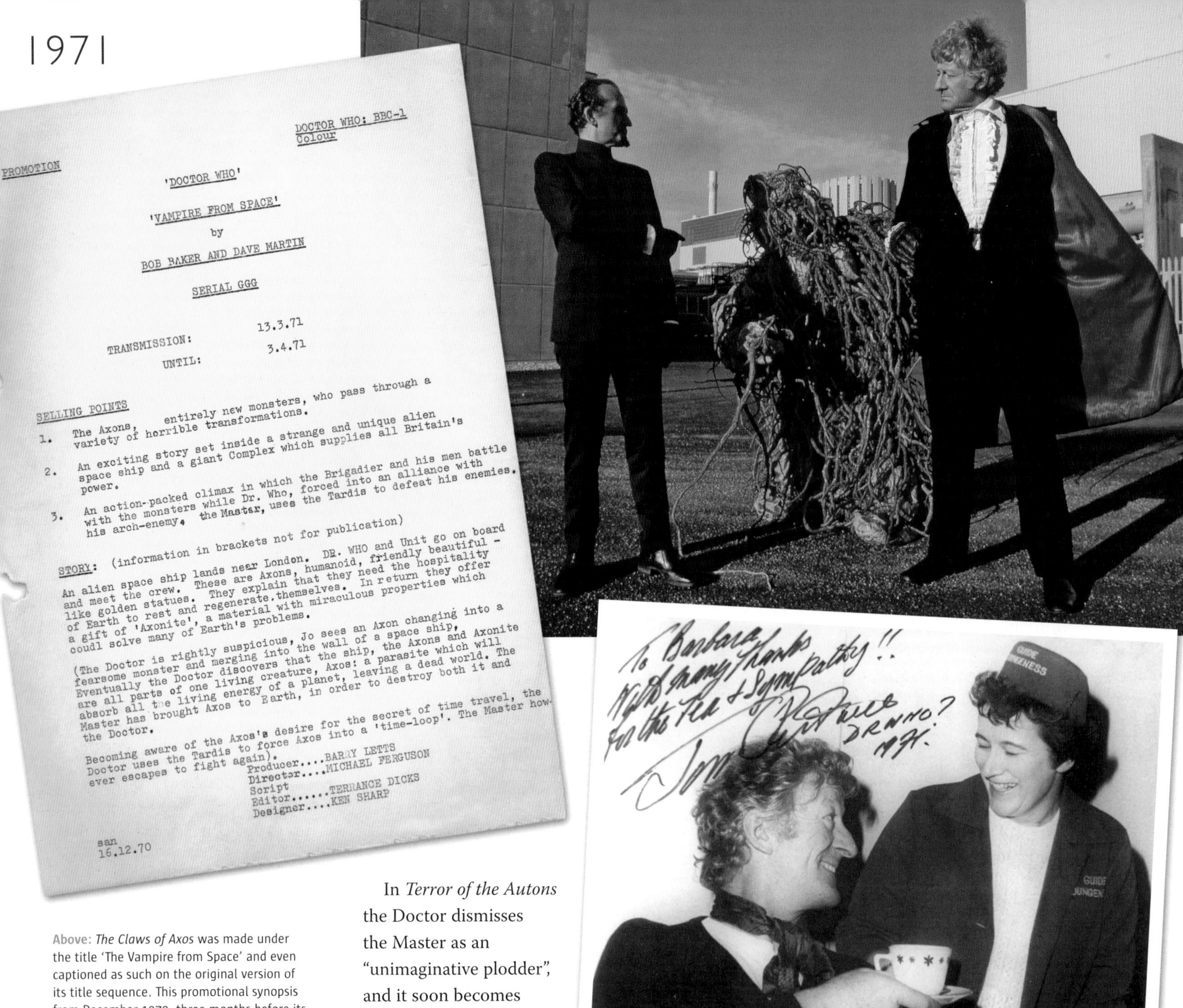

PROMOTION

DOCTOR WHO: BBC-1
Colour

'DOCTOR WHO'

'VAMPIRE FROM SPACE'

by

BOB BAKER AND DAVE MARTIN

SERIAL GGG

TRANSMISSION: 13.3.71

UNTIL: 3.4.71

SELLING POINTS

1. The Axons, entirely new monsters, who pass through a variety of horrible transformations.

2. An exciting story set inside a strange and unique alien space ship and a giant Complex which supplies all Britain's power.

3. An action-packed climax in which the Brigadier and his men battle with the monsters while Dr. Who, forced into an alliance with his arch-enemy, the Master, uses the Tardis to defeat his enemies.

STORY: (information in brackets not for publication)

An alien space ship lands near London. DR. WHO and Unit go on board and meet the crew. These are Axons, humanoid, friendly beautiful - like golden statues. They explain that they need the hospitality of Earth to rest and regenerate themselves. In return they offer a gift of 'Axonite', a material with miraculous properties which coudl solve many of Earth's problems.

(The Doctor is rightly suspicious, Jo sees an Axon changing into a fearsome monster and merging into the wall of a space ship. Eventually the Doctor discovers that the ship, the Axons and Axonite are all parts of one living creature, Axos: a parasite which will absorb all the living energy of a planet, leaving a dead world. The Master has brought Axos to Earth, in order to destroy both it and the Doctor.

Becoming aware of the Axos's desire for the secret of time travel, the Doctor uses the Tardis to force Axos into a 'time-loop'. The Master however escapes to fight again).

Producer....BARRY LETTS
Director....MICHAEL FERGUSON
Script Editor......TERRANCE DICKS
Designer....KEN SHARP

san
16.12.70

Above: *The Claws of Axos* was made under the title 'The Vampire from Space' and even captioned as such on the original version of its title sequence. This promotional synopsis from December 1970, three months before its transmission, features the working title.

Top right: Roger Delgado, Jon Pertwee and an Axon at Dungeness Nuclear Power Station, the principal filming location for *The Claws of Axos*, on 8 January 1971.

Centre: Jon Pertwee signed this photograph for one of the Dungeness staff.

In *Terror of the Autons* the Doctor dismisses the Master as an "unimaginative plodder", and it soon becomes clear that the Master suffers from a staggering lack of foresight. In *Terror of the Autons* and its immediate follow-ups *The Mind of Evil* and *The Claws of Axos* he resorts to forging an alliance with the Doctor when his plans misfire. The ensuing bargaining between the two renegade Time Lords creates the impression that theirs is a feud built on intellectual – if essentially child-like – one-upmanship.

The Master featured in every story broadcast in 1971 but, sensing that the character had become over-exposed, Letts and Dicks used him more sparingly thereafter. Fearing that his now infrequent association with *Doctor Who* could jeopardise his career, Delgado asked Letts if he could be written out of the series. Pertwee was getting similarly restless, so in 1973 Letts and writer Robert Sloman came up with a story that would end the Third Doctor's era with a fatal confrontation. Remarkably, *The Final Game* would exploit the Freudian concept that the Doctor and the Master were more than rival Time Lords, but diametrically opposed aspects of the *same* Time Lord.

"The Master was the id to the Doctor's ego," revealed Sloman. "They are the same character. That would have explained all those dodgy bits of plot, like, 'I could kill you now, but I'm not going to until I've explained something.' It's a terrible, deliberate pinch from *Forbidden Planet* (1956), but the explanation is that he's the same man, divided."

Delgado's untimely death in 1973 put paid to those plans, leaving him with a sadly ignominious final scene in Episode Six of *Frontier in Space*. It would be another three years before the character returned in *The Deadly Assassin* – now played by

Above left: The renewed popularity of *Doctor Who* brought a fresh wave of merchandise in 1971, the year this 100-piece jigsaw by Michael Stanfield Holdings was originally issued. The picture of Bessie was doctored to obscure the car's legitimate number plate, MTR 5.

Above: This iron-on transfer is a rare item of Dalek merchandise from the period.

Below left: The Master featured prominently on these Nestlé chocolate bars, which were the first *Doctor Who* confectionery produced since the mid-1960s. The back of each wrapper featured an instalment of the 15-part story *Doctor Who Fights Masterplan "Q"*.

WOOEEEEEEEEEEE
WHAT'S HAPPENING DOCTOR?
IT MUST BE THE TIME LORDS.
A PLANET WITH AN ATMOSPHERE — THAT MUST BE OUR DESTINATION.
THE TARDIS IS BEING OPERATED BY REMOTE CONTROL — WE'RE TRAVELLING THROUGH SPACE AND TIME!
WE'LL JUST HAVE A LOOK ROUND, THEN I'LL TRY AND GET YOU BACK TO EARTH.
DOCTOR! THE PETALS. THEY'RE DIFFERENT COLOURS!
???!!!!!

MEANWHILE A FEW YARDS

Dr Who zooms off into time again

HOLIDAY SPECIAL ON BBC1 NEW DR WHO STORY ON SATURDAY

DOCTOR WHO in Colony in SPACE

In his last adventure 'Vampire from Space,' Dr Who used his space/time machine the Tardis to save Earth from the alien parasite creature Axos by forcing it into a time-loop. But Dr Who is afraid that the Master, who sent Axos to destroy Earth, has escaped and will fight again . . . now read on

I'VE DONE IT! I'VE MADE A NEW DEMATERIALISATION CIRCUIT FOR THE TARDIS!
BUT WILL IT WORK? THE TIME LORDS MEAN TO KEEP ME HERE ON EARTH...
MEANWHILE SOMEWHERE IN SPACE AND TIME...
THE REPORT ON THE DOOMSDAY WEAPON IS MISSING. THE MASTER MUST HAVE TAKEN IT!
WHY NOT USE THE DOCTOR TO DEAL WITH HIM? BUT WAIT —
WE'VE IMMOBILISED HIS TARDIS.
THEN LET US RESTORE HIS FREEDOM — FOR AS LONG AS IT SERVES OUR PURPOSE...
... AND BACK ON EARTH.
NOW WE'LL TRY OUT THE NEW CIRCUIT.
DOCTOR! THE DOORS HAVE CLOSED!

Top: This piece, by renowned comic strip artist Frank Bellamy, appeared in the 8 April issue of the *Radio Times*. It was commissioned by art editor David Driver. "I'd always wanted Frank Bellamy to do work associated with *Doctor Who* because of the connection I felt it had with the strip story," he remembers.

Above: The Alien Priest (Roy Heymann) in Episode Six of *Colony in Space*. The priest's mask was based on a design by visual effects assistant Ian Scoones and sculpted by John Friedlander.

Peter Pratt and unrecognisable as a virtual corpse, clinging to life at the end of his regenerative cycle. The story was produced by Philip Hinchcliffe and written by Robert Holmes. "Philip wanted the story to be set on the Time Lords' own planet, and we both felt sufficient time had elapsed since Roger Delgado's death for us to be able to re-introduce the Master," remembered Holmes. "However, we didn't want to tie our successors to a particular actor – by this time we knew that our time with the programme was coming to an end – so I got the idea that he was in the terminal stage of his existence. This led me to the story – the Master was back on Gallifrey to try to steal himself a new supply of the Time Lord life essence."

This mellifluous but decrepit Master was reprised by Geoffrey Beevers in *The Keeper of Traken* (1981). At the end of this story he stole another body and, now played by Anthony Ainley, would plague the Fourth, Fifth, Sixth and Seventh Doctors until the end of the decade.

Although Ainley superficially resembled Delgado, his performance was significantly broader. The latest Master chuckled through a number of over-elaborate schemes, finding himself hoist by his own petard in *Time-Flight* (1982), *Planet of Fire* (1984) and *Survival* (1989). The latter was Ainley's most affecting performance, his final duel with the Doctor (Sylvester McCoy) acting as a fitting conclusion to the series' original run.

This Master's desperation lent credibility to a character whose exact motivation for galactic dominance remained unclear. The trend continued in the *Doctor Who* TV movie (1996), where the apparently executed Master (Eric Roberts) returns from the dead, appropriates another body then attempts to steal the Doctor's remaining regenerations.

Resurrected by the Time Lords to fight on their behalf in the Time War, the Master returned with a new regenerative cycle in *Utopia* (2007). This story marked the beginning of his most spectacular comeback to date. Derek Jacobi chillingly evoked the quiet menace of the Roger Delgado years before his elderly Master regenerated into a younger model (John Simm). The nature of his rivalry with the Doctor was now all the more poignant, as they remained the last two Time Lords in existence.

Originally portrayed by Delgado in the style of a cold and calculating Bond villain, Simm's more casually dressed Master is an unstable psychopath with a rampaging ego. In *The End of Time* (2009-10) he creates a short-lived 'Master race' by repopulating the Earth in his image.

Like Sherlock Holmes and James Bond, the Master is a character who is frequently reinvented to reflect the audience's prevailing expectations. In that respect he has represented the dark side of the Doctor all along.

Above left: A publicity still from *The Daemons*, featuring the Doctor (Jon Pertwee), Jo Grant (Katy Manning) and the animated gargoyle Bok (Stanley Mason).

Above: Boys' comic *Countdown* was launched in February 1971 and maintained close links with *Doctor Who*. This cover picture was taken during the making of *The Daemons* on 29 April.

Below left: Roger Delgado, John Levene and director Christopher Barry rehearse a scene from Episode Five of *The Daemons* on location in Aldbourne during April.

SEASON EIGHT
Producer: Barry Letts
Script editor: Terrance Dicks

EEE *Terror of the Autons*
17 September – 24 October 1970
written by Robert Holmes
directed by Barry Letts (uncredited)

Episode One	2 January
Episode Two	9 January
Episode Three	16 January
Episode Four	23 January

FFF *The Mind of Evil*
26 October – 12 December 1970
written by Don Houghton
directed by Timothy Combe

Episode One	30 January
Episode Two	6 February
Episode Three	13 February
Episode Four	20 February
Episode Five	27 February
Episode Six	6 March

GGG *The Claws of Axos*
22 December 1970 – 6 February 1971
written by Bob Baker & Dave Martin
directed by Michael Ferguson

Episode One	13 March
Episode Two	20 March
Episode Three	27 March
Episode Four	3 April

HHH *Colony in Space*
10 February – 3 April
Written by Malcolm Hulke
Directed by Michael E Briant

Episode One	10 April
Episode Two	17 April
Episode Three	24 April
Episode Four	1 May
Episode Five	8 May
Episode Six	15 May

JJJ *The Daemons*
19 April – 26 May
written by Guy Leopold (pseudonym for Barry Letts and Robert Sloman)
directed by Christopher Barry

Episode One	22 May
Episode Two	29 May
Episode Three	5 June
Episode Four	12 June
Episode Five	19 June

1972

The Daleks had not made a major appearance in *Doctor Who* since 1967, but Terry Nation was now happy to allow their return. He didn't have time to write a story himself, however, so Terrance Dicks grafted the Daleks onto an existing Louis Marks script at that stage entitled *Years of Doom*. *Day of the Daleks* opened the ninth season of *Doctor Who* on New Year's Day 1972.

The last episodes of the previous year had been broadcast at 6.10, but the production team were sensitive to criticisms that *Terror of the Autons* had upset some younger viewers. For 1972 more subtle ways were found to appeal to the show's burgeoning adult audience. *The Curse of Peladon* revolved around a planet's efforts to join a galactic federation, a plot which older viewers may have equated with Britain's long-debated entry to the European Common Market. Once again the Doctor was given special dispensation to leave Earth, and it was intriguing to see the story's Martian delegation, the Ice Warriors, revealed as allies.

Doctor Who's six-month transmission seasons afforded Barry Letts the luxury of advance recording dates, which in turn gave him the opportunity to screen serials out of production order. *The Sea Devils*, a nautical sequel to *Doctor Who and the Silurians*, was made before *The Curse of Peladon* and, like the previous year's *The Mind of Evil*, benefited greatly from Letts' good relationship with the Ministry of Defence. Location filming in Portsmouth enjoyed the full co-operation of the Royal Navy.

The Doctor's latest mission for the Time Lords took place in *The Mutants*, a disturbing tale that was the most complete expression yet of Letts' and Dicks' thematic approach to the series. Thirtieth century Solos was the setting for a thinly veiled allegory about contemporary racism and the decline of the British Empire.

Last seen in *The Sea Devils*, the Master returned for the season finale. *The Time Monster* was an epic scenario that once again addressed science and spirituality, although it was rather less accomplished than Letts' and Sloman's previous collaboration, *The Daemons*.

Despite the uneven quality of this season's stories the popularity of *Doctor Who* remained high. There was considerable confidence in the future of the show, but as the tenth anniversary approached thoughts inevitably turned to acknowledging the past.

MAKING HISTORY

Radio Times WIN A DALEK COMPETITION 1972

THIS CERTIFICATE
HAS BEEN AWARDED TO

FOR THE ENTRY IN THE
RADIO TIMES WIN A DALEK COMPETITION
WHICH WAS DISPLAYED IN
A SPECIAL EXHIBITION IN LONDON
MARCH–APRIL 1972

Above: In January 1972 *Countdown* merged with *TV Action* but continued to include *Doctor Who*. This issue from August featured the latest instalment of *The Ugrakks*, a strip featuring elephant-like aliens designed by competition winner Ian Fairnington.

Above right: Terry Nation helps to judge the 7,000 entries to the *Radio Times*' Win a Dalek competition at the magazine's Marylebone High Street offices in February. The first prize was one of the 'Mark 7' toy Daleks pictured here.

Right: A Frank Bellamy illustration adorned the winners' certificates.

"I think we've seen the last of the Doctor," crows the Master in *The Time Monster*, "buried for all time under the ruins of Atlantis!" The destruction of the legendary city helped to bring Season Nine to a momentous conclusion. Unfortunately *Doctor Who* had demolished Atlantis twice already – demonstrably in *The Underwater Menace* (1967) and reportedly in *The Daemons* (1971). The latter story had been broadcast only the year before *The Time Monster*, and was the work of the same writers – Barry Letts and Robert Sloman.

Mistakes such as these were easy to commit in the age before home video and – crucially, in the case of *Doctor Who* – published research for the show's professional custodians to draw upon.

The first official attempt to chronicle the series had been made back in 1965. *The History of Doctor Who* was compiled by outgoing story editor Dennis Spooner as part of the handover notes to his replacement Donald Tosh and new producer John Wiles. Intended for internal use only, *The History of Doctor Who* identified the first 19 stories by their production code and author, offering a brief synopsis for each one. For example, Serial G was described as: "A 6 part story with the Sensorites. The gimmick was telepathy – thought transference, and involved the captured Earth ship and its crew."

Even at this early stage in the programme's life, Spooner was aware that it was creating a complex framework of references that had to remain consistent. He wrote: "I think it is a point to bear in mind that any stories that are commissioned and are set in the future will have to be checked from their date point of view. Serials G [*The Sensorites*], K [*The Dalek Invasion of Earth*], L [*The Rescue*] involve the Earth in some way, so any given date must not clash with these. The Dalek serials have also to be watched with this in mind, as in the first Dalek serial (Serial B) Doctor Who did in fact wipe out the Dalek race. With a time machine at his disposal this is not as

Left: A BBC publicity postcard featuring a climactic scene from Episode Four of *Day of the Daleks*.

Below: This February issue of *Countdown for TV Action* featured a reversed cover shot from the same episode.

Below left: Barry Letts and Jon Pertwee attend a Young Observers presentation at the London Planetarium on 21 December 1971. The gold Chief Dalek would make its first appearance in *Day of the Daleks* on New Year's Day 1972.

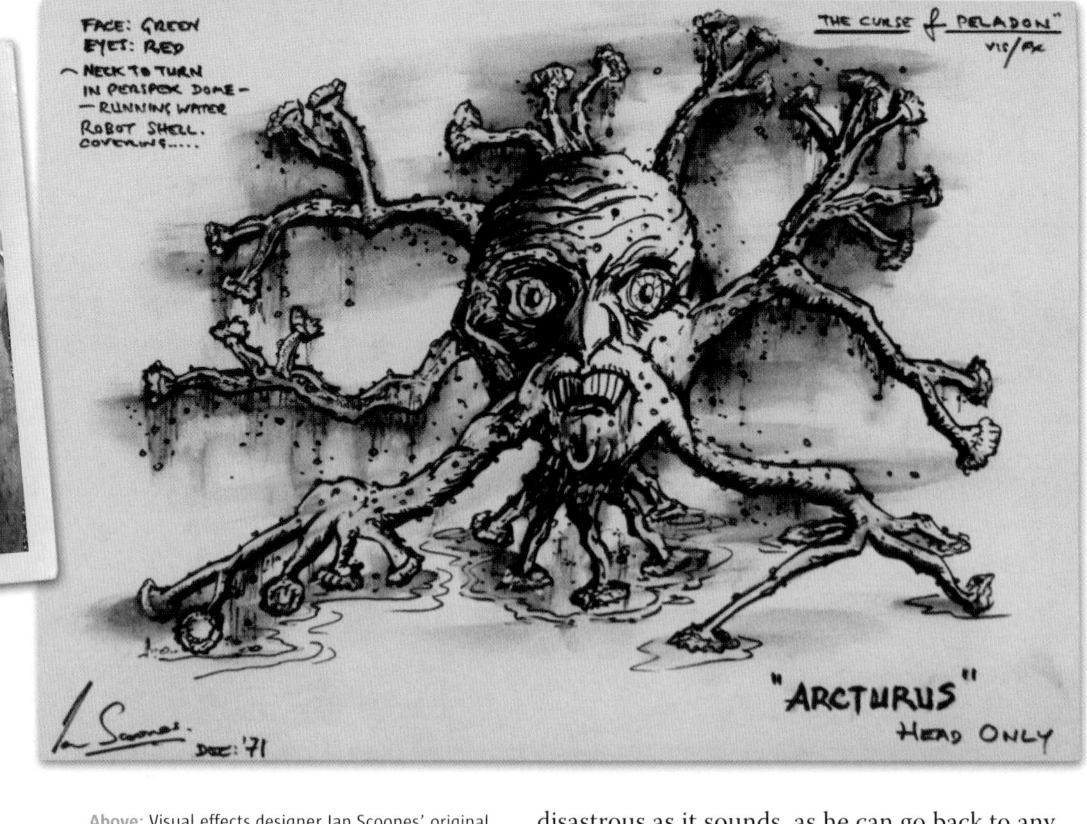

Above: Visual effects designer Ian Scoones' original sketch for Arcturus in *The Curse of Peladon*.

Above left: Arcturus (Murphy Grumbar) and the Ice Lord Izlyr (Alan Bennion).

Left: The hermaphrodite hexapod Alpha Centauri (Stuart Fell), part of the Galactic Federation's Committee of Assessment in *The Curse of Peladon*. The character was voiced by Ysanne Churchman.

Below: Ian Scoones' concept painting for King Peladon's citadel was redolent of his work on the classic Hammer horror films.

disastrous as it sounds, as he can go back to any point in their history; but one has to be careful in Serial B, K, R [*The Chase*] and in the Dalek to come [*The Daleks' Master Plan*] that they are true to the Dalek history calendar."

By the time Terrance Dicks joined as script editor in 1968 this three-page document had been consigned to the BBC archive. "In my days

as script editor, *Who* never really had a 'Bible', a book of rules and facts in which continuity was set in stone," wrote Dicks in his foreword to the 1995 reference book *The Discontinuity Guide*. "*Who* continuity was rather like the *1066 And All That* definition of history – What You Can Remember. It was what I could remember about my predecessors' shows, what my successors could remember of mine."

While Dicks was struggling with continuity in the production office his occasional writing partner Malcolm Hulke approached him with a proposition. Hulke's editor at Pan Books had recalled the 1968 paperback *The Making of Star Trek* and suggested that *Doctor Who* could be a candidate for the same approach. Hulke asked Dicks to collaborate on the volume, and *The Making of Doctor Who* was published in April 1972.

The book was predominantly written by Hulke and pitched at a younger readership. Nevertheless it was rich in detail about the programme's background and current production methods. Sydney Newman and Donald Wilson were credited as the show's originators, and one chapter was dedicated to the creation of the Daleks and the Cybermen. Every writer, producer, director and story/script editor was listed, with 18 pages following Hulke's *The Sea Devils* from script to screen. There was even a theological interlude contributed by the Rev John D Beckwith.

The book's greatest innovation, however, was the section entitled 'The Travels of Doctor Who'.

Below: The Doctor uses his Sonic Screwdriver to detonate a minefield in Episode Four of *The Sea Devils*.

Bottom: The Sonic Screwdriver prop used in the series from *The Sea Devils* to *Carnival of Monsters*.

Bottom left: The first edition of *The Making of Doctor Who*. 'This fascinating book tells you everything about the Doctor and his adventures,' promised the back cover.

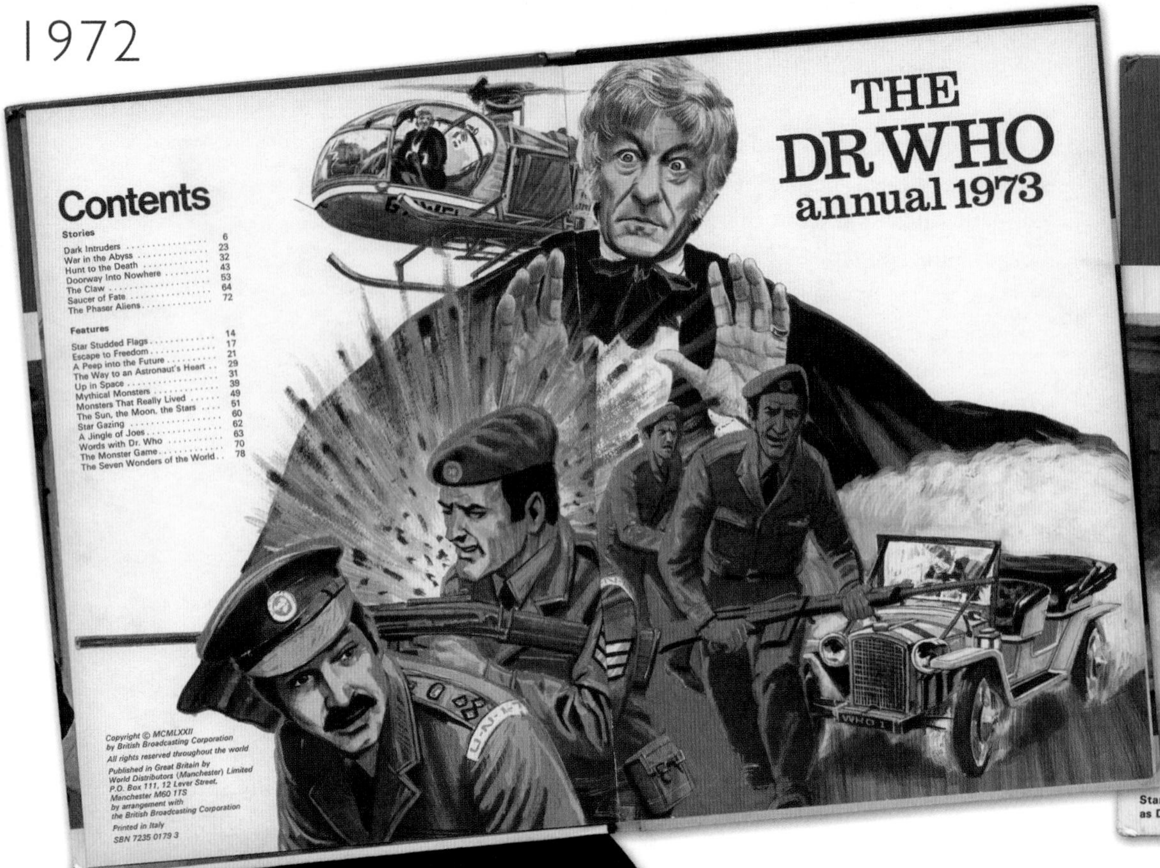

This was a summary of every television story, initially presented by the counsel for the defence during the Doctor's trial in *The War Games*. The Doctor's earthbound adventures, beginning with *Spearhead from Space* and ending with *The Sea Devils*, were outlined in the form of a memorandum by Brigadier Lethbridge-Stewart to UNIT HQ in Geneva. The framing devices may have been unorthodox, but there is no disguising the fact that this was the first professionally published episode guide to *Doctor Who*.

The Making of Doctor Who was the first step towards a consensus on the programme's byzantine history. In 1973 there would be another, when David Driver, the art editor of the *Radio Times*, suggested that the magazine should publish a special issue to celebrate *Doctor Who*'s tenth anniversary. "Everyone at *Radio Times* thought it was a joke, because they knew I was such a fan of *Doctor Who*," he remembers. "They all fell about, saying 'There's no point, it's got no following,' and so on. I said, 'Well, looking at the audience figures, I don't think that's the case, and besides, I just know that it's going to go on forever.'"

Top: The contents pages and front cover of the *Doctor Who Annual* published in September 1972. Poor sales of the 1970 edition meant that there hadn't been a book in 1971.

Above: Jon Pertwee's *Who is the Doctor* was released on Deep Purple's label, Purple Records, on 10 November.

Right: Now showing its age, this Sea Devil head is carefully retained by the BBC.

Left, above and below: A Draconian, a Sea Devil, an Ogron and a Dalek were among the *Doctor Who* attractions at the BBC TV Special Effects Exhibition which opened at London's Science Museum in December. Exhibits also included a recreation of the TARDIS control room.

Below left: A souvenir badge from the exhibition.

Driver relished the project, working with Jack Lundin to create a visually distinctive souvenir that included interviews, specially commissioned photography, a new story by Terry Nation and instructions on how to build a Dalek. The lengthy episode guide far surpassed *The Making of Doctor Who* and, unlike the previous book, attributed titles to each of the Doctor's adventures. For the serials made prior to *The Savages*, just the titles of first episodes were given. Hence Story A became *An Unearthly Child*, Story B became *The Dead Planet*, Story C became *The Edge of Destruction* and so on. "I did try so hard to ensure it was accurate, but I do remember that all the [early] scripts had the individual episode titles, not the full story titles, on them," says Driver. "I didn't discover this until it was too late. It was very frustrating; I didn't want to get that sort of thing wrong because it was so easy to get it right."

It would be churlish to criticise Hulke and Dicks for describing the robots in *The Chase* as 'Mechons', or Driver and Lundin for referring to *The Chase* as *The Executioners*; both claims were made before the plethora of more accurate works they inspired.

Leading this reference revolution was *The Doctor Who Programme Guide*, written by Jean-Marc Lofficier and published by Target Books in 1981. In his foreword, Barry Letts tacitly admitted the errors made in less enlightened times when he pointed out that "in various *Doctor Who* stories there have been three entirely different and incompatible versions of the destruction of Atlantis..."

SEASON NINE

Producer: Barry Letts
Script editor: Terrance Dicks

KKK *Day of the Daleks*
7 September – 19 October 1971
written by Louis Marks
directed by Paul Bernard

Episode One	1 January
Episode Two	8 January
Episode Three	15 January
Episode Four	22 January

MMM *The Curse of Peladon*
15 December 1971 – 1 February 1972
written by Brian Hayles
directed by Lennie Mayne

Episode One	29 January
Episode Two	5 February
Episode Three	12 February
Episode Four	19 February

LLL *The Sea Devils*
21 October – 14 December 1971
written by Malcolm Hulke
directed by Michael E Briant

Episode One	26 February
Episode Two	4 March
Episode Three	11 March
Episode Four	18 March
Episode Five	25 March
Episode Six	1 April

NNN *The Mutants*
7 February – 28 March
written by Bob Baker and Dave Martin
directed by Christopher Barry

Episode One	8 April
Episode Two	15 April
Episode Three	22 April
Episode Four	29 April
Episode Five	6 May
Episode Six	13 May

OOO *The Time Monster*
29 March – 24 May
written by Robert Sloman
directed by Paul Bernard

Episode One	20 May
Episode Two	27 May
Episode Three	3 June
Episode Four	10 June
Episode Five	17 June
Episode Six	24 June

SEASON TEN

Producer: Barry Letts
Script editor: Terrance Dicks

RRR *The Three Doctors*
6 November – 12 December 1972
written by Bob Baker and Dave Martin
directed by Lennie Mayne

Episode One	30 December 1972

1973

Television was still relatively young, and the tenth anniversary of *Doctor Who* was a milestone in a medium that had only reached a mass audience 20 years before. Barry Letts was determined to prove it was a medium that was still evolving; he continued to experiment with the sequence of shooting within stories and commissioned Bernard Lodge to create impressive new titles using the slit-scan photography technique.

The programme was moving further away from the stifling production techniques of the 1960s, but the anniversary was an opportunity to celebrate the best of the previous decade. Patrick Troughton and a sadly incapacitated William Hartnell joined Jon Pertwee for *The Three Doctors*, the serial that opened the tenth season at the end of December 1972.

As well as looking over its shoulder, *The Three Doctors* pointed a way to the future. As a reward for saving the Time Lords, the Third Doctor's exile was lifted. His intergalactic travels would resume immediately, making assignments for UNIT increasingly rare.

Carnival of Monsters' postmodern premise saw the Doctor and Jo miniaturised inside an extraterrestrial peepshow. Held over from the end of Season Nine's recording block, this was the story in which Robert Holmes emerged as one of *Doctor Who*'s wittiest and most imaginative writers. The Draconians in *Frontier in Space* would become Jon Pertwee's favourite aliens during his time on the series, but the six-episode space opera in which they appeared was less inventive. It is now chiefly notable for featuring the last appearance of Roger Delgado's Master.

Frontier in Space led directly into *Planet of the Daleks*, the first Terry Nation script for *Doctor Who* since 1965. It was a shameless return to the *Boy's Own* style (and some of the plot elements) he had last employed during that era, but matched the eccentric audacity of Holmes' ideas with some characteristically epic aspirations. Visual effects had moved on since the mid-1960s, but Nation's request to see an army of 10,000 Daleks remained over-ambitious.

As was now traditional, the season ended with a serial by Robert Sloman and Barry Letts. *The Green Death* was arguably their most successful collaboration – a story that combined political satire and contemporary ecological concerns with a rarely seen emotional literacy. The poignancy of the Doctor's farewell to Jo in the final episode was, in its unassuming way, probably the most startling innovation of the anniversary year.

ON TARGET

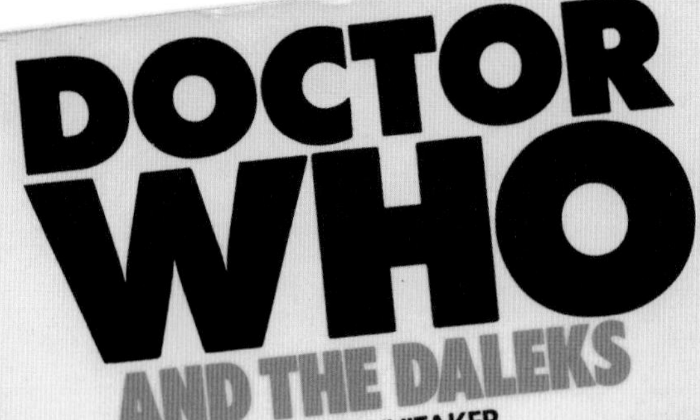

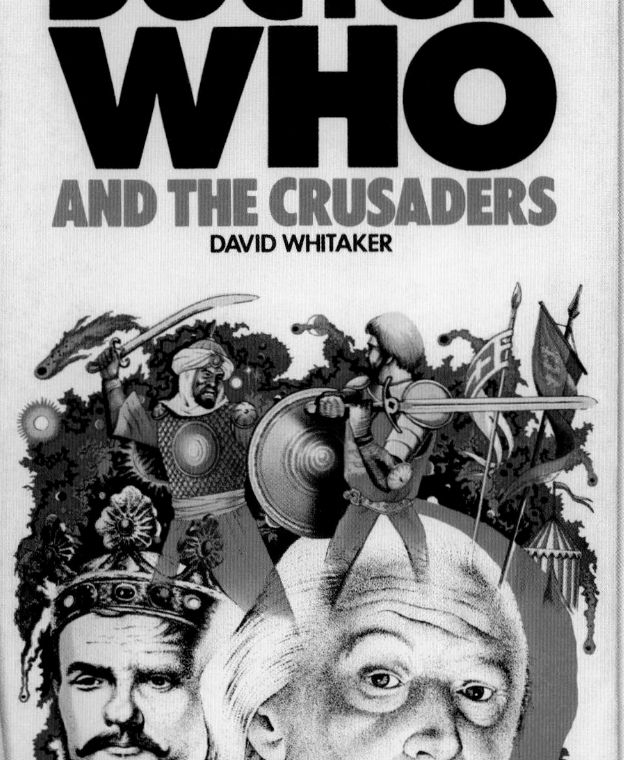

Top left: The cover of the first Target catalogue, which included the imprint's initial list of 12 books. As well as the three *Doctor Who* titles there were such diverse publications as *The Nightmare Rally* by Pierre Castex, *Investigating UFOs* by Larry Kettelkamp and *Fishing* by JH Elliott.

Top right: Two of the paperback *Doctor Who* reprints that launched the range.

Above: Chris Achilleos, the original cover artist for Target's *Doctor Who* range.

In the 1960s and early '70s there were very few ways to experience *Doctor Who* when the show wasn't on the air. Before the advent of affordable home video, extra-curricular *Doctor Who* mainly comprised its annual, a gift publication of variable quality that, as the 1970s progressed, maintained only a tenuous link to the programme it was based on.

Things began to improve with the launch of Target Books, a children's imprint with an initial list that included reprints of three *Doctor Who* novelisations from the 1960s: *Doctor Who in an exciting adventure with the Daleks*, *Doctor Who and the Zarbi* and *Doctor Who and the Crusaders*. These out-of-print titles were selected on a hunch by Target editor Richard Henwood, even though he was barely acquainted with the programme and didn't own a television.

The three paperbacks were published in May 1973, with print runs estimated at 20,000 copies each. Initial feedback had been so good that Henwood had already approached the BBC for the rights to publish new adaptations. With a deal in place and his first three titles selling fast, he realised he had to find some new books quickly. "He fetched up in the *Doctor Who* office with me and Barry [Letts]," remembers Terrance Dicks. "He said 'What I desperately need now is some more *Doctor Who* books. Who is going to write them?' I immediately leapt up and said, 'I will!'"

In summer 1973 Dicks' ambition to become an author was fulfilled by his commission to adapt the 1970 story *Spearhead from Space*. Shortly afterwards Malcolm Hulke was asked to adapt *Doctor Who and the Silurians*. Henwood disliked the title *Spearhead from Space* so named Dicks'

Left: The first cover illustration that Chris Achilleos completed for Target was this piece for Bill Strutton's book *Doctor Who and the Zarbi*. Achilleos' original draft had featured giant ants of his own invention, but the BBC insisted they should be changed to resemble those from the source serial, *The Web Planet*.

Below: The 7" single of the *Doctor Who* theme was reissued in a picture sleeve for the first time in April. The cover was designed by Andrew Prewett.

Bottom: A close-up of the Plesiosaurus from *Carnival of Monsters*.

book *The Auton Invasion*. Hulke's book was similarly christened *The Cave-Monsters*. Much to Dicks' consternation, Target also insisted on prefixing every story title with 'Doctor Who and the' – a habit that would take years to break.

Dicks' economical style would become familiar to readers who enjoyed identifying his standard descriptions of the TARDIS' "wheezing, groaning sound" and the Third Doctor's "shock of white hair" amongst many others. "My own style of writing, or at least what I aim at, is simplicity, clarity and pace," he says unaffectedly.

Hulke, on the other hand, tended to extrapolate more than Dicks; in *The Cave-Monsters* various Silurian characters are given names, and the book opens with an evocative prologue describing the prehistoric events prior to the creatures' hibernation.

Right: In January the *Countdown* name was dropped from *TV Action*. Issue 123 was published in June, with cover artwork by Gerry Haylock.

Below: Haylock's original artwork for the cover of *TV Action* Issue 113, published in April. His reference photos of Daleks had obviously been reversed. The story *The Threat From Beneath* was illustrated by Haylock from a script by Dick O'Neill.

With no idea that the Target *Doctor Who* range would eventually encompass almost every televised story, the authors strove to make sure their books were self-contained. *The Doomsday Weapon*, Hulke's 1974 novelisation of *Colony in Space*, was the first to feature companion Jo Grant. Hulke introduced her in the second chapter, creating the impression that this had also been her debut in the series. Dicks' novelisation of *Terror of the Autons*, which *had* been Jo's first story, was published just over a year later, in May 1975.

As Target became better established there was a greater adherence to on-screen titles, and any story changes were generally only made in an effort to improve on the original. In the television *Terror of the Autons* the Master has a sudden change of heart about allowing the Nestene Consciousness to invade Earth, but the novelisation has a more plausible ending when the Brigadier forces him to make the decision at gunpoint. Dicks' 1978 novelisation of *The Android Invasion* was also re-edited. "*The Android Invasion* was, as a script, a bit of a mess," says Dicks. "When you got to the end there were several loose ends which were not tied up. I just felt they needed to be tied up, so I provided a slightly more coherent ending." His 1976 adaptation of *Pyramids of Mars* had gone further, adding a charming epilogue featuring Sarah Jane Smith and set some years after the events of the original story.

Dicks would write a staggering 64 *Doctor Who* books for Target. The next most prolific author was actor Ian Marter, who had appeared in *Carnival of Monsters* (1973) and later returned to the show as companion Harry Sullivan. He wrote nine books for the range, his occasionally lurid prose stretching the limits of acceptability in children's fiction of the time. Marter was presumably trying to nudge the books towards a teenage readership with his controversial use of the word 'bastard' in his 1981 adaptation of *The Enemy of the World*. His 1985 novelisation of *The Invasion* was, in places, also rather extreme: "Routledge remained standing like a waxen dummy for several seconds. Then he vomited a stream of blood and pitched forward onto his face at Vaughn's feet."

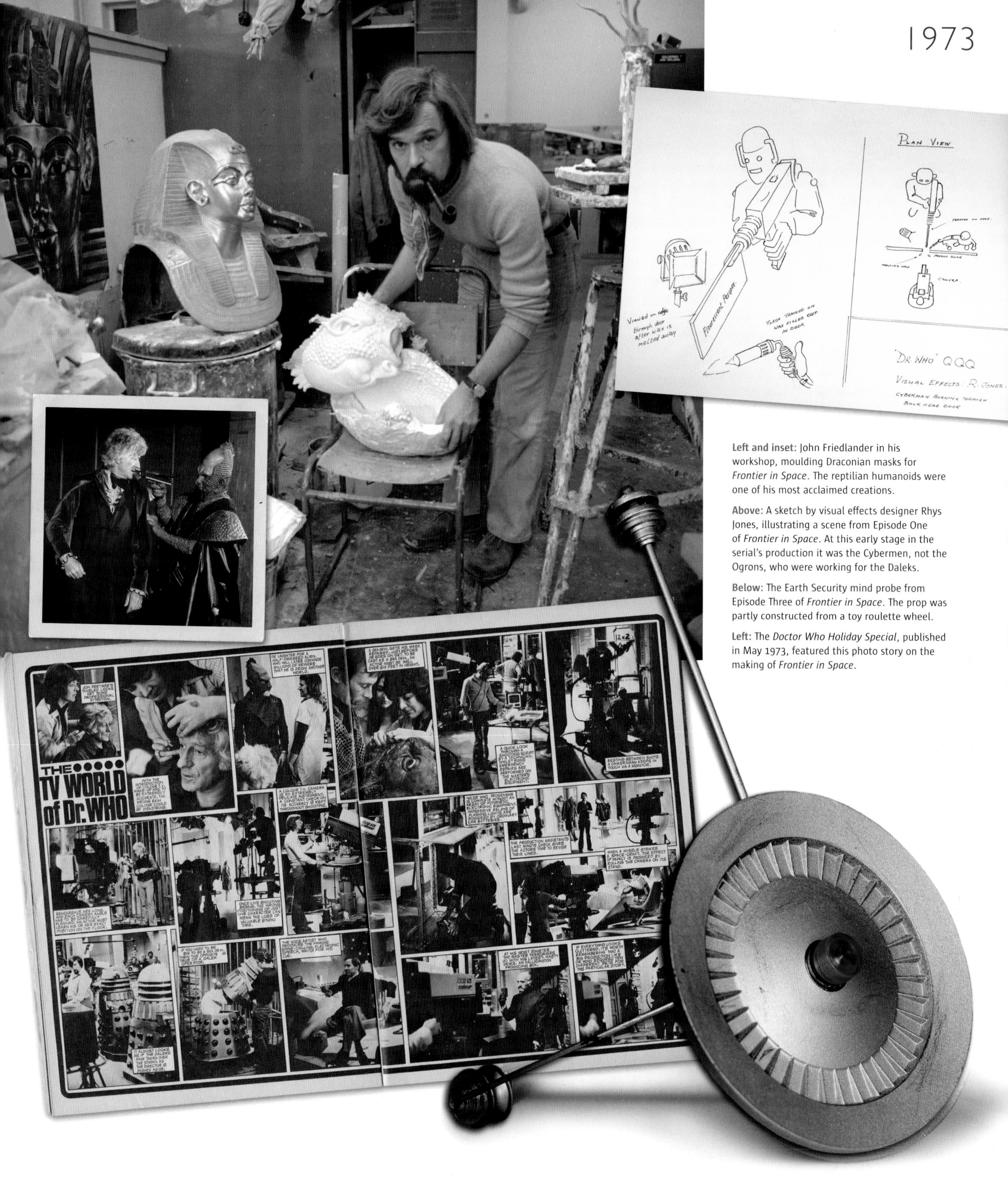

Left and inset: John Friedlander in his workshop, moulding Draconian masks for *Frontier in Space*. The reptilian humanoids were one of his most acclaimed creations.

Above: A sketch by visual effects designer Rhys Jones, illustrating a scene from Episode One of *Frontier in Space*. At this early stage in the serial's production it was the Cybermen, not the Ogrons, who were working for the Daleks.

Below: The Earth Security mind probe from Episode Three of *Frontier in Space*. The prop was partly constructed from a toy roulette wheel.

Left: The *Doctor Who Holiday Special*, published in May 1973, featured this photo story on the making of *Frontier in Space*.

Above: One of Terry Nation's movie props was adapted for use in *Planet of the Daleks*.

Above right: Director David Maloney (centre) with Jon Pertwee and the crew of *Planet of the Daleks*, on location at Beachfields Quarry in Redhill, Surrey, in early January.

Below: A picture from *The Green Death* featured on this jigsaw puzzle from Whitman Publishing.

Below right: The cover of the *Doctor Who Annual* published in September.

By the 1980s Dicks was using videos as well as scripts to compile his adaptations. In 1985 Marter tried to do the same thing. "I remember when he was going to write *The Invasion* and he came round to see a very dud copy I had of it," said Nicholas Courtney in 2007. "It was very hard for him to elicit any information from this ropey old copy. There was a Russian air base in the story, and he gave this air base the name Nykortney. I thought that was rather sweet."

The Target *Doctor Who* books are fondly remembered for their iconic artwork, which only came about in 1973 because Henwood was reluctant to pay Jon Pertwee for the rights to reproduce his image as a photograph. The original

Right: The *Radio Times* special celebrating *Doctor Who*'s tenth anniversary was published in November. Terry Nation was among the interviewees, and used the opportunity to admit that his oft-quoted inspiration for the Daleks' name was a fabrication. "The fact is that no encyclopedia in print covers those letters DAL-LEK. Anyone checking the facts would have found me out."

Below: Stewart Bevan (Clifford Jones in *The Green Death*) and Katy Manning in one of the specially posed photographs commissioned by the magazine.

artist is still the one most associated with the books – Cypriot illustrator Chris Achilleos contributed 28 covers, traditionally presenting a stippled black and white portrait of the Doctor as part of a colour montage. His most notorious illustration was for *The Dinosaur Invasion*, Malcolm Hulke's 1976 adaptation of his story *Invasion of the Dinosaurs.* Achilleos' cover showed the Third Doctor retreating from a pterodactyl. Behind its snapping jaws is the comic-style exclamation "Kklak!" The illustration was only grudgingly accepted by Target but now enjoys cult status among *Doctor Who* fans.

The popularity of home video and an exhaustion of stories put paid to Target Books in 1994. Its 156 *Doctor Who* titles had sold more than 13 million copies.

Doctor Who novelist and scriptwriter Gareth Roberts is quick to acknowledge his debt to the company, and its most prolific author. "For me, and for thousands of other people of my generation, Terrance is an absolutely central figure in *Doctor Who.* He was the voice of *Doctor Who* to us. He was our video recorder. He was the person that got us reading, and in some cases writing."

SEASON TEN (continued)
Producer: Barry Letts
Script editor: Terrance Dicks

RRR ***The Three Doctors*** (continued)
Episode Two 6 January
Episode Three 13 January
Episode Four 20 January

PPP ***Carnival of Monsters***
30 May – 4 July 1972
written by Robert Holmes
directed by Barry Letts
Episode One 27 January
Episode Two 3 February
Episode Three 10 February
Episode Four 17 February

QQQ ***Frontier in Space***
14 August – 1 November 1972, 22 January 1973
written by Malcolm Hulke
directed by Paul Bernard
Episode One 24 February
Episode Two 3 March
Episode Three 10 March
Episode Four 17 March
Episode Five 24 March
Episode Six 31 March

SSS ***Planet of the Daleks***
2 January – 20 February
written by Terry Nation
directed by David Maloney
Episode One 7 April
Episode Two 14 April
Episode Three 21 April
Episode Four 28 April
Episode Five 5 May
Episode Six 12 May

TTT ***The Green Death***
12 February – 30 April
written by Robert Slomandirected by Michael E Briant
Episode One 19 May
Episode Two 26 May
Episode Three 2 June
Episode Four 9 June
Episode Five 16 June
Episode Six 23 June

SEASON 11
Producer: Barry Letts
Script editor: Terrance Dicks

UUU ***The Time Warrior***
7 May – 12 June 1973
written by Robert Holmes
directed by Alan Bromly
Part One 15 December
Part Two 22 December
Part Three 29 December

1974

The first story of the 11th broadcast season had been produced at the end of the recording block for Season Ten. *The Time Warrior* unveiled Elisabeth Sladen as the Doctor's new assistant, investigative journalist Sarah Jane Smith. Privately, Jon Pertwee was still struggling to get over the departure of Katy Manning. He would be further traumatised by the death of Roger Delgado, who was killed in a car accident on 18 June 1973.

The only significant pseudo-historical of the Pertwee years, *The Time Warrior* introduced the Sontarans – a race of stout, aggressive clones, in this instance represented by the ruthless Commander Linx (Kevin Lindsay). It is during an interrogation by Linx in Part Two that the Doctor casually reveals the name of his home planet: Gallifrey.

The TARDIS returned the Doctor and Sarah to present-day London for *Invasion of the Dinosaurs.* Malcolm Hulke's final script for the series was also his most accomplished. It's therefore unfortunate that it relied on some of the most woefully inadequate visual effects in the show's history. In this story the Doctor swapped Bessie for a futuristic flying car described in the scripts as the 'Whomobile'. The familiar UNIT repertory company further dissipated with the revelation that Captain Yates had turned traitor.

Demoralised by the fragmentation of the old team, the death of Delgado and the news that Barry Letts and Terrance Dicks intended to move on, Jon Pertwee decided to stay with the programme only if he received an increased salary. His request was denied, and on 9 February various newspapers announced his intention to leave. A new Doctor, the relatively unknown Tom Baker, had already been recommended to Letts by Bill Slater, the BBC's Head of Drama Serials.

The eleventh season continued with *Death to the Daleks* and *The Monster of Peladon,* the latter of which featured the last appearance of the Ice Warriors for 39 years. It fell to Letts and Robert Sloman to close the Third Doctor's tenure in memorable style. In *Planet of the Spiders* Letts made his most explicit statement yet about his spiritual beliefs with a parable on Buddhist meditation.

This thoughtful story was a fitting bookend to five years of consistently popular television that never patronised its audience. But this is only part of Barry Letts' legacy; his decision to cast Tom Baker would take *Doctor Who* to even greater heights.

AS SEEN ON TV

Above: A poster advertising *Seven Keys to Doomsday* at the Adelphi Theatre.

Above right: Jimmy (James Mathews), Jenny (Wendy Padbury) and the Doctor (Trevor Martin) wrestle a Clawrantular on stage at the Adelphi. Three Clawrantulars were built for the show by model-maker Allister Bowtell. The set design was by John Napier, who went on to win Tony Awards for his work on *Cats*, *Starlight Express* and *Les Misérables*.

Doctor Who was created for television, and any reinterpretations of the concept for other media have inevitably suffered comparisons to the original. The only professional stage productions of *Doctor Who* have strong links to the Jon Pertwee years, and they number amongst the programme's most inventive spin-offs.

Producers Robert De Wynter and Anthony Pye-Jeary secured a licence to stage the first *Doctor Who* play in London's West End from December 1974 to January 1975. A previous proposal had floundered following the salary demands of Jon Pertwee's agent, and Tom Baker, Pertwee's successor in the role of the Doctor, was committed to recording Season 12 during the period the play was due to run. "So it was a bit like *Hamlet* without the Prince of Denmark!" remembers Terrance Dicks, who was commissioned to write the script. "I said all along that we needed either Jon or the Daleks, and they eventually managed to get the Daleks. For the Doctor they got in Trevor Martin, who was actually pretty good."

Dicks' script was called 'Doctor Who and the Daleks in *Seven Keys to Doomsday*', the title a reference to the segmented Crystals of Power that serve as the catalyst to the Daleks' 'Ultimate

Weapon'. The opening of the first act included ingenious introductions to both the new Doctor and stage *Doctor Who*. A white-haired Doctor staggers out of the TARDIS and makes a call for help which is answered by two youngsters sitting in the front row of the stalls. Jimmy Forbes and Jenny Wilson rush onto the stage and help the injured man inside his ship. As images of the previous television Doctors are projected on a big screen the Doctor's features are obscured by Jimmy and Jenny. He emerges 'rejuvenated', and his mission for the Time Lords continues with the help of his new friends.

The programme on sale at the Adelphi Theatre pointed out that Trevor Martin and his four children were avid fans of *Doctor Who*. It didn't mention that he had played a Time Lord before – as part of the prosecuting council in *The War Games*. Wendy Padbury, the actress second-billed as Jenny, was certainly familiar to *Doctor Who* fans as Zoe in the later Patrick Troughton stories. James Mathews, who played Jimmy, was making his West End debut.

The play opened on 16 December. The following day Irving Wardle, *The Times'* theatre critic, betrayed his ignorance of the television series when he highlighted "a moment in *Seven Keys to Doomsday* when Terrance Dicks allows

Far left: This version of the TARDIS key was introduced in *The Time Warrior*.

Left: Kevin Lindsay, who played the Sontaran Linx, pictured during location filming for *The Time Warrior* at Peckforton Castle, Cheshire in May 1973.

Below: Jon Pertwee joins Elisabeth Sladen for her Sarah Jane Smith photocall at Television Centre on 26 June 1973.

Bottom left: Linx was based on a sketch by costume designer James Acheson and realised by several departments. John Friedlander made the finished mask and three-fingered gloves, while Allister Bowtell were responsible for the space helmet and collar.

1974

Below: In 1973 Pertwee took delivery of 'The Alien', as the Whomobile was originally called, from designer Peter Farries. The car appeared in the 1974 stories *Invasion of the Dinosaurs* and *Planet of the Spiders*.

Right: The Whomobile was powered by a Chrysler engine that gave it a top speed of 105 mph. In 1974 Pertwee allowed the car's use in one of the company's marketing campaigns.

Bottom: The first *Doctor Who* titles written specifically for the Target range were published on 17 January. The imprint also produced badges featuring its logo.

authorship to go to his head, and permits the Doctor to open a Dalek like a hinged biscuit tin and scrape out its occupant while the rest of the cast avert their eyes in horror."

Dicks wasn't about to take this lying down. Three days later *The Times* ran the following correction: "Mr Wardle chides me for assuming that the outer casing of a Dalek conceals a living creature. Yet such is, in fact, the case. The Daleks are not, and never have been, any kind of robot – a fact clearly established on television many times, since the Daleks were created by Terry Nation well over ten years ago."

Seven Keys to Doomsday was an impressive production ultimately compromised by more than the absence of a television Doctor. The play's four-week run closely followed IRA bombs in Guildford and Birmingham, and a discarded attaché case caused a scare at the Adelphi on the first night. Families stayed away from London over Christmas and when the production closed on 11 January 1975 it had failed to recoup its estimated £35,000 cost. The size of its sets made it impractical to transfer to smaller venues.

Although *Seven Keys to Doomsday* was the first *Doctor Who* stage show, elements from the series had previously appeared in the West End. *The Curse of the Daleks* was a whodunnit set on Skaro 50 years after the creatures' defeat by mankind.

Based on a script credited to David Whitaker and Terry Nation, the play ran in matinees at Wyndham's Theatre from 21 December 1965 to 15 January 1966 while *Maigret and the Lady* was performed in the evenings.

The Curse of the Daleks and *Seven Keys to Doomsday* failed to capitalise on the strong popularity of the television series during their respective runs. Ironically, the longest-running *Doctor Who* stage production began when the series was at one of its lowest ebbs in terms of ratings. *The Ultimate Adventure* was written by Terrance Dicks and staged in 1989. A rejection of the mature, more complex style of contemporary episodes, *The Ultimate Adventure* was an accessible, occasionally tongue-in-cheek escapade that was nominally about the Daleks' latest attempt to conquer Earth. The supporting cast included Cybermen, Margaret Thatcher and a furry alien called Zog. The proceedings were temporarily halted for a number of musical interludes, which Dicks was pleased to reduce as the run continued.

The role of the Doctor was played by the 69-year-old Jon Pertwee when the production opened at London's Wimbledon Theatre on 23 March 1989. The play toured around the country, with Sixth Doctor Colin Baker taking over when it reached Oxford in June. His more dynamic portrayal of the character better suited the knockabout action of the play, but despite an enthusiastic response from fans the production

Above left: This year's *Doctor Who Annual*, with a main cover photo from *Planet of the Daleks*.

Above: Original artwork by Walter Howarth from *The House That Jack Built*, a text story from the 1975 annual. It was written by teenager Keith Miller, the head of the Doctor Who Fan Club.

Bottom left: One of the surviving puppets designed by Ian Scoones for *Planet of the Spiders*.

Below: A publicity shot of Sarah and the Queen.

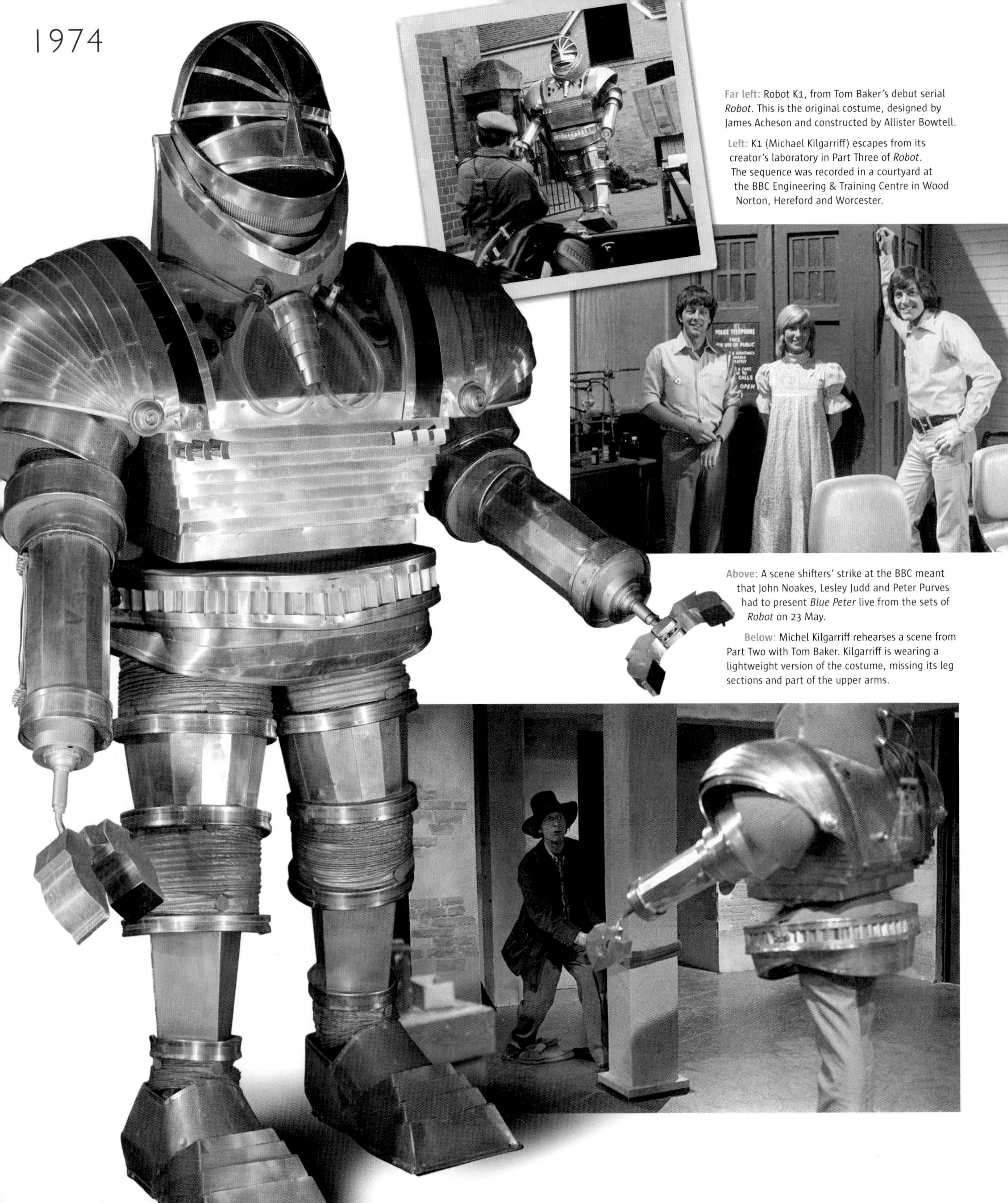

Far left: Robot K1, from Tom Baker's debut serial *Robot*. This is the original costume, designed by James Acheson and constructed by Allister Bowtell.

Left: K1 (Michael Kilgarriff) escapes from its creator's laboratory in Part Three of *Robot*. The sequence was recorded in a courtyard at the BBC Engineering & Training Centre in Wood Norton, Hereford and Worcester.

Above: A scene shifters' strike at the BBC meant that John Noakes, Lesley Judd and Peter Purves had to present *Blue Peter* live from the sets of *Robot* on 23 May.

Below: Michel Kilgarriff rehearses a scene from Part Two with Tom Baker. Kilgarriff is wearing a lightweight version of the costume, missing its leg sections and part of the upper arms.

Seen in the Stable Yard
LONGLEAT, WARMINSTER, WILTSHIRE.
Every day Easter to October

We hope that you have enjoyed your visit to this BBC Exhibition. This memento is offered with the compliments of BBC Enterprises.

An exciting new single is available from BBC Records featuring the television series theme music from "Dr. Who". The B-side of this disc is an intriguing example of the fascinating new and trendy sounds emanating from the BBC's Radiophonic Workshop. This 45rpm stereo record no. RESL 11 is modestly priced and is available at the Exhibition and at all record shops.

Left: Official *Doctor Who* exhibitions opened in Blackpool on 10 April and in the grounds of Longleat House two days later. This souvenir postcard was available from Longleat.

Below left: An iron-on transfer licensed by BBC Enterprises in 1974.

Below: This award from the Writers' Guild of Great Britain remained in the *Doctor Who* production office until the original run of the series ended in 1989.

couldn't attract big enough audiences to last beyond its final performance at Eastbourne on 19 August.

Outside the realms of pantomime – where Doctors William Hartnell, Peter Davison, Colin Baker and Sylvester McCoy have all acknowledged their television alter egos – the most successful *Doctor Who* stage production was not a play in the accepted sense. A sequel to the Jon Pertwee story *Carnival of Monsters*, *Doctor Who Live* was performed in nine arenas across the UK from October to November 2010. Subtitled *The Monsters are Coming!*, the show was hosted by the flamboyant Vorgenson (Nigel Planer), who unleashed an alien menagerie including Judoon, Silurians, Weeping Angels and Cybermen from within his 'Minimiser'. Many of the monsters roamed amongst the audience, the Eleventh Doctor (Matt Smith), Amy (Karen Gillan) and Rory (Arthur Darvill) appeared on screen in specially recorded footage and in the final act the Supreme Dalek appeared to hover above the stage. The show ended with a highly convincing video projection of the Doctor waving goodbye from the TARDIS threshold. *The Independent* responded with a review that cheekily asked "Who can fill Wembley without turning up?"

Doctor Who Live was written by Gareth Roberts (one of the show's television scriptwriters) and director Will Brenton. The event maintained the frenetic pace of the series by playing more like a concert, with Vorgenson popping up as the MC between a variety of performances. *Doctor Who* had finally been turned into successful theatre – but only through the rejection of a conventional theatrical narrative.

iron-ons
an Imagine product made in England
DOCTOR WHO and the DALEKS
E.X.T.E.R.M.I.N.A.T.E.
©BBC 1974

THE WRITERS' GUILD OF GREAT BRITAIN
ANTE OMNIA VERBUM
Awarded to THE WRITERS of "DR WHO" THE BEST BRITISH TV CHILDREN'S ORIGINAL DRAMA SCRIPT 1974

SEASON 11
(continued)
Producer: Barry Letts
Script editor: Terrance Dicks

UUU *The Time Warrior* (continued)

Part Four	5 January

WWW *Invasion of the Dinosaurs*
2 September – 30 October 1973
written by Malcolm Hulke
directed by Paddy Russell

Part One	12 January
Part Two	19 January
Part Three	26 January
Part Four	2 February
Part Five	9 February
Part Six	16 February

XXX *Death to the Daleks*
13 November – 18 December 1973
written by Terry Nation
directed by Michael E Briant

Part One	23 February
Part Two	2 March
Part Three	9 March
Part Four	16 March

YYY *The Monster of Peladon*
14 January – 27 February
written by Brian Hayles
directed by Lennie Mayne

Part One	23 March
Part Two	30 March
Part Three	6 April
Part Four	13 April
Part Five	20 April
Part Six	27 April

ZZZ *Planet of the Spiders*
22 February – 1 May
written by Robert Sloman
directed by Barry Letts

Part One	4 May
Part Two	11 May
Part Three	18 May
Part Four	25 May
Part Five	1 June
Part Six	8 June

SEASON 12
Producer: Philip Hinchcliffe
(unless otherwise stated)
Script editor: Robert Holmes

4A *Robot*
28 April – 7 June
written by Terrance Dicks
directed by Christopher Barry
Producer: Barry Letts

Part One	28 December

1975

Tom Baker's mercurial Doctor made an immediate impact in *Robot*, a story that also welcomed UNIT's medical officer Harry Sullivan (Ian Marter) as a new travelling companion.

Baker's electrifying performance, some ingenious set design and a disturbing central concept helped attract more than 11 million viewers to *The Ark in Space*. This was where new producer Philip Hinchcliffe and script editor Robert Holmes set out their stall for the reinvention of *Doctor Who* as a darker, Gothic horror show.

The Sontaran Experiment was made entirely on location in Dartmoor, recorded with the same lightweight video cameras that had been used on *Robot*. During the production of this grimly sadistic tale Tom Baker fell and broke his collar bone.

The epoch-making *Genesis of the Daleks* introduced Davros, the crippled psychopath who created a supposed master race from the mutated Kaleds. Michael Wisher's detailed performance and John Friedlander's prosthetics helped create one of the most memorable villains in the show's history.

Revenge of the Cybermen was screened at the end of the season, but had been recorded directly after *The Ark in Space* in order to recycle many of its sets.

The death of William Hartnell on 23 April was announced as part of the BBC's television news. Hartnell's health had been failing for some years, and *The Three Doctors* proved to be his final job.

Season 13 began in the Scottish Highlands with *Terror of the Zygons*, a story held back from the end of the previous recording block. At the end of Part Four Harry shunned the TARDIS to take more conventional transport home.

Planet of Evil and *Pyramids of Mars* were clearly inspired by classic horror films, as well as Gothic fiction. The former's retelling of *Dr Jekyll and Mister Hyde* and *Forbidden Planet* was enhanced by a jungle setting that numbered among the greatest sets ever seen in the programme.

Harry made a brief comeback in *The Android Invasion*, a Terry Nation script that wrung considerable intrigue from juxtaposing *Invasion of the Body Snatchers*, a rural English setting and race of rhinoceros-like aliens called Kraals.

In September, Tom Baker, Elisabeth Sladen and Ian Marter were cheered by huge crowds when they switched on the Blackpool Illuminations. Many years later, their adventures are still considered a golden age of *Doctor Who*.

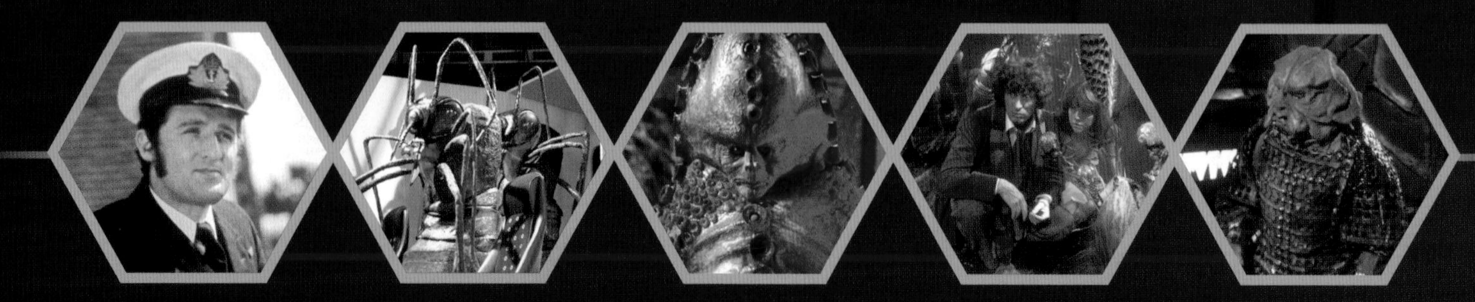

SUSPENDING DISBELIEF

Left: One of the chest unit props from *Revenge of the Cybermen*.

Above: The First Cyberman (Melville Jones, centre) reports to the Cyberleader (Christopher Robbie, right) aboard their spaceship in Part Two of *Revenge of the Cybermen*.

Top: A *Doctor Who* poster magazine, issued by Legend Publishing in May.

Doctor Who's schedule and budget have always conspired against the quality of its visual effects. During the show's original run the effects were sometimes exceptional, the earliest example being the title sequence filmed in August 1963. Generally speaking, however, the look of *Doctor Who* was often found wanting in comparison to the Supermarionation series produced by Gerry Anderson and American productions such as *Star Trek* and *Star Wars. Doctor Who* traditionally dealt with the challenge by sidestepping it, emphasising its quirky or more serious qualities as a safeguard against bigger-budgeted rivals.

The BBC's Visual Effects Department (so named to distinguish it from 'special effects' that were also required for sound) was not directly associated with the early years of *Doctor Who*. Department heads Jack Kine and Bernard Wilkie feared that the programme's demands would overstretch their

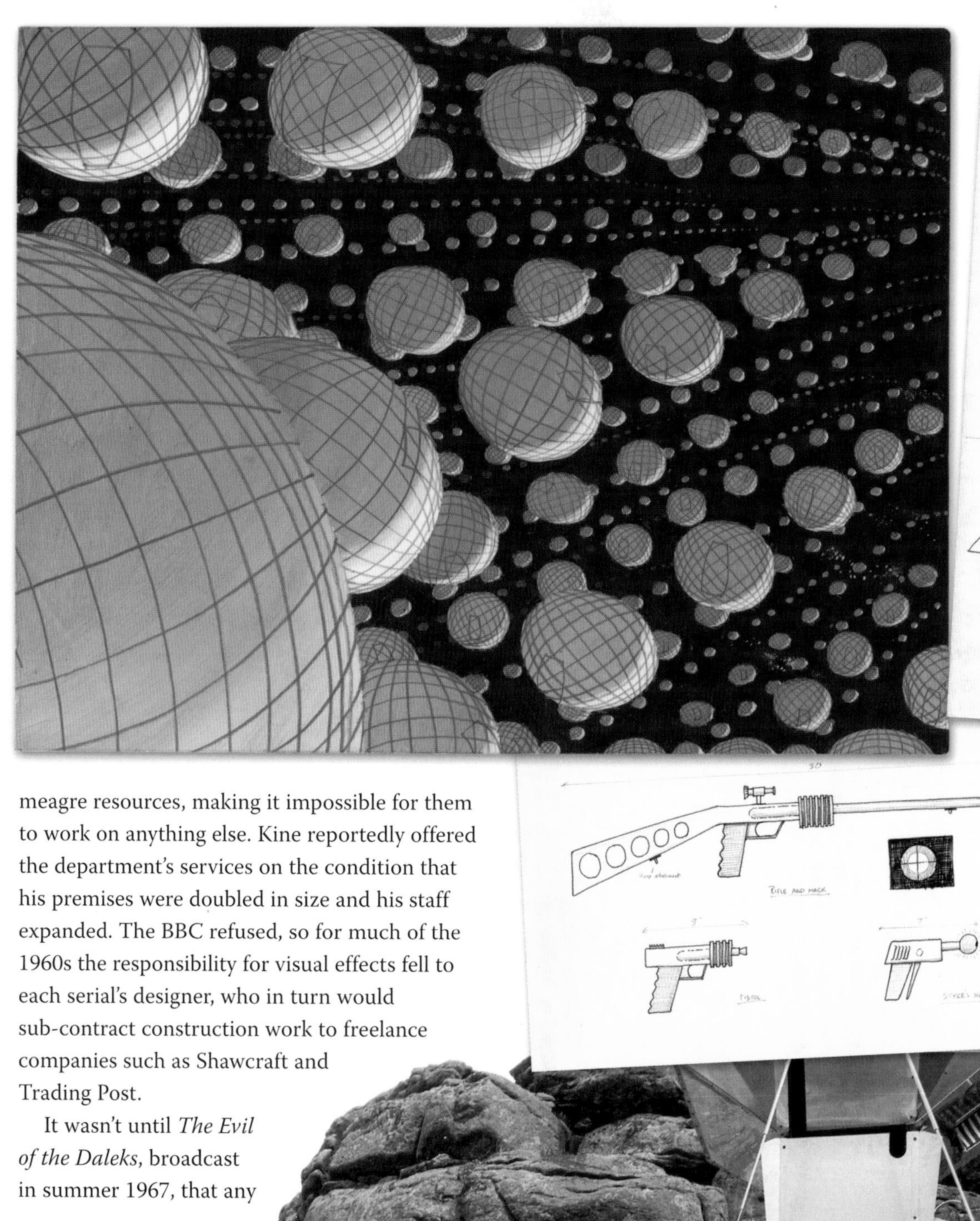

meagre resources, making it impossible for them to work on anything else. Kine reportedly offered the department's services on the condition that his premises were doubled in size and his staff expanded. The BBC refused, so for much of the 1960s the responsibility for visual effects fell to each serial's designer, who in turn would sub-contract construction work to freelance companies such as Shawcraft and Trading Post.

It wasn't until *The Evil of the Daleks*, broadcast in summer 1967, that any

Top left: Assistant floor manager Russ Karel suggested this depiction of the 'invasion warships' during post-production of *The Sontaran Experiment*. The waiting ships would have been shown on Styre's screen at the end of Part Two.

Above: Visual effects designer Tony Oxley made these sketches for the 'reflectors', the 'pressure bar', various firearms and Styre's robot for *The Sontaran Experiment*.

Left: "The female of the species..." Field Major Styre (Kevin Lindsay) assesses Sarah (Elisabeth Sladen) and Roth (Peter Rutherford) in Part Two of *The Sontaran Experiment*.

members of the Visual Effects Department were credited on *Doctor Who,* and from that point onwards the department became increasingly involved. The Pertwee years had their fair share of visual effects disasters, but there were numerous highlights such as the destruction of the church in *The Daemons* (supervised by Peter Day), the Drashigs in *Carnival of Monsters* (John Horton) and the giant maggots in *The Green Death* (built by Colin Mapson for supervisor Ron Oates).

One of the BBC's leading visual effects designers of the era was Ian Scoones, a graduate of Hammer Film Productions and Gerry Anderson's Century 21 studio before his first *Doctor Who* assignment, contributing to the outstanding miniature effects on *The Space Pirates* (1969). Scoones' enthusiasm for horror films had influenced his designs for *The Curse of Peladon* (1972) and *Planet of the Spiders* (1974), both of which were reined in by Barry Letts. By 1975, however, *Doctor Who* had a new ethos and a new producer. "Philip Hinchcliffe was a totally different kettle of fish," remembered Scoones. "He wanted to make it much more of an adult series. My favourite of all has to be *Pyramids of Mars,* which was a total spoof on Hammer's *The Mummy.*"

The effects on *Pyramids of Mars* were, in common with many *Doctor Who* stories, a collaborative effort between departments. Although Scoones was the effects supervisor, the mask worn by Sutekh (Gabriel Woolf) was made by John Friedlander, the prosthetics expert who also contributed to *The Ark in Space, The Sontaran Experiment* and *Genesis of the Daleks* the same year. The bandaged mummy robots were the responsibility of costume designer Barbara Kidd, although their skeletal internal mechanisms were created by Scoones.

As the holder of a pyrotechnics licence, Scoones was also in charge of the fire that ended the story in classic Hammer style. "It was the biggest the fire prevention

Above: The front and back of a Dalek's Death Ray wrapper, and a shop display for the chocolate mint lolly produced by Wall's.

Above right: A publicity postcard featuring an image from the previous year's *Robot* and a printed version of Tom Baker's signature.

Below: The silver version of Palitoy Bradgate's Talking Dalek, which spoke one of four messages when the button on top was pressed. The Dalek was also available in red.

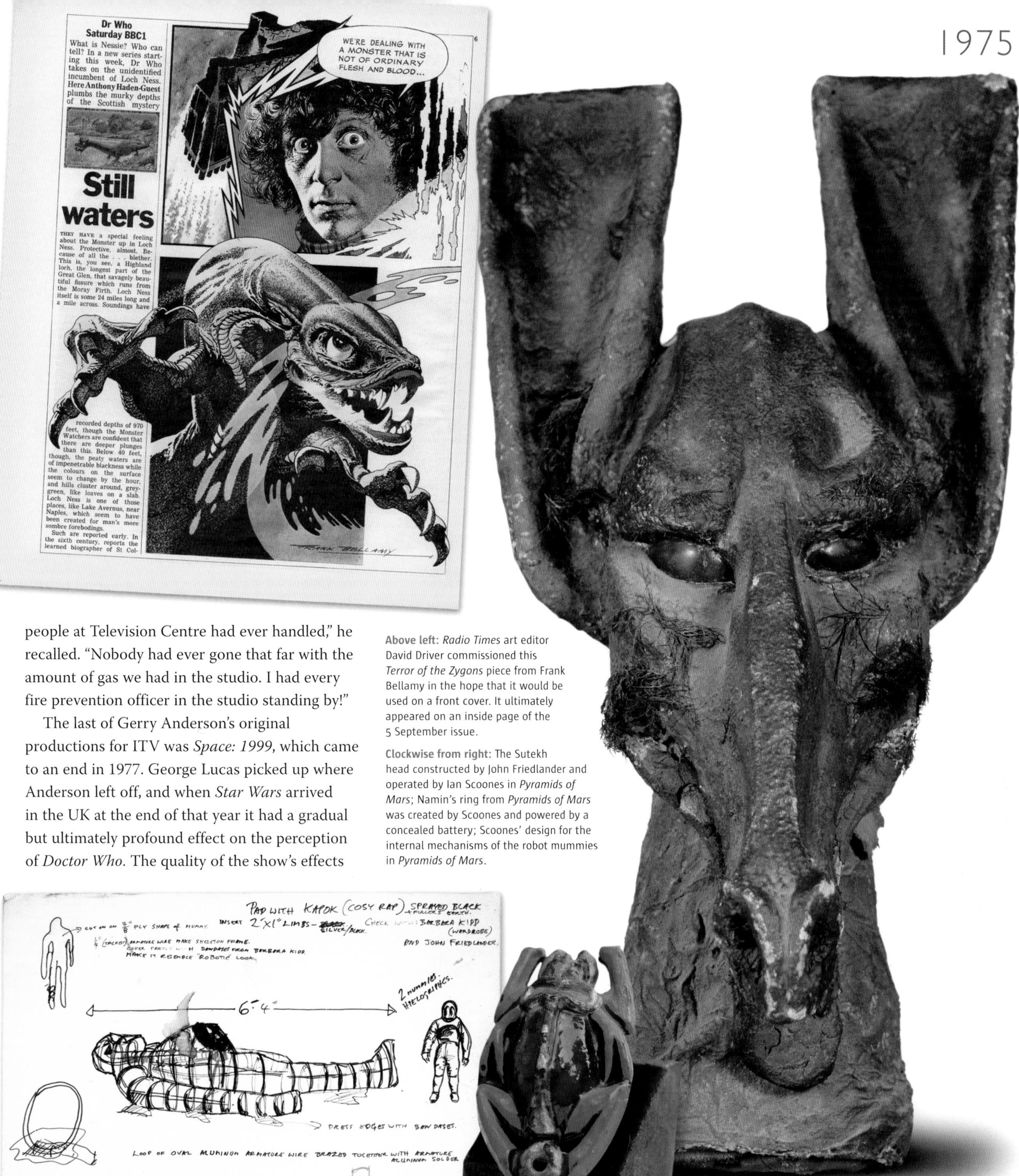

Dr Who
Saturday BBC1

What is Nessie? Who can tell? In a new series starting this week, Dr Who takes on the unidentified incumbent of Loch Ness. Here **Anthony Haden-Guest** plumbs the murky depths of the Scottish mystery

Still waters

THEY HAVE a special feeling about the Monster up in Loch Ness. Protective, almost. Because of all the . . . blether. This is, you see, a Highland loch, the longest part of the Great Glen, that savagely beautiful fissure which runs from the Moray Firth. Loch Ness itself is some 24 miles long and a mile across. Soundings have recorded depths of 970 feet, though the Monster Watchers are confident that there are deeper plunges than this. Below 40 feet, though, the peaty waters are of impenetrable blackness while the colours on the surface seem to change by the hour, and hills cluster around, grey-green, like loaves on a slab. Loch Ness is one of those places, like Lake Avernus, near Naples, which seem to have been created for man's more sombre forebodings.

Such are reported early. In the sixth century, reports the learned biographer of St Col-

people at Television Centre had ever handled," he recalled. "Nobody had ever gone that far with the amount of gas we had in the studio. I had every fire prevention officer in the studio standing by!"

The last of Gerry Anderson's original productions for ITV was *Space: 1999*, which came to an end in 1977. George Lucas picked up where Anderson left off, and when *Star Wars* arrived in the UK at the end of that year it had a gradual but ultimately profound effect on the perception of *Doctor Who*. The quality of the show's effects

Above left: *Radio Times* art editor David Driver commissioned this *Terror of the Zygons* piece from Frank Bellamy in the hope that it would be used on a front cover. It ultimately appeared on an inside page of the 5 September issue.

Clockwise from right: The Sutekh head constructed by John Friedlander and operated by Ian Scoones in *Pyramids of Mars*; Namin's ring from *Pyramids of Mars* was created by Scoones and powered by a concealed battery; Scoones' design for the internal mechanisms of the robot mummies in *Pyramids of Mars*.

Right: A selection of the *Doctor Who* cards found inside packets of Weetabix between April and June. The illustrations were by Gordon Archer.

Below: Larger packets featured this cut-out TARDIS model on the reverse. Smaller sizes included a variety of 'action settings' on which to display the cards.

Inset: A *Doctor Who* badge, issued by BBC Enterprises.

Bottom: The anti-matter monster costume was superimposed on this publicity still of Ponti (Louis Mahoney) and De Haan (Graham Weston) from Part Two of *Planet of Evil*.

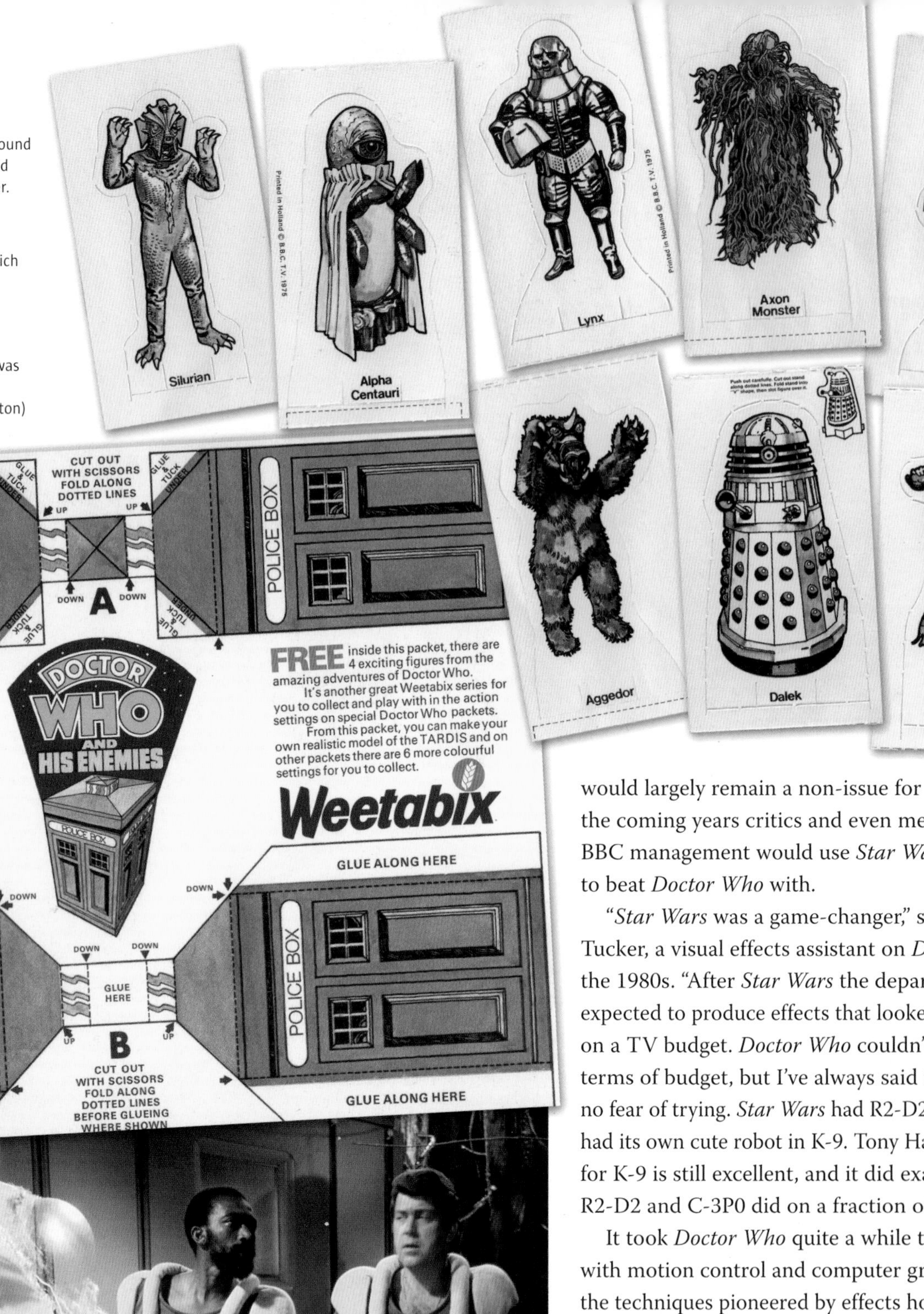

would largely remain a non-issue for fans, but over the coming years critics and even members of BBC management would use *Star Wars* as a stick to beat *Doctor Who* with.

"*Star Wars* was a game-changer," says Mike Tucker, a visual effects assistant on *Doctor Who* in the 1980s. "After *Star Wars* the department was expected to produce effects that looked just as good on a TV budget. *Doctor Who* couldn't compete in terms of budget, but I've always said the BBC had no fear of trying. *Star Wars* had R2-D2, *Doctor Who* had its own cute robot in K-9. Tony Harding's design for K-9 is still excellent, and it did exactly what R2-D2 and C-3P0 did on a fraction of the budget."

It took *Doctor Who* quite a while to catch up with motion control and computer graphics, two of the techniques pioneered by effects house Industrial Light & Magic in *Star Wars*. "Motion control was becoming affordable when I started on the show with *The Trial of a Time Lord* in 1986," says Tucker. "Computer graphics started when a company called CAL Video worked on Sylvester McCoy's first story, *Time and the Rani*, the year after. *Doctor Who* was taken off just as CG was coming down to a point where the BBC could have afforded to do it."

From 2005 to 2013 the London branch of post-production and visual effects company The Mill supervised visual effects on the revived

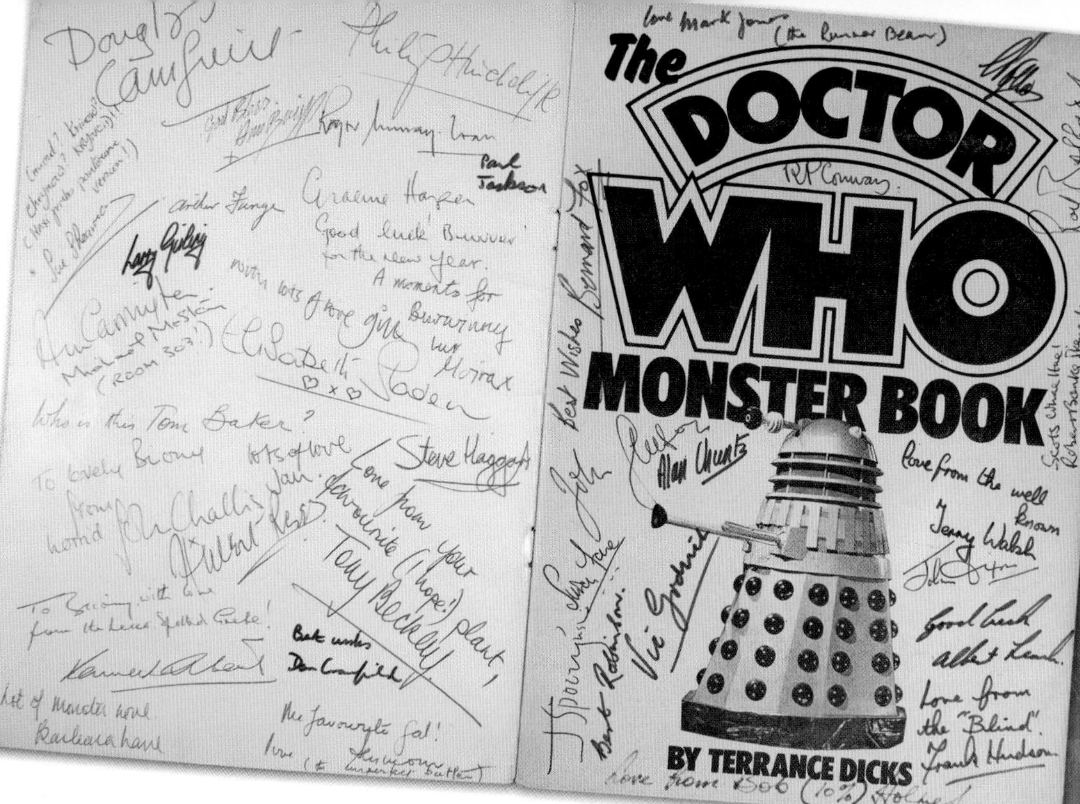

Left: From October to December *Seeds of Doom* crew member Briony Brown asked her colleagues to sign this copy of *The Doctor Who Monster Book*.

Below: The striking cover of the *Doctor Who Annual* published in September. The same photograph was used in a different setting for a Dutch edition of the book published at the same time.

programme. "I think something we bring to *Doctor Who* is that sense of scale that possibly it didn't have in the '60s, '70s and '80s when it originally aired," said Dave Houghton, *Doctor Who*'s visual effects supervisor, in 2011. "We can bring some scale to the proceedings, we can bring a sense of the wonder of alien planets."

Although visual effects are now predominantly digital, there has occasionally been room for traditional techniques. Mike Tucker returned to the series as a model unit supervisor on Series One in 2005 and Series Seven in 2013. He uses an example from Series One's *The Parting of the Ways* to illustrate how various disciplines now complement each other. "For the scenes showing the Dalek Emperor you had actors in the foreground shot against a green screen, a miniature Dalek Emperor in mid-ground, CG Daleks flying around it, a 2D matte painting sitting behind that and an animatronic creature created by Neill Gorton in the middle of it all. So that was a fantastic example of every single method under the sun, including model work that was exactly the same as it would have been in the 1960s."

Whatever the technique – from the relatively crude in-camera effects of the early 1960s to The Mill's award-winning digital vistas – the aim of visual effects has always been the same: to enable the audience to suspend its disbelief for the duration of the shot, the scene, the sequence or the episode. And the majority of *Doctor Who*'s visual effects have achieved that aim in admirable style.

"I get a bit impatient when people say, 'I loved watching *Doctor Who* because of the shaky sets,'" says the Sixth Doctor, Colin Baker. "No you didn't, you liar. You loved watching because you believed it and you were scared."

SEASON 12
(continued)
Producer: Philip Hinchcliffe
(unless otherwise stated)
Script editor: Robert Holmes

4A *Robot* (continued)
Producer: Barry Letts

Part Two	4 January
Part Three	11 January
Part Four	18 January

4C *The Ark in Space*
16 October – 12 November 1974
written by Robert Holmes
directed by Rodney Bennett

Part One	25 January
Part Two	1 February
Part Three	8 February
Part Four	15 February

4B *The Sontaran Experiment*
26 September – 2 October 1974
written by Bob Baker and Dave Martin
directed by Rodney Bennett

Part One	22 February
Part Two	1 March

4E *Genesis of the Daleks*
6 January – 25 February
written by Terry Nation
directed by David Maloney

Part One	8 March
Part Two	15 March
Part Three	22 March
Part Four	29 March
Part Five	5 April
Part Six	12 April

4D *Revenge of the Cybermen*
12 November 1974 – 17 December 1974, 10 February 1975
written by Gerry Davis
directed by Michael E Briant

Part One	19 April
Part Two	26 April
Part Three	3 May
Part Four	10 May

SEASON 13
Producer: Philip Hinchcliffe
Script editor: Robert Holmes

4F *Terror of the Zygons*
17 March – 23 April
written by Robert Banks Stewart
directed by Douglas Camfield

Part One	30 August
Part Two	6 September
Part Three	13 September
Part Four	20 September

4H *Planet of Evil* 11 June – 15 July
written by Louis Marks
directed by David Maloney

Part One	27 September
Part Two	4 October
Part Three	11 October
Part Four	18 October

4G *Pyramids of Mars* 29 April – 3 June
written by Stephen Harris (pseudonym for Lewis Greifer and Robert Holmes)
directed by Paddy Russell

Part One	25 October
Part Two	1 November
Part Three	8 November
Part Four	15 November

4J *The Android Invasion*
21 July – 25 August
written by Terry Nation
directed by Barry Letts

Part One	22 November
Part Two	29 November
Part Three	6 December
Part Four	13 December

1976

Season 13 reached its Gothic apotheosis with *The Brain of Morbius*; the blackest comedy in the show's history borrowed heavily from classic horror films in its depiction of a mad surgeon (Philip Madoc) assembling a body to house the brain of an executed Time Lord. Terrance Dicks was so angered by the changes made to his script by Robert Holmes that he insisted the serial was credited to "some bland pseudonym."

The season closed with *The Seeds of Doom*, six episodes of body horror that paid homage to *The Thing from Another World* (1951) and *The Quatermass Experiment* (1953). Its explosive finale brought the *Doctor Who* career of Douglas Camfield to a memorable end.

Holmes and producer Philip Hinchcliffe continued to exploit the theme of possession in Season 14, which opened with *The Masque of Mandragora*. The North Wales village of Portmeirion convincingly doubled as 15th century Italy, and even the TARDIS was given a Gothic interior with a wood-panelled secondary control room built around an Edwardian-style console.

It was in this room that the Doctor made an abrupt farewell to Sarah Jane Smith at the end of *The Hand of Fear*, a serial that saw her survive numerous perils only to be deposited back on Earth when the Doctor was summoned to his home planet.

The Deadly Assassin, the tale that followed, was the most audacious since *The War Games* had explained the Doctor's background seven years earlier. The first story to feature the Doctor working without a regular companion or assistant, *The Deadly Assassin* revolved around him being framed for the murder of the President at the Panopticon on Gallifrey.

Holmes scandalised many *Doctor Who* fans with his sardonic deconstruction of Time Lord society and his radical reinvention of the emaciated Master (Peter Pratt), desperately trying to extend his life beyond its natural length. This was the story that revealed only 12 regenerations were possible before Time Lords die.

The Deadly Assassin also proved divisive outside the realms of *Doctor Who* fandom, with the drowning sequence at the end of Part Three prompting the most serious complaint yet by pressure group the National Viewers' and Listeners' Association.

By the end of the year it proved impossible for the BBC to ignore the mounting calls for a less controversial style...

TEA-TIME BRUTALITY

Above: The generic folder cover for BBC Enterprises sales brochures offering new episodes of *Doctor Who* to foreign broadcasters.

Top right: "Don't you recognise me? I made you!" Solon (Philip Madoc) pleads with the deranged Morbius (Stuart Fell) in Part Four of *The Brain of Morbius*.

Above right: *Doctor Who and the Dinosaur Invasion* was published on 19 February. Target initially rejected Chris Achilleos' cover illustration, but the artist refused to remove the word 'Kklak!'

Doctor Who has freely exploited the visual grammar of the horror film ever since a lone policeman patrolled a fog-shrouded street in the opening moments of *An Unearthly Child*.

The Doctor had met Dracula and Frankenstein's Monster in *The Chase* (1965), and *The Tomb of the Cybermen* (1967) was clearly a homage to *The Mummy*. But in the mid-1970s, Seasons 12 to 14 brought a new reliance on both the iconography and the central tenets of Gothic horror. "I had a passion for the Gothic – the genuine Gothic," says producer Philip Hinchcliffe. "Bob Holmes, my script editor, had a natural predilection for the Gothic anyway."

The success of this approach was reflected not only in the viewing figures but in the enduring regard for these episodes by critics and fans. The series has often been at its most successful when horror is its dominant genre, partly because of the perception shared by many that the staging of such stories has been more competent.

"For me, *Doctor Who* at its best is a horror series," says visual effects designer Mike Tucker. "The stories I love are *Pyramids of Mars* (1975) and *The Brain of Morbius* (1976). *Doctor Who* lets itself down when it tries to do full-on outer

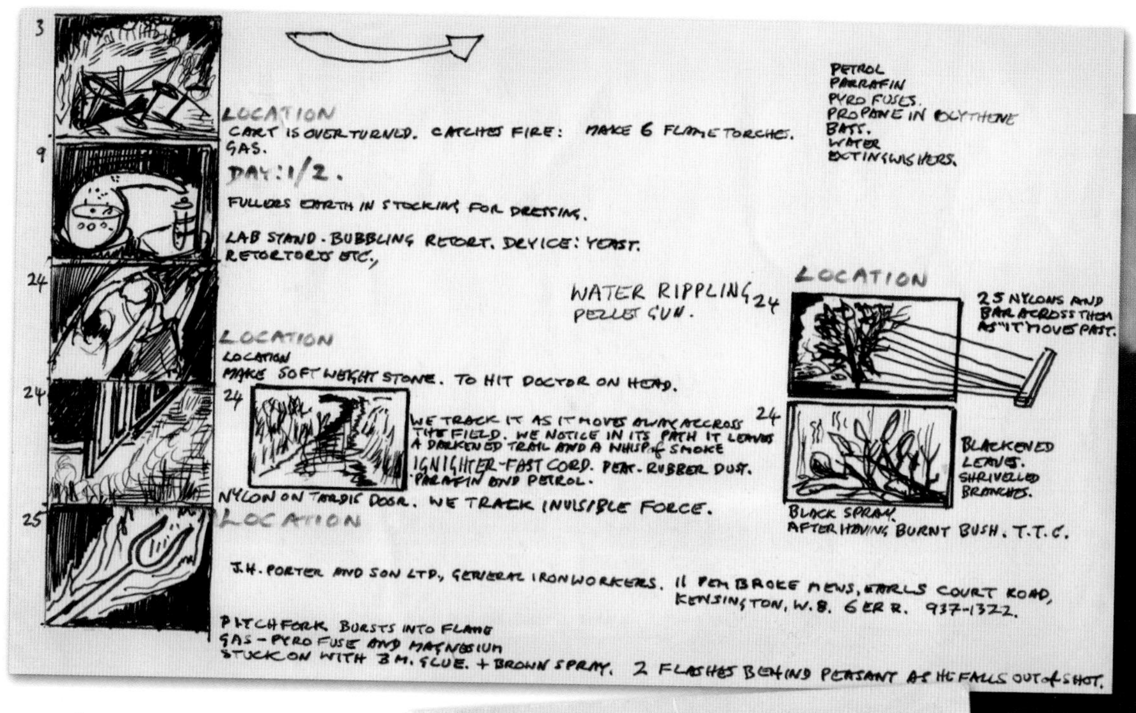

space, because not everyone gets how to do that sort of thing. I think the design teams responded best when they were presented with something they intrinsically understood. If the brief was to create the backdrop for a Gothic horror story set in ancient Egypt and a spooky old house then they got it immediately."

During these years the show presented its own take on the sometimes violent and disturbing narratives associated with Gothic horror, and from *The Ark in Space* onwards Hinchcliffe and the BBC's Head of Drama Serials Bill Slater acted as their own censors, curbing some of Robert Holmes' excesses before they reached the screen. "Bob used to sit there chuckling at the back," says Hinchcliffe. "Barry [Letts] was very serious, and liked to do themes like popular ecological problems and all this sort of thing. Bob was a bit of a devil and used to say, 'Let's scare the buggers!'"

Unfortunately for the programme, this approach came at a time when self-appointed moral guardian Mary Whitehouse and her pressure group the National Viewers' and Listeners' Association were at the height of their powers. Whitehouse had been a thorn in the BBC's side since the 1960s, but for a time

Top: Ian Scoones' visual effects storyboards for Part One of *The Masque of Mandragora*.

Top right: Scoones prepares to film the Mandragora Helix effect using a camera positioned over a Perspex cone containing a thin wallpaper paste solution. The whirlpool effect seen on the TARDIS scanner in Part One was achieved by spinning the solution with an electric drill motor.

Above: A publicity shot of the Doctor and Sarah taken at Athelhampton House in Dorset, during location recording for *The Seeds of Doom* in late 1975.

Right: This Krynoid pod prop was made by visual effects designer Richard Conway and used in Part One of *The Seeds of Doom*.

1976

Below: Paul Crompton's original artwork for the back and front covers of the *Doctor Who Annual* published in September. The latest edition was much larger than previous books in the series.

Bottom: This Dalek bubble bath was produced by The Water Margin and featured a screw cap disguised as the Dalek's dome.

she briefly diverted her attention from controversial screenwriters such as Johnny Speight and Dennis Potter, instead launching a sustained attack on *Doctor Who.* Her regular letters to BBC executives on the subject express some of her most notorious prejudices. Whitehouse criticised *The Sontaran Experiment*'s depiction of "helpless adults in a state of terror," before claiming that *Genesis of the Daleks* had turned *Doctor Who* into "tea-time brutality for tots. This series has moved from fantasy to real-life violence with cruelty, corpses, poison gas and Nazi-type stormtroopers, not to mention revolting experiments in human genetics." In her opinion, the programme now belonged after the nine o' clock watershed.

Some time later, *Daily Express* columnist Jean Rook recalled *Pyramids of Mars* with a shudder: "While I was frying his fish fingers, my child was alone in a room with a programme which could have screwed up and permanently crunched his nerve with one mummified hand," she wrote. "*Dr Who* is no longer suitable for children... it has grown out of a rubber monster show into a full, scaly,

Above: A window sticker advertising Ty-Phoo's *Doctor Who* promotion, which ran from July to September.

Left: Some of the cards that could be collected in boxes of Ty-Phoo tea bags and (far left) the cover of the hardback book available by mail order from the company. The book included a poster with spaces to attach all 12 cards.

Below: The cuddly owl retained by Elisabeth Sladen as a memento of Sarah's final scenes in *The Hand of Fear*.

unknown horror programme. Compared with it, an old Hammer movie wouldn't crack toffee."

Mary Whitehouse resumed her campaign in the new year. On 25 January 1976 she wrote to BBC Chairman Sir Michael Swann, describing *The Brain of Morbius* as containing "some of the sickest and most horrific material ever seen on children's television." Several weeks later she objected to *The Seeds of Doom*: "Strangulation – by hand, by claw, by obscene vegetable matter – is the latest gimmick. And, just for a little variety, [they] show the children how to make a Molotov Cocktail."

A misconception shared by these detractors was that *Doctor Who* was a children's programme. It had never been produced by the BBC's Children's Department and had never been aimed squarely at a junior audience. But it was undeniably watched by a significant number of children, and it was the show's potential influence on them that allowed Mrs Whitehouse her ultimate triumph.

Following the Part Three cliffhanger of *The Deadly Assassin*, Whitehouse pointed out that the BBC's own guidelines had been contravened by the sequence where the Doctor was apparently drowned by having his head forced underwater. She quoted a five-year-old boy of her acquaintance who, having seen the episode, told his mother what he intended to do to his younger brother next time he misbehaved: "I shall hold his head under the bath water until he's still, like the man did with Doctor Who..."

In this instance, the NVALA's complaint was at least partly upheld by the BBC's Director General, Sir Charles Curran. He wrote back to Mrs Whitehouse, assuring her that Bill Slater had already edited the sequence before transmission, "but would have liked to have cut out just a few more frames of the action than he did." Whitehouse wrote a victorious letter to *The Daily Telegraph*, highlighting "this acknowledgement that a mistake of judgement has been made." By the time the serial was repeated the following

Top: The Doctor leaves a note and a crude decoy in the TARDIS control room in Part One of *The Deadly Assassin*.

Above: When *TV Comic* was relaunched in a tabloid format in September, a separate *Mighty Midget Doctor Who Comic* was included as a free gift.

Left: The Time Lord robes for *The Deadly Assassin* were designed by James Acheson and Joan Ellacott. The main gowns were made by the Costume Department, while the headpieces and collars were produced by Allister Bowtell. Acheson and director David Maloney chose different colours to denote rank or class on Gallifrey (inset).

August its master tape had been permanently edited.

Hinchcliffe later dismissed the *Deadly Assassin* controversy, arguing that the moment in question was clearly established as part of a dream sequence. Nevertheless, when Graham Williams became the next producer in 1977 he was given strict instructions to tone down the show's more visceral content and its transmission time was moved to a later slot.

The debate over violence in the programme resumed in 1985, but it wasn't until 2005 that the series gained a new strain of Gothic horror. Steven Moffat's *The Empty Child* featured disturbing imagery in the shape of zombie-like assailants with gas masks fused to their faces. "The gas masks are as far as we push it," said showrunner Russell T Davies at the time. "I'm glad we pushed it, because they're absolutely terrifying, and fantasy-like enough not to really disturb very young kids too much."

Perhaps the greatest of Moffat's nightmare scenarios was the award-winning *Blink* (2007), which introduced the Weeping Angels – Gothic stone statues that only strike when their victims aren't looking.

The scripts by Moffat and fellow horror advocates are calculated to frighten, but have attracted none of the controversy of the Hinchcliffe/Holmes years. Of course audiences are now more sophisticated, even desensitised, but a change in the programme's format has helped.

Life-and-death cliffhangers between episodes are now relatively rare, so the audience no longer has to wait a week to discover whether their heroes will survive. The happy ending, and the rational explanation of the supernatural, generally occur at the end of each self-contained story.

Doctor Who continues to borrow the iconography of Gothic horror, but now wakes from its nightmares in the final act.

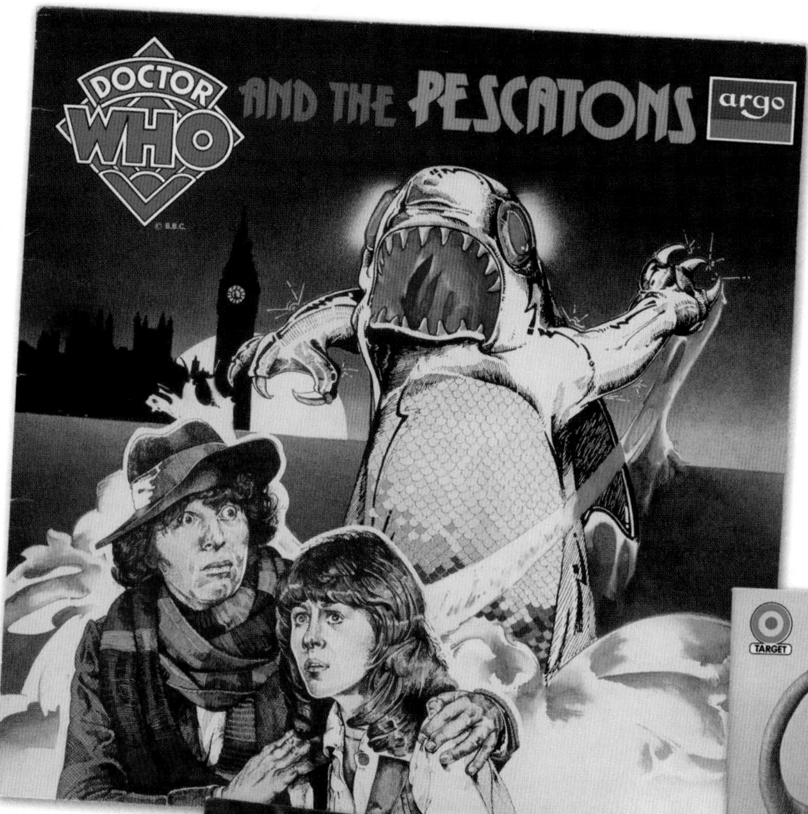

Left: *Doctor Who and the Pescatons* was the first commercially available audio production based on the series. The two-part adventure was written by Victor Pemberton and starred Tom Baker, Elisabeth Sladen and Bill Mitchell. It was released as an LP record and cassette by the Decca label Argo in August.

Below: Written by Terrance Dicks and illustrated by George Underwood, *The Doctor Who Dinosaur Book* was published on 16 December and proved so successful that it spawned a number of other educational titles over the next few years.

Below left: The latest 7" reissue of the *Doctor Who* theme was packaged in this sleeve.

SEASON 13 (continued)
Producer: Philip Hinchcliffe
Script editor: Robert Holmes

4K *The Brain of Morbius*
6 – 24 October 1975
written by Robin Bland (pseudonym for Terrance Dicks and Robert Holmes)
directed by Christopher Barry

Part One	3 January
Part Two	10 January
Part Three	17 January
Part Four	24 January

4L *The Seeds of Doom*
30 October – 19 December 1975
written by Robert Banks Stewart
directed by Douglas Camfield

Part One	31 January
Part Two	7 February
Part Three	14 February
Part Four	21 February
Part Five	28 February
Part Six	6 March

SEASON 14
Producer: Philip Hinchcliffe
Script editor: Robert Holmes

4M *The Masque of Mandragora*
28 April – 8 June
written by Louis Marks
directed by Rodney Bennett

Part One	4 September
Part Two	11 September
Part Three	18 September
Part Four	25 September

4N *The Hand of Fear* 14 June – 20 July
written by Bob Baker and Dave Martin
directed by Lennie Mayne

Part One	2 October
Part Two	9 October
Part Three	16 October
Part Four	23 October

4P *The Deadly Assassin*
26 July – 2 September
written by Robert Holmes
directed by David Maloney

Part One	30 October
Part Two	6 November
Part Three	13 November
Part Four	20 November

1977

The introduction of the scantily clad savage Leela (Louise Jameson) as a new companion must have done little to allay fears that the series was becoming too violent. In an effort to lighten the mood Tom Baker added a scene to Part One of *The Face of Evil* where the Doctor threatens to kill an opponent with one of his trademark jelly babies.

The Robots of Death was an Agatha Christie-style whodunnit set on a futuristic Sandminer. Kenneth Sharp's Art Deco design for the vessel extended to the crew's flamboyant costumes and, most impressively, the murderous robots themselves.

Philip Hinchcliffe's final production was *The Talons of Weng-Chiang*. Robert Holmes created a rich Gothic soup from Sherlock Holmes, *The Phantom of the Opera* and Fu Manchu in his tale of a ruthless time fugitive lurking beneath Victorian London. Holmes' favourite story – and one of the show's acknowledged classics – *The Talons of Weng-Chiang* was the focus of the first *Doctor Who* documentary when BBC2's *The Lively Arts* followed its production.

Season 15 got off to a tricky start for new producer Graham Williams when Terrance Dicks' *The Vampire Mutation* was suddenly cancelled. Head of Drama Serials Graeme McDonald argued it could be perceived as a parody of the BBC's forthcoming, and highly prestigious, adaptation of *Dracula*. Dicks hastily assembled a substitute in *Horror of Fang Rock*, which placed the occupants of an early 20th century lighthouse at the mercy of a shape-shifting Rutan.

The Invisible Enemy was the first indication that the direction of the series would change under Williams' guidance. A bizarre science fiction adventure in the style of the 1966 film *Fantastic Voyage*, the story's appeal to younger viewers was confirmed in Part Four when the Doctor and Leela were joined on their travels by a mobile computer called K-9.

The mechanical dog was confined to the TARDIS for the duration of the atmospheric *Image of the Fendahl*, but emerged to prove himself by Leela's side in *The Sun Makers*. Robert Holmes' arch condemnation of a cruel plutocratic society incorporated a number of satirical swipes at the Inland Revenue.

From *The Sun Makers* onwards Williams worked alongside Anthony Read, Holmes' replacement in the *Doctor Who* office. Before they could realise Williams' grand design for the series, they would be forced to collaborate on another last-minute script...

MUTUAL APPRECIATION

Centre: The control deck of the Sandminer, pictured during recording of *The Robots of Death*.

Below: An original Voc mask from *The Robots of Death*. Designer Kenneth Sharp realised the finished look of the masks in collaboration with sculptor Rose Garrard.

On Saturday 6 August 1977 a small piece of social history was made in Battersea, south London, when Broomwood Church Hall became the venue for the world's first *Doctor Who* convention. The guests included Tom Baker, his co-star Louise Jameson, Jon Pertwee, Graham Williams and visual effects designer Mat Irvine. None of them were paid for attending, and they mingled freely with the 200 attendees.

The event was organised by the Doctor Who Appreciation Society, but this was far from the beginning of organised *Doctor Who* fandom. The William Hartnell Fan Club had been established in Stoke-on-Trent in the mid-1960s, giving way to the Doctor Who Fan Club when Patrick Troughton assumed the role.

In 1971, 13-year-old Scottish fan Keith Miller contacted the *Doctor Who* production office to ask about running the club. With the support and encouragement of Barry Letts' secretary Sarah Newman, Miller took over at the end of the year. Miller duplicated *The Doctor Who Fan Club Monthly* using equipment at his school, but as his mailing list expanded Newman offered to handle the duplication and even the postage at the BBC.

Under the patronage of Letts and Newman, and with the support of Jon Pertwee, Roger Delgado and other members of the cast, Miller's Doctor Who Fan Club provided a relatively basic but unique service. It was superseded in 1976, when Stephen Payne and Jan-Vincent Rudzki founded the Doctor Who Appreciation Society on the suggestion of Tom Baker.

During the late 1970s improved print technology, including the advent of economical photocopying, made it possible for the society to distribute an array of material to members, including its newsletter *Celestial Toyroom*, fanzines *TARDIS* and *Cosmic Masque*, and a variety of synopses and other material from its Reference Department.

In the interview sessions for the *Lively Arts* documentary *Whose Doctor Who*, broadcast by BBC2 in 1977, Philip Hinchcliffe discussed the role the society played in helping the production team. "The first I knew about it was when a computer printout landed on my desk which had every detail about the programme since the very first transmission; who'd written it, who'd directed it, etc. And so I rang these people up and found out

Top: John Bloomfield's costume designs for the Doctor and Leela in *The Talons of Weng-Chiang*.

Above: Working from initial designs by Michaeljohn Harris, visual effects assistant George Reed created this reference illustration of the trionic lattice – the key to Magnus Greel's Time Cabinet in *The Talons of Weng-Chiang*.

Right: Louise Jameson as Leela, wearing a costume designed by John Bloomfield for *The Face of Evil*.

Below: Leela's knife, and the blowpipe she used in *The Talons of Weng-Chiang*.

Below: Pages from a Denys Fisher catalogue showing prototypes of the *Doctor Who* toys issued by the company in September.

Bottom: Two of the action figures from the range, and packaging for the Giant Robot. Following an accident during the preparation of the Doctor figure, Denys Fisher substituted a head based on *New Avengers* star Gareth Hunt.

who they were and we've had quite a good working relationship ever since. We use them sometimes. We say, 'Can you tell us what happened ten years ago with such-and-such a character?' Because they're hot on continuity and they keep records. They have a historian, and we get into deep water with them if we do something which they think destroys the continuity of the programme."

The BBC's support of the DWAS may have been offered in the hope that it would act as a conduit for fans. As producer Graham Williams remembered, in the late 1970s the amount of fan mail received by the production office was becoming a problem. "We took a tremendous amount of time out of the actual programme to take care of it. My secretary, Ann Rickard, and I instigated a system whereby all the fan mail was replied to – we spent a fortune on photographs – and made sure that everything we could do was done. In fact, the interest was such that in my last year I asked if I could have some back-up in the form of an associate producer and/or another secretary. They said 'no' to both, yet in the last six months I was criticised for spending too much time on the Enterprises side of things and too much time with the fans!"

The DWAS became the officially recognised fan organisation, approved by the production office and tolerated by BBC Enterprises in exchange for not encouraging the sale or distribution of copyright material such as audio recordings of episodes.

The Holy Grail for many fans was the opportunity to *see* old episodes, and

they got the chance when *An Unearthly Child* was screened at the second DWAS convention, PanoptiCon '78. The event was once again headlined by Tom Baker and Jon Pertwee, with other guests including Nicholas Courtney, Terrance Dicks, Robert Holmes and William Hartnell's widow Heather.

Gareth Roberts, who many years later would become one of *Doctor Who*'s television writers, was in the audience when the society showed the whole of *100,000 BC* at PanoptiCon '79. "I was only

Top: Two strips of Weetabix cards, illustrated by Gordon Archer and issued as part of the cereal's second *Doctor Who* promotion from March to May.

Top right: The packets featured *Doctor Who* games which could be played with the cards and these cut-out tokens and spinners.

Above: Artwork from the Weetabix sales reps' promotional campaign.

Right: Elements of the original artwork for the front cover of the *Dalek Annual* published in September. Four such annuals were published, cover-dated 1976 to 1979.

Below right: Tom Baker, Louise Jameson and K-9 outside the BBC's North Acton rehearsal rooms during production of Season 15.

Below: For *The Sun Makers*, costume designer Christine Rawlins reworked John Bloomfield's original design for Leela's costume in a lighter chamois leather, adding details such as lace-up armbands.

11 so my mum didn't want me going into London on my own. We asked the DWAS if she could accompany me and they said, 'Only if she joins the DWAS.' So we'd get two issues every month of their *Celestial Toyroom*, one addressed to me and one to my mum. I don't think the interview with Gerry Davis particularly gripped her."

He was already rather jaded by the experience of being a DWAS member. "I thought it was going to be people talking about how jolly *Doctor Who* was," he says. "Instead, it was people complaining that Graham Williams was rubbish, and I didn't even know who he was!"

Williams and his script editor Douglas Adams had certainly added a light-hearted veneer to many of their episodes, partly as a result of the backlash against the perceived horror and violence of the Philip Hinchcliffe/Robert Holmes years. While the show's popular appeal was undimmed, this superficial element of undergraduate humour didn't click with the earnest undergraduates of the DWAS. Organised fandom was giving *Doctor Who* fans a reputation for being cynical about their favourite show.

Despite this, membership of the DWAS continued to rise, boosted in no small part by its presence at the 20th anniversary celebration organised by the BBC in 1983. The most renowned convention ever held in the UK – if not necessarily the best organised – the Longleat weekend attracted more than 35,000 visitors on its first day alone.

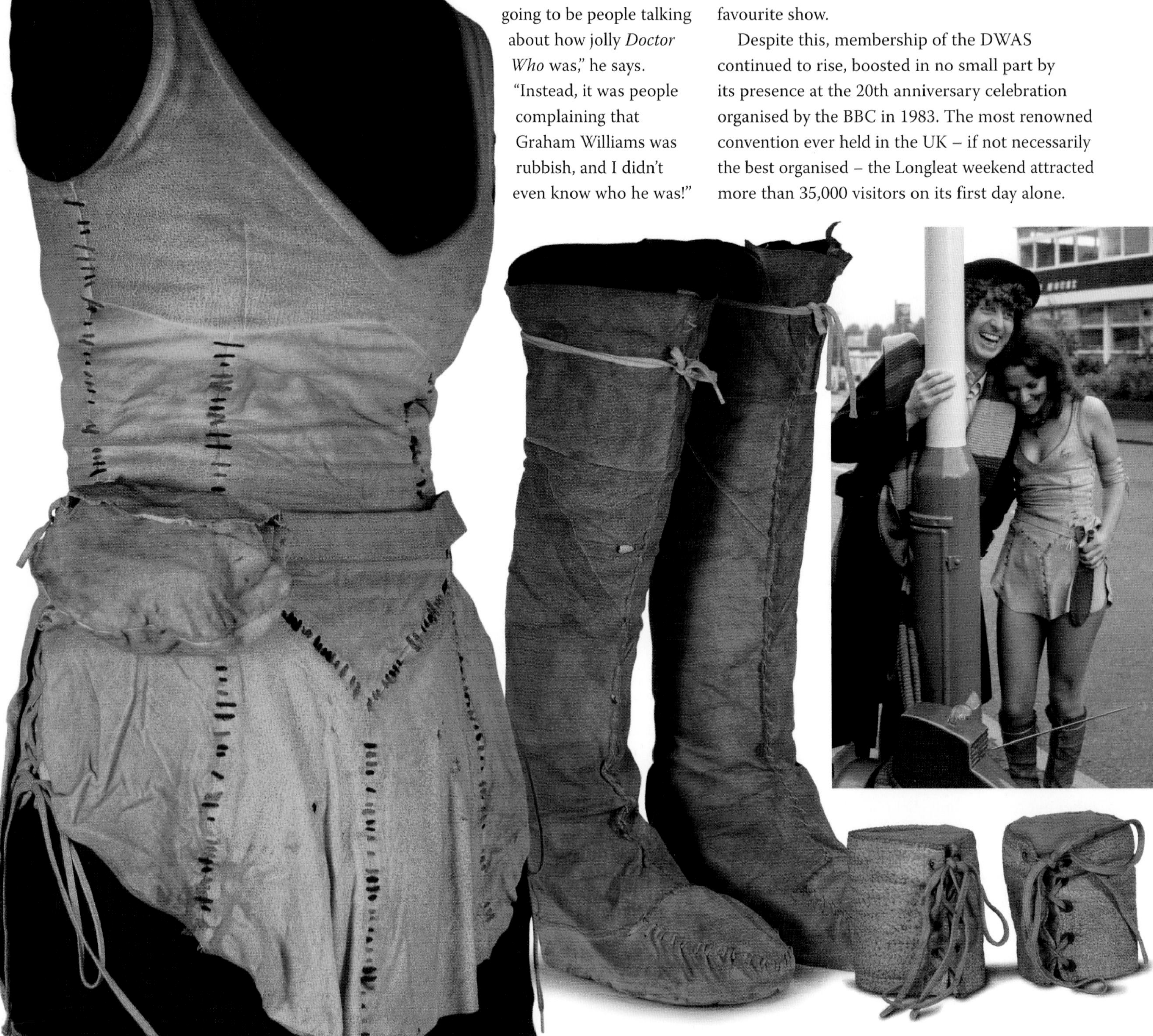

Left: A publicity picture of Wanda Ventham as the possessed Thea Ransome, from Part Four of *Image of the Fendahl*.

Below: The K-9 prop, which made its debut in *The Invisible Enemy*, was based on designs by the Visual Effects Department's Tony Harding.

Right: An illustration by Paul Crompton featured on the cover of the *Doctor Who Annual* published in September.

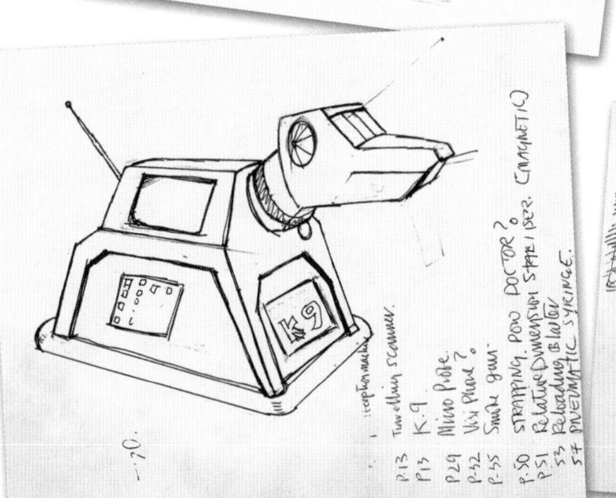

By the middle of the decade membership of the DWAS had peaked at more than 3,000. The subsequent cancellation of the programme, the specialised nature of the official *Doctor Who Magazine* and the rise of the internet all diminished the necessity for a formal fan organisation.

The DWAS is still the pre-eminent *Doctor Who* fan club, however, and the convention circuit continues to thrive. Like Longleat, the biggest events are now once again sanctioned by the BBC. What started in a church hall in 1977 is now significantly more than a cottage industry, providing a valuable income to many licensees and *Doctor Who* alumni.

"When I left the series, nearly 20 years ago, I never imagined that *Doctor Who* would always remain such a large part of my life," said Jon Pertwee in 1993. "Mind you, I've always been utterly aware of the effect science fiction has on its followers... I did a series about a talking scarecrow called *Worzel Gummidge*. It had 12-and-a-half million viewers a week, but if I held a *Worzel Gummidge* convention about four people would turn up. That's the difference!"

SEASON 14 (continued)
Producer: Philip Hinchcliffe
Script editor: Robert Holmes

4Q *The Face of Evil*
20 September – 26 October 1976
written by Chris Boucher
directed by Pennant Roberts

Part One	1 January
Part Two	8 January
Part Three	15 January
Part Four	22 January

4R *The Robots of Death*
2 November – 7 December 1976
written by Chris Boucher
directed by Michael E Briant

Part One	29 January
Part Two	5 February
Part Three	12 February
Part Four	19 February

4S *The Talons of Weng-Chiang*
13 December 1976 – 10 February 1977
written by Robert Holmes
directed by David Maloney

Part One	26 February
Part Two	5 March
Part Three	12 March
Part Four	19 March
Part Five	26 March
Part Six	2 April

SEASON 15
Producer: Graham Williams
Script editor: Robert Holmes
(unless otherwise stated)

4V *Horror of Fang Rock* 26 April – 9 June
written by Terrance Dicks
directed by Paddy Russell

Part One	3 September
Part Two	10 September
Part Three	17 September
Part Four	24 September

4T *The Invisible Enemy*
22 March – 26 April
written by Bob Baker and Dave Martin
directed by Derrick Goodwin

Part One	1 October
Part Two	8 October
Part Three	15 October
Part Four	22 October

4X *Image of the Fendahl*
1 August – 6 September
written by Chris Boucher
directed by George Spenton-Foster

Part One	29 October
Part Two	5 November
Part Three	12 November
Part Four	19 November

4W *The Sun Makers* 13 June – 19 July
written by Robert Holmes
directed by Pennant Roberts
Script editor: Anthony Read

Part One	26 November
Part Two	3 December
Part Three	10 December
Part Four	17 December

1978

Graham Williams was beset by bad luck during his turbulent tenure as producer. An overspend on Season 15's penultimate serial *Underworld* forced him to superimpose actors in front of models instead of building the missing sets. The production was saved, but the results looked especially impoverished in comparison with the recently released *Star Wars*.

Williams aimed to expand on *The Deadly Assassin*'s depiction of Time Lord society in the season finale, which was commissioned under the title *Killers of the Dark*. Unfortunately writer David Weir's scenario was far too ambitious for the limited budget, so Williams and Anthony Read hastily drafted *The Invasion of Time* as a replacement. Shortly afterwards, the story narrowly averted cancellation as a consequence of industrial action at the BBC.

In Part Six the Sontarans' convoluted invasion plan was thwarted and Leela made the surprising decision to remain on Gallifrey with her new love, Chancellery guard Andred (Chris Tranchell). K-9 stayed with them.

In Season 16 Williams realised his long-held ambition to connect every story with a linking narrative. The Doctor's mission to identify and recover the six segments of the Key to Time encompassed *The Ribos Operation*, *The Pirate Planet* (written by newcomer Douglas Adams), *The Stones of Blood*, *The Androids of Tara*, *The Power of Kroll* and, in the new year, *The Armageddon Factor*. The Doctor was assisted by a new version of K-9 (still voiced by John Leeson) and the elegant Time Lady Romana (Mary Tamm).

The Stones of Blood was *Doctor's Who*'s 100th story and coincided with the 15th anniversary, so Tom Baker, Mary Tamm and John Leeson improvised a scene where the Doctor was given a birthday cake. Williams was invited to a rehearsal and immediately vetoed the idea; he had already been advised by Graeme McDonald that humour was getting out of hand. For his part, Baker was developing his own light-hearted script for a *Doctor Who* film, and becoming frustrated by what he considered to be his lack of input to the series.

During the recording of *The Power of Kroll* Baker threatened to resign unless he was given approval on scripts, casting and directors. For the put-upon Williams, a temperamental star was just the latest in a long line of problems.

CONTINUING DRAMA

In terms of its narrative structure, Season 16 was quite unlike any previous series of *Doctor Who*. Each of its six stories was linked by the Doctor's mission to locate and reassemble the fragmented sections of the Key to Time, a perfect cube that maintained the equilibrium of the universe.

Quests for keys had appeared in *Doctor Who* before, notably in Season One's *The Keys of Marinus* and more recently in the stage play *Seven Keys to Doomsday*. In fact the quest had been something of an archetype during the early years of the show, with the First Doctor trying to get Ian and Barbara home again and often struggling to retrieve, or return to, his TARDIS.

With hindsight, umbrella themes can also be discerned in the series' early history. It could be argued that developments from the first episode of *100,000 BC* to the final episode of *The War Games* comprise an epic narrative establishing the Doctor's freedom and concluding with his exile. The events in *Spearhead from Space* through to *The Three Doctors* led to a reversal of this situation.

However, during the 1960s and early '70s continuity was the preserve of the TARDIS crew, their affiliates and adversaries – it was characters, not plots, that interlinked stories. It wasn't until 1974 that a producer and script editor devised a loose narrative interlinking several stories within a series. Philip Hinchcliffe

Top: Tom Baker and friends outside the US Embassy in London's Grosvenor Square on 14 February. This photocall celebrated the sale of Baker's first four seasons to American television.

Above: K-9 wears the Crown and Sash of Rassilon, watched by Andred (Chris Tranchell) and the Doctor in Part Four of *The Invasion of Time*.

Left: *Doctor Who Discovers Strange and Mysterious Creatures* was published by Target on 20 April, with a cover by Jeff Cummins. This was the last in the short series of non-fiction *Doctor Who Discovers* titles.

Right: Derek Deadman in full make-up as Stor and Stuart Fell, minus Sontaran mask and collar, wait for their cue during location recording for *The Invasion of Time*.

Left: The Target novelisation of *Death to the Daleks* was published in July, with a cover illustration by Roy Knipe. *The Time Warrior* was published in June, with a cover by Jeff Cummins.

Below: The first segment of the Key to Time is revealed in Part Four of *The Ribos Operation*.

Bottom: One of the original Key to Time props designed by Dave Harvard and the Tracer used to locate each piece. The interlocking segments are cast in resin.

and Robert Holmes ordered much of Season 13 around their justification for reusing the Nerva Beacon sets from *The Ark in Space* in *Revenge of the Cybermen.*

In 1976, when Graham Williams pitched to become the next producer of *Doctor Who*, he suggested an umbrella theme that would encompass an entire season. His proposal to Graeme McDonald, the BBC's Head of Serials, was dated 30 November and outlined *Doctor Who*'s first pre-meditated story arc: "Over 26 episodes, telling six individual stories, [the Doctor] will recover the Sections which form the Key to Time... [He] will be seen to have six independent ventures, all linked by the common theme... Those who wish to join him in episode one and follow him through to episode twenty-six will gain the momentum and bonus of following the story through. Those who choose to watch only one venture will enjoy it for its own sake – the scope in each venture is as wide and as free-ranging as ever – but should be encouraged as far as possible to see what happens next."

Williams assumed his duties in early 1977, but he wasn't able to put the first of his Key to Time stories into production until spring the

1978

ROMANA
DR. WHO
BBC

Above: The BBC's publicity postcard for Romana, as played by Mary Tamm.

Right: June Hudson's costume design for Romana in *The Ribos Operation*.

Below: The flamboyant gown was so eye-catching that, as Mary Tamm recalled, Tom Baker resented its scene-stealing properties.

POLICE PUBLIC CALL BOX

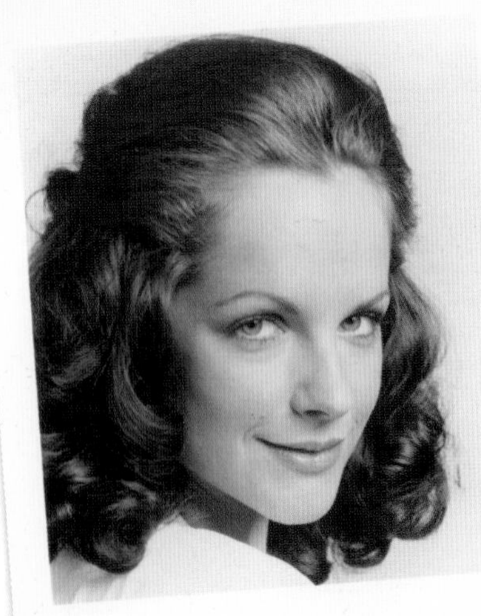

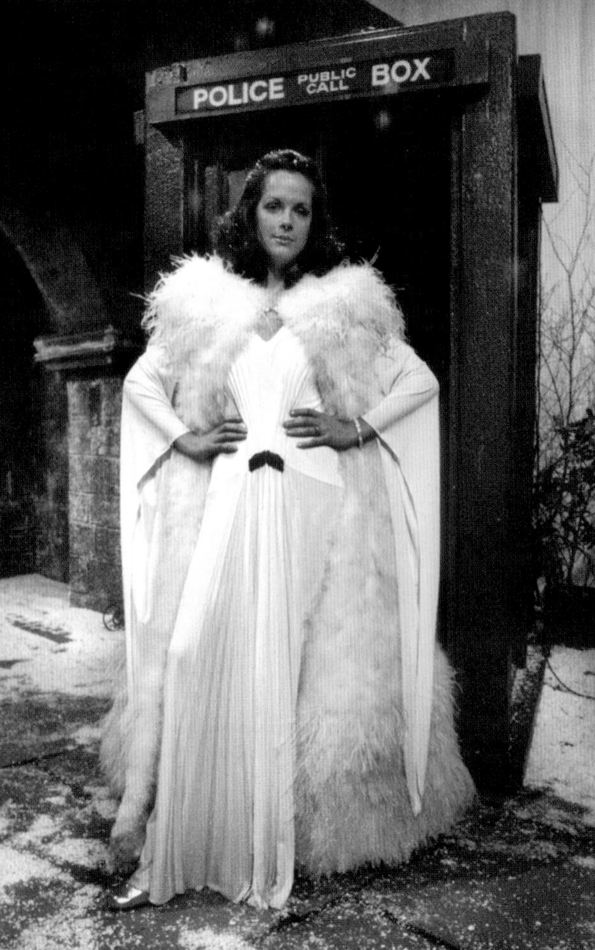

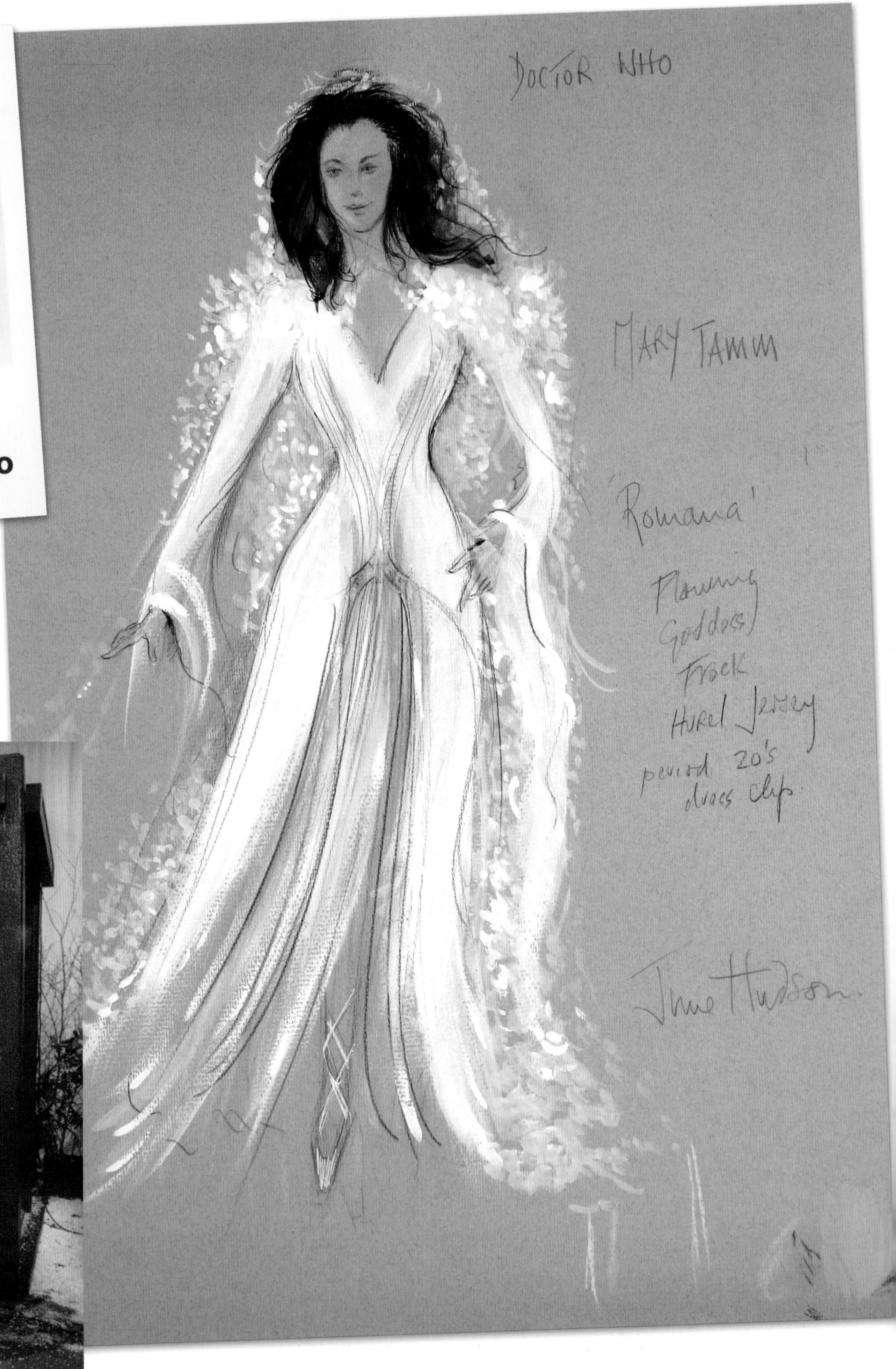
DOCTOR WHO
MARY TAMM
'Romana'
Flowing Goddess Frock
Hurel Jersey
period 20's dress clip
June Hudson.

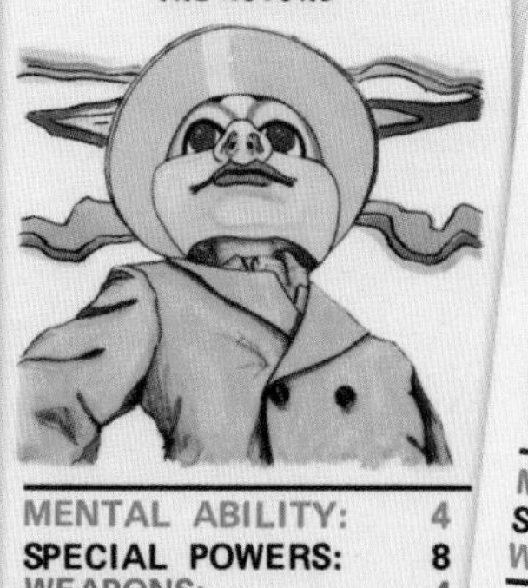

following year. *The Ribos Operation*, the opening instalment of Season 16, was written by Robert Holmes as a standalone serial. The elements establishing the Key to Time were added by Williams and his script editor Anthony Read.

In Part One the Doctor is approached by the White Guardian (Cyril Luckham), who asks him to recover the scattered fragments of the Key to Time "before the universe is plunged into eternal chaos." The White Guardian gives the Doctor an assistant (the Time Lady Romana, played by Mary Tamm) and tells him to beware his evil counterpart, the Black Guardian (Valentine Dyall). The Guardians, mythical beings who exerted incredible power, were also described by Williams in 1976: "Perhaps the Guardians are Time Lords, advanced to a higher degree along their own paths. Perhaps this is a test as to whether the Doctor is to qualify for advancement... Whatever the outcome, the Doctor will arrive at his own conclusions and decisions in his own fashion."

Williams' plan chimed with Read's own ambition to stretch *Doctor Who*'s format, but it soon became clear that the Key to Time would lead to difficulties. In 1984 Williams remembered that the umbrella narrative "seemed like enough of a stricture, but when we came to actually do it, it was very much worse than that. Usually, we were able to juggle the transmission order of the first three or four stories we made, so that if we had horrific problems on one, we could shove it down the line and bring another one forward. But, having made our Key to Time, we had to

Above: In Jotastar's Doctor Who Trump Card Game aliens from the series challenged the historical figures of 'Doctor Who and the Legendary Legion' in a contest of mental ability, special powers and weapons. Unfortunately the illustrations of the Sea Devils and Ogrons were transposed.

Below left: Visual effects designer Mat Irvine, Tom Baker, Mary Tamm and Beatrix Lehmann (playing Professor Rumford) during location recording for *The Stones of Blood* at the Rollright Stones, Oxfordshire, on 13 June.

Below: Manufactured by Shortman Trading, the TARDIS Tuner was essentially a medium wave radio with added sound effects and lights.

transmit them exactly in the order we made them; if a script turned out very weak, then we had to make it strong before we started shooting. That caused a few sleepless nights!"

Variety was maintained by disguising the various segments in increasingly inventive ways, and the experiment was ultimately deemed a success. The unit production manager for Season 16 was John Nathan-Turner, and when he took over from Graham Williams as producer many of his episodes were linked by story arcs: the E-Space trilogy in Season 18 was followed by a trilogy of stories featuring the Black Guardian (again played by Valentine Dyall) in Season 20. The whole of Season 23 would comprise a 14-episode adventure called *The Trial of a Time Lord*.

With the exception of the 1996 TV Movie, *Doctor Who* was off the air between 1990 and 2004. During that time the increased dominance of continuing dramas (ie, soap operas) and the rise of home entertainment viewing (where whole seasons of television shows would be marketed as box sets) had a major impact on television. In common with American productions such as *The X-Files* (1993-2002), every series of *Doctor Who* since 2005 has been governed to a lesser or greater degree by at least one story arc.

Never has this been more pronounced than in the episodes overseen by current showrunner Steven Moffat. "The controlled release of information – which is more or less what storytelling is – is really, really hard," he says. "Especially on *Doctor Who*, where you do need quite a lot going on, because everything is new. It's not like you've got a big regular cast, or one set that you're going to see every week. In *Doctor Who*, you're lucky if there's one prop you see every week! So you need to release quite a lot of information. It's a whole new world and you have to know what a normal day is like there, so you can twist it... Right up until the last minute, you're constantly changing your mind about the release

Top: Tom Baker and Beatrix Lehmann on location for *The Stones of Blood*.

Above: The Pirate Captain (Bruce Purchase) and his parrot-like robot, the Polyphase Avatron, in *The Pirate Planet*.

Left: One of two Polyphase Avatron props, this articulated puppet was operated by cables hidden behind Bruce Purchase's shoulder.

of information in a story. There's a fine line between mysterious and confusing. And I've often drifted over it!"

As well as creating carefully plotted themes, Moffat has found it impossible to resist retrospectively engineering story arcs in the style of *The War Games*. In 2013 *The Name of the Doctor* rewrote history by showing us a fresh aspect of the character's origin. In the process we were reminded that in a series about time travel it's always possible to create the greatest story arc of all.

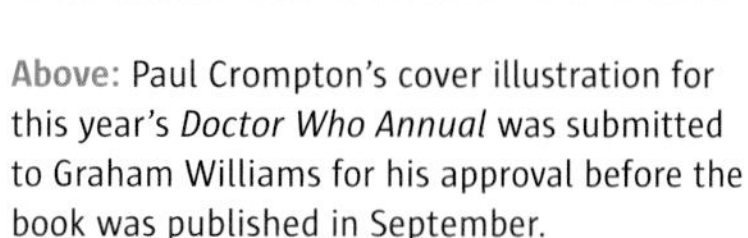

Above: Paul Crompton's cover illustration for this year's *Doctor Who Annual* was submitted to Graham Williams for his approval before the book was published in September.

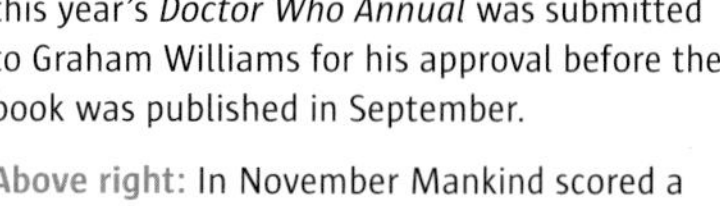

Above right: In November Mankind scored a number 25 hit with their version of the *Doctor Who* theme. This is an early pressing of the 12" on blue vinyl, and the 7" record sleeve.

Right: Number 19 in the BBC Sound Effects series, this album featured 30 tracks from *Doctor Who*. The cover was designed by Andrew Prewett and the run-out groove on Side A included the sly etching '(alias Luke Skywalker)'.

Below left: Romana's costume in *The Androids of Tara* was suggested by Mary Tamm's own design.

SEASON 15 (continued)
Producer: Graham Williams
Script editor: Anthony Read

4Y *Underworld*
3 October – 18 October 1977
written by Bob Baker and Dave Martin
directed by Norman Stewart

Part One	7 January
Part Two	14 January
Part Three	21 January
Part Four	28 January

4Z *The Invasion of Time*
1 November – 16 December 1977
written by David Agnew (pseudonym for Graham Williams and Anthony Read)
directed by Gerald Blake

Part One	4 February
Part Two	11 February
Part Three	18 February
Part Four	25 February

Part Five	4 March
Part Six	11 March

SEASON 16
Producer: Graham Williams
Script editor: Anthony Read

5A *The Ribos Operation* 9 April – 25 April
written by Robert Holmes
directed by George Spenton-Foster

Part One	2 September
Part Two	9 September
Part Three	16 September
Part Four	23 September

5B *The Pirate Planet* 1 May – 5 June
written by Douglas Adams
directed by Pennant Roberts

Part One	30 September
Part Two	7 October
Part Three	14 October
Part Four	21 October

5C *The Stones of Blood* 12 June – 18 July
written by David Fisher
directed by Darrol Blake

Part One	28 October
Part Two	4 November
Part Three	11 November
Part Four	18 November

5D *The Androids of Tara* 24 July – 29 August
written by David Fisher
directed by Michael Hayes

Part One	25 November
Part Two	2 December
Part Three	9 December
Part Four	16 December

5E *The Power of Kroll*
18 September – 20 October
written by Robert Holmes
directed by Norman Stewart

Part One	23 December
Part Two	30 December

1979

Tom Baker was placated without the granting of any unworkable concessions, but despite Graham Williams' best efforts it proved impossible to persuade Mary Tamm to stay for another season.

When *Doctor Who* returned in September for its 17th series Williams and his incoming script editor Douglas Adams dealt with the problem in typically light-hearted style – Romana initiates a voluntary regeneration at the beginning of *Destiny of the Daleks*, trying several different bodies for size before adopting a new model based on Princess Astra from *The Armageddon Factor*. Lalla Ward, who had played the princess, was offered the role of Romana partly because of her compatibility with Baker. The two got on so well that they would eventually marry in December 1980.

Destiny of the Daleks was Terry Nation's final script for *Doctor Who*, and brought the Daleks back to Skaro in an effort to recruit Davros (David Gooderson) as a strategist in their war with the robotic Movellans. What the production lacked in budget director Ken Grieve made up for with inventive camera work and a tangible sense of mounting dread.

City of Death was built around a location shoot in Paris, the series' first overseas filming, but also boasted some outstanding visual effects and a strong guest cast led by Julian Glover as Scaroth, the last of the Jagaroth. With ITV off the air owing to industrial action, *City of Death* attracted the biggest audience in *Doctor Who*'s history, peaking at 16.1 million for its final episode.

The remaining serials in Season 17 – *The Creature from the Pit*, *Nightmare of Eden* and *The Horns of Nimon* – had intelligent stories but were less distinguished productions. Williams had been keeping his powder dry for *Shada*, a six-parter named after the Time Lords' prison planet. Location filming took place in Cambridge, where writer Douglas Adams had spent his student days, but studio recording was curtailed by more of the industrial action that had previously jeopardised *The Invasion of Time* and *The Armageddon Factor*. On this occasion, no amount of rescheduling could save the project and *Shada* was abandoned. The disillusioned Williams resolved to quit the programme, and Adams left to develop a film adaptation of his radio serial and novel *The Hitchhiker's Guide to the Galaxy*.

Doctor Who was ready for a fresh start.

HAPPY TIMES AND PLACES

Above: Pages from the dummy issue of *Doctor Who Weekly* that Dez Skinn presented to BBC Enterprises. A licence was granted to Skinn's publisher Marvel Comics in August.

Above right: The first issue included a free set of transfers and *Doctor Who and the Iron Legion*, a strip story written by Pat Mills and John Wagner and illustrated by Dave Gibbons.

There had been a *Doctor Who* comic strip since the spoof *Doctor What and his Time Clock* appeared in the pages of *Boys' World* in May 1964. From November a more conventional *Doctor Who* strip began in *TV Comic* and continued across various publications until summer 1979.

In October that year the strip attained a new sophistication and a permanent home in the pages of the first regular magazine dedicated to the series. *Doctor Who Weekly* has grown up with the show's many followers, evolving through several formats and name changes to become the bedrock of fandom.

"It *was* a good idea," says the magazine's creator Dez Skinn, "but in those days good ideas were easy!"

Skinn had first applied his innovative combination of features and comic strips to a publication celebrating Hammer horror. In the late 1970s *House of Hammer* struggled in the absence of any new films from the production company, but Skinn recognised that the format he had devised could be easily translated to other, more prominent fantasy subjects.

"BBC Enterprises told me there was a two-page strip in *TV Comic*, and so they had the rights to *Doctor Who*," he says. "I considered that a waste. So I watched the progress of that comic strip, and

as soon as the publishers, Polystyle Publications, stopped doing the *Doctor Who* strip in *TV Comic* I shot along to BBC Enterprises. At the time I happened to be working at Marvel Comics."

Skinn called Marvel's forthcoming publication *Doctor Who Weekly*, "because I couldn't decide whether it was a comic or a magazine." In August 1979 he handed the mock-up for Issue 1 to Tom Baker and Graham Williams at the Worldcon convention in Brighton. Skinn and Baker hit it off, and Baker agreed to undertake a national publicity tour of newsagents and other venues to support the launch of the magazine. "Tom didn't get a penny for it," says Skinn. "He was promoting his livelihood. He did tell me, however, that he wouldn't have time to write an editorial every week, so I practised copying how he wrote 'The Doctor' so I would be able to scribble it at the bottom of the 'Letter From The Doctor' each week."

The first such letter said: "I hope you all enjoy the first issue of my new weekly as much as I did last time I was in town next week... By the way, watch out for issue 879 – it really was a beauty." It ended with what would become the usual sign-off: "Happy times and places, The Doctor."

Williams was another enthusiastic supporter of *Doctor Who Weekly*, although, as Skinn recalls, he had no desire to take an active role in contributing to, or even approving, the content. "Graham was never very hands-on with the comic. In fact, no one even told us that Terry Nation had the copyright on the Daleks. We just went ahead and did what we wanted. I would go and see BBC Enterprises now and then, give them a bunch of copies, and we'd get the whisky out."

Top left: From 10 to 13 October Tom Baker undertook a publicity tour to launch *Doctor Who Weekly*. The tour visited Leeds, Manchester, Birmingham, Wolverhampton, Woodford Green, Wandsworth and Liverpool. In Coventry he signed free copies of the magazine for pupils of Annie Osborn Primary School.

Top right: An example of the promotional material that helped *Doctor Who Weekly* achieve huge sales when it launched in October.

Above: The cover of the second issue, and the free transfers that came with this and Issue 3. *Doctor Who and the Iron Legion* continued until Issue 8.

Left: Tom Baker and Lalla Ward in Paris for *City of Death*. Director Michael Hayes conducted four days of location filming from 30 April to 3 May.

Above: The mask designed by Ian Scoones for Scaroth (Julian Glover) in *City of Death* was sculpted by John Friedlander in his last assignment for *Doctor Who*.

Below: Scoones' design for Scaroth's spaceship was based on a spider and included a mechanism that retracted the ship's legs after take-off. Scoones conducted the model filming at Bray Studios, where he had served his apprenticeship with Hammer Film Productions.

Below left: An annotated page from Hayes' camera script.

Skinn insists he never aimed *Doctor Who Weekly* at diehard fans, on the grounds that he expected them to buy it anyway. The comic strip was there to lure casual purchasers, and benefited from the cream of British talent; the earliest strips were written by Pat Mills and John Wagner, and illustrated by Dave Gibbons. To cater for those curious about the history of the Doctor's television adventures, Skinn recruited Jeremy Bentham, the head of the Doctor Who Appreciation Society's Reference Department.

Home video was still not widely adopted, so Bentham's principal reference source for his articles were episode synopses prepared by BBC Enterprises. "When the DWAS started in 1976 pretty much the only thing they had to go on were the Enterprises synopses," says Bentham. "There was the *Radio Times* tenth anniversary special and *The Making of Doctor Who* book,

Far left: June Hudson's costume design for Agella (Suzanne Danielle) in *Destiny of the Daleks*.

Top: Agella, Commander Sharrel (Peter Straker) and the unconscious Romana (Lalla Ward) in Episode Four.

Above: *The Adventures of K9 and Other Mechanical Creatures* was published by Target on 17 September.

Left: One of the surviving Mandrel costumes from *Nightmare of Eden*.

Below: K-9 defends Romana and Della (Jennifer Lonsdale) from a Mandrel in Part Four.

but there was no access to production files, directors or producers."

Skinn regarded the comic strip as central to *Doctor Who Weekly*, and the features accordingly received a much smaller percentage of the meagre budget. Bentham was paid £15 for a synopsis of an old TV story, and £7 for a biographical article about an actor or actress. Before long, his job was made even harder by the dwindling supply of pictures. "The BBC gave you photographs of the current series they wanted you to promote – *City of Death, Destiny of the Daleks, Creature from the Pit* and so on," says Bentham. "If you said, 'What about doing a Hartnell cover?' it was a different matter. The search and acquisition fees at the BBC photo library were very expensive."

Thirty-four years later, the comic strip, archive features and rare photographs are still among the selling points of a publication now entitled *Doctor Who Magazine*. Estimated to be the world's biggest-selling science fiction

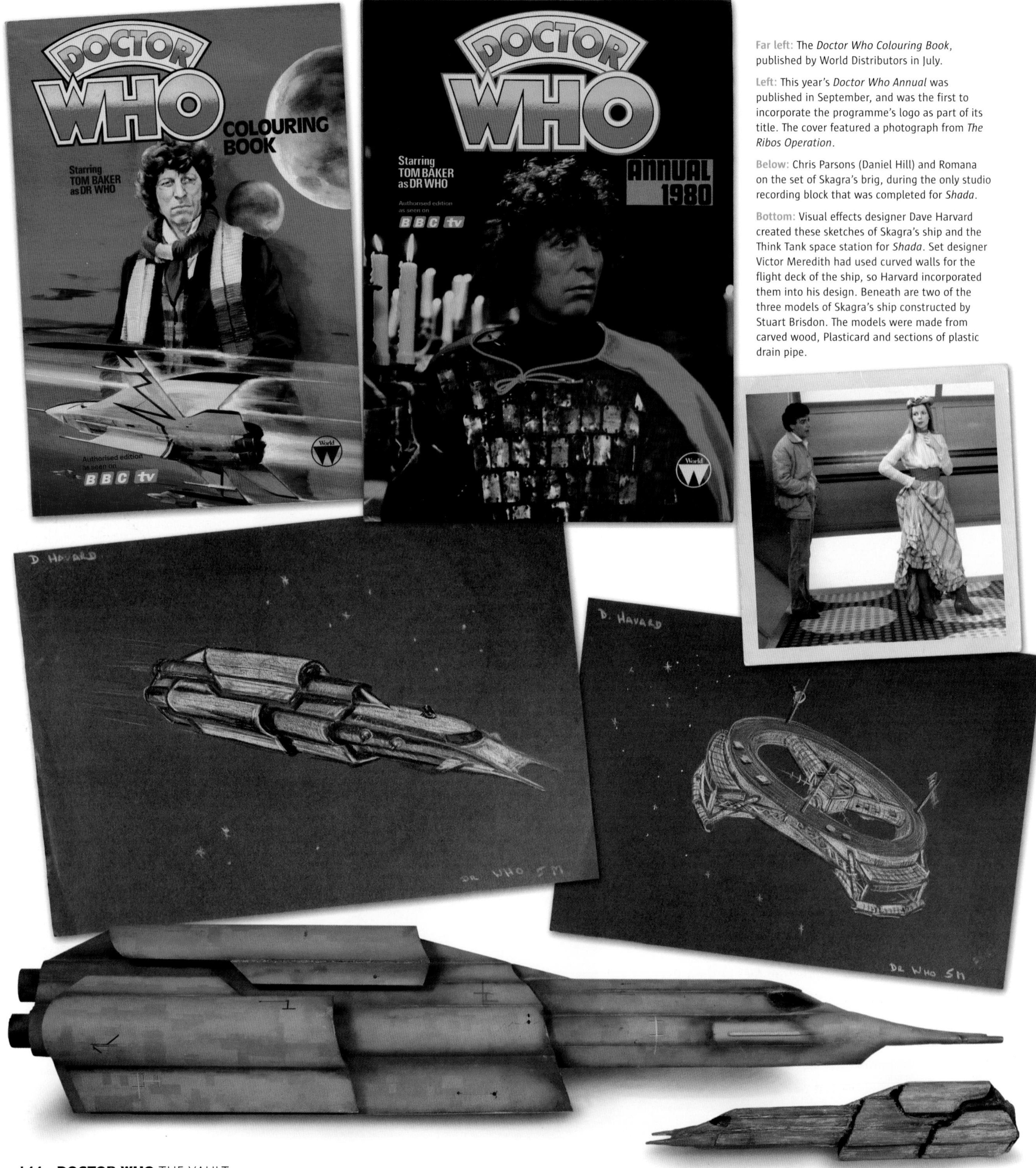

Far left: The *Doctor Who Colouring Book*, published by World Distributors in July.

Left: This year's *Doctor Who Annual* was published in September, and was the first to incorporate the programme's logo as part of its title. The cover featured a photograph from *The Ribos Operation*.

Below: Chris Parsons (Daniel Hill) and Romana on the set of Skagra's brig, during the only studio recording block that was completed for *Shada*.

Bottom: Visual effects designer Dave Harvard created these sketches of Skagra's ship and the Think Tank space station for *Shada*. Set designer Victor Meredith had used curved walls for the flight deck of the ship, so Harvard incorporated them into his design. Beneath are two of the three models of Skagra's ship constructed by Stuart Brisdon. The models were made from carved wood, Plasticard and sections of plastic drain pipe.

magazine, DWM has chronicled every corner of the show's history and in the world of licensed publishing is as much of an institution as the programme it's based on.

The current editor, Tom Spilsbury, is also the longest serving. "The comic strip is probably the most expensive and laboured-over part of the magazine," he says. "I'm aware that to some people it may be anachronistic, but I'm glad it's still there. Everything else in the magazine is *about Doctor Who* – it's responding to it in some way – whereas the comic strip is the magazine's opportunity to actually *be Doctor Who*."

Doctor Who Adventures, launched by BBC Magazines in 2006, now caters for the junior market while *Doctor Who Magazine* offers more sophisticated analysis of the show. "It's more specialised now than it was in 1979, partly because it had to survive during the years when the show wasn't on television," says Spilsbury. "By the early 1980s it was already very different from Dez's original conception, and in the 1990s it became even more in-depth. We didn't want to change all that when the series came back in 2005, but we also tried to bring in as many new fans as we could. It's a balancing act. Hopefully it still appeals to all those readers who have stayed with it for decades, but equally I hope it isn't totally incomprehensible to anyone who's just seen a few episodes on a Saturday night."

Spilsbury is optimistic about the magazine's prospects in the digital age. "I think DWM has a better future in the print marketplace than many other titles. A decent proportion of our readers keep each issue after they've read it and add it to their collection. Lifestyle magazines or fashion magazines are usually read and discarded. Having a digital version of magazines like that is almost a logical extension because most people aren't planning to keep them anyway. I'm sure there will be a digital version of *Doctor Who Magazine* in the not too distant future, but I think it would be a mistake to consider that a replacement for the print edition.

"I hope *Doctor Who Magazine* will continue for as long as the show itself, if not longer. It's already done it once before!"

Above: In September Tom Baker took part in a photo shoot for licensed merchandise produced by Denis Alan Print. These two greetings cards were part of the ensuing range.

Right: On 5 September BBC Records released an abridged version of *Genesis of the Daleks* on LP and cassette. A new narration, written by Derek Goom and recorded by Tom Baker, linked scenes from the original television serial. The cover featured a coloured photo montage by Mario Moscardini.

SEASON 16 (continued)
Producer: Graham Williams
Script editor: Anthony Read

5E *The Power of Kroll* (continued)
Part Three 6 January
Part Four 13 January

5F *The Armageddon Factor*
27 October – 5 December 1978
written by Bob Baker and Dave Martin
directed by Michael Hayes
Part One 20 January
Part Two 27 January
Part Three 3 February
Part Four 10 February
Part Five 17 February
Part Six 24 February

SEASON 17
Producer: Graham Williams
Script editor: Douglas Adams

5J *Destiny of the Daleks*
11 June – 17 July
written by Terry Nation
directed by Ken Grieve
Episode One 1 September
Episode Two 8 September
Episode Three 15 September
Episode Four 22 September

5H *City of Death*
30 April – 5 June
written by David Agnew (pseudonym for Douglas Adams, Graham Williams and David Fisher)
directed by Michael Hayes
Part One 29 September
Part Two 6 October
Part Three 13 October
Part Four 20 October

5G *The Creature from the Pit*
21 March – 24 April
written by David Fisher
directed by Christopher Barry
Part One 27 October
Part Two 3 November
Part Three 10 November
Part Four 17 November

5K *Nightmare of Eden*
12 August – 28 August
written by Bob Baker
directed by Alan Bromly (and Graham Williams, uncredited)
Part One 24 November
Part Two 1 December
Part Three 8 December
Part Four 15 December

5L *The Horns of Nimon*
24 September – 9 October
written by Anthony Read
directed by Kenny McBain
Part One 22 December
Part Two 29 December
Part Three 5 January 1980
Part Four 12 January 1980

5M *Shada* 15 October – 16 November
written by Douglas Adams
directed by Pennant Roberts
Not completed or transmitted

1980

John Nathan-Turner, *Doctor Who*'s ninth producer, made an immediate mark on the programme by instituting a new logo and title sequence. The familiar theme tune was replaced by a radically different interpretation played on synthesisers. Incidental music was now similarly electronic, courtesy of the Radiophonic Workshop. Even Tom Baker received a make-over, his famous multi-coloured scarf now matching the shades of burgundy in his redesigned costume.

Season 18's *The Leisure Hive* was a startling overture to this modernity, delivered with panache by director Lovett Bickford. Nathan-Turner's reimagining of the programme felt more superficial in lesser efforts such as the follow-up, *Meglos*. This story did, at least, contribute the unforgettable image of a cactoid Doctor. Madame Tussaud's were particularly impressed, and unveiled a waxwork replica alongside their more conventional statue of the Time Lord in August.

With Nathan-Turner's endorsement, script editor Christopher H Bidmead initiated a shift away from the jokiness of recent Graham Williams/Douglas Adams episodes, towards more earnest science fiction. *Full Circle* was the first part of a trilogy of stories set in the 'pocket universe' of E-Space. When the TARDIS left the planet Alzarius, the Doctor, Romana and K-9 were unaware that the young Adric (Matthew Waterhouse) had stowed away.

The talent involved in *Full Circle* made it something of a watershed. Matthew Waterhouse and the serial's writer Andrew Smith were both 18 years old, and both members of the Doctor Who Appreciation Society. They were the most conspicuous examples so far of fans influencing the course of their favourite show.

In contrast, Terrance Dicks – the writer of the next E-Space story, *State of Decay* – was well on the way to becoming one of *Doctor Who*'s elder statesmen. *State of Decay* had begun life as *The Vampire Mutation* in 1977, and still bore many of the Gothic hallmarks of the era.

On 24 October, shortly before the broadcast of *Full Circle*, it had been announced that Tom Baker would be leaving *Doctor Who* in 1981. Interviewed by *The Times*, he hinted that his replacement could be a woman.

The rumour ensured that publicity for the series remained strong. But after months of disappointing ratings, a more pressing challenge was how to regain the interest of viewers...

STAY TUNED

22. The Costume Designer arrives.

There is a knock at the door. It is June Hudson, the programme's Costume Designer. June is one of *Doctor Who*'s greatest fans, and has worked on the series for four years. She describes the programme as 'a designer's paradise', although it takes all her imagination and energy to think up the costumes for the aliens and monsters in each series.

'This series is very hard work,' she says, 'but the results are worth it.'

23. John discusses new costume designs.

June opens her portfolio, and lays out her design drawings on the floor. John has asked June to come to the office to discuss the design for the Doctor's new costume. This is part of the updated look which he has decided to give to the programme.

June is holding a drawing of the outfit that John decided on: a plum-coloured Regency-style coat, with a matching hat and long scarf.

~~Barry~~ Letts

With the compliments
of
DOCTOR WHO

BBC tv

John Nathan-Turner

Above: *A Day With a TV Producer* was written by Graham Rickard and issued by Wayland Publishers in November. The book was tailored towards primary school libraries and followed John Nathan-Turner during production of his first story, *The Leisure Hive*.

Top right: The book devoted a section to June Hudson, who was shown presenting her designs for the Doctor's new costume.

Centre: A compliments slip from the production office, as used by Nathan-Turner and executive producer Barry Letts.

Graham Williams was asked to stay with *Doctor Who* for a fourth year, but sheer exhaustion forced him to decline. His replacement as producer became one of the most influential and divisive figures in *Doctor Who*'s history.

John Nathan-Turner began his long association with the programme when he served as a floor assistant during recording of *The Space Pirates* at Lime Grove in 1969. After several further junior assignments on early 1970s stories he became the unit production manager on all three of Graham Williams' series. Efficient at rationing the show's meagre budgets, it was Nathan-Turner's careful planning that had enabled the Paris location shoot for *City of Death*.

Despite this acumen, his lack of experience at such a senior level meant that Barry Letts was recalled as an executive producer for Nathan-Turner's first series. According to Letts, he was primarily a sounding board and his input diminished as Nathan-Turner's confidence grew.

John Nathan-Turner, or 'JNT' as he liked to be known, was a very different character from the studious Letts, or indeed any of his predecessors. A flamboyant showman with a penchant for Hawaiian shirts, he became almost as familiar to fans as the stars of his show. He even had his own catchphrases: "surprised and delighted" was a standard response to acclaim for his work, "no hanky panky in the TARDIS" was often included in his denials of rumours about the Doctor's love life and, most popular of all, "stay tuned" deflected queries about future storylines.

His showy personality and passion for musical theatre suggested he would have been more at

Left: Nathan-Turner asked Hudson to devise a more uniform look for the Doctor in Season 18. Her initial black and white sketch was refined in a number of vivid colour paintings.

Above: In March, Tom Baker, Lalla Ward and K-9 visited Brighton to film the opening sequence of *The Leisure Hive*.

Below: The Doctor's Season 18 hat was made by Herbert Johnson of St James, London.

Right: Hudson provided Nathan-Turner with this illustration of Mena (Adrienne Corri) and her fellow Argolins following production of *The Leisure Hive*. Mena's costume was complemented with antique lace, denoting her authority and grandeur.

Below: The wig created for the Argolin Pangol in *The Leisure Hive*.

Below right: David Haig in full costume as Pangol, pictured during recording on the Great Hall set in April.

home producing light entertainment; he would direct several *Doctor Who*-themed pantomimes and relished his cabaret appearances at American conventions. Primarily, however, he is best remembered as *Doctor Who*'s longest serving producer, a post he actively held from 1980 until the original run of the series ended in 1989.

"I believe Graham Williams had never shown much interest in the DWAS, and Season 17 was certainly unpopular with members," says Gary Russell, who was editing the fanzine *Shada* when he first met JNT in 1980. "JNT was much more attentive to the fans, and initially they thought he was wonderful. Season 18 was very well received, partly because it seemed so different from Season 17 but largely because it was very good. *The Leisure Hive*, in particular, was an incredible opening salvo."

As well as the glossy production values, fans welcomed the frequent nods to the past which began with the casting of Jacqueline Hill as a guest star in *Meglos*, the second story of Season 18. Later in that season the revitalised Master (Anthony Ainley) became a semi-regular character. In Season 19 the Cybermen made a spectacular comeback in *Earthshock*, and the nostalgic Season 20 culminated in the anniversary special *The Five Doctors*. So far so good, but behind the scenes JNT made enemies. "I never got on with him," says Alan McKenzie, who edited Marvel's *Doctor Who Monthly* from 1981 to 1985. "I didn't care for his approach to the show, which I thought was way too camp, and I didn't like the high level of interference he tried to bring to the magazine."

Richard Marson was one of McKenzie's writers who joined the BBC as a floor assistant in 1988. "JNT was always suspicious of fans who worked at the BBC. He called them 'spies'. A lot of the other fans he rather cruelly referred to as 'barkers' – because they were barking mad. The phone would ring for him and you'd hear his secretary say, 'It's a barker – he wants to interview you for his fanzine.'"

Between 1984 and 1990 Gary Russell worked in the BBC press office, making him a target for suspicion. "The moment anything leaked out about the programme John assumed it came from one of us," he says. "The chances are it didn't – none of us were that stupid. Nine times out of ten things would actually leak out because John opened his mouth when a fanzine editor came to interview him."

Above: *Doctor Who Weekly* treated some aspects of the series as if they were real. From Issue 26 the magazine offered readers membership of UNIT.

Below left: The last of the Zolfa-Thurans assumes the appearance of the Doctor in Part Two of *Meglos*.

Below: Jacqueline Hill as Lexa, on the set of *Meglos* with Lalla Ward. Hill had played Barbara in the first and second seasons of *Doctor Who*.

Bottom: A Tigellan gun from *Meglos*, based on a design by Steven Drewett and vac-formed in silicone by Roger Perkins.

Above: The Doctor and K-9 hide as they watch the Marshmen rise from the water. Location filming for this sequence in *Full Circle* took place at Black Park in Buckinghamshire on 25 July.

Top right: *The Curse of Peladon* novelisation received a belated hardback publication in July. The cover artwork by Bill Donohoe was unique to this edition.

Right: This TARDIS-shaped container could be used as a money box, and was produced by Avon Tin. The following year Avon revised the design to feature the Fifth Doctor.

The 20th anniversary was the peak of 1980s popularity for both *Doctor Who* and JNT. A number of his decisions over the next few series proved controversial, including his insistence that Sixth Doctor Colin Baker wore a garishly-coloured costume that Baker himself despised. In later years JNT conceded it had been a misjudgement. In 1985 *Doctor Who* gained a reputation as a series in decline and fell victim to a cancellation attempt by BBC1 Controller Michael Grade.

"There are utterly opposing views about John's approach, and I think they're both true, frankly," says Andrew Cartmel, who became JNT's final script editor in 1987. "He brought *Doctor Who* to its knees; the show Michael Grade wanted to cancel was entirely the show that JNT saw fit to produce. But at the same time, when he got a creative team that included me, [Seventh Doctor] Sylvester McCoy and [companion] Sophie Aldred, and we were successfully climbing out of the pit, he was the man responsible for approving that team and giving us the backing to do what we needed to do."

Gary Russell counted JNT as a friend, as well as a colleague. "I think *Doctor Who* destroyed his career, because he left the BBC with effectively one line on his CV. It wasn't his fault – he was trying to leave the show from 1984 onwards, but he was on staff and you didn't leave unless you had something else to go to. And they never gave him anything else to go to. When Colin Baker left in 1986 they told John that if he also left the programme they'd end it. He desperately didn't want to be known as the last producer of *Doctor Who* so he stayed, always hoping that someone else would come up through the ranks and take over."

It is sadly ironic that JNT earned exactly the reputation he most feared. When he was finally succeeded in 1996 the show's first revival came too late to lift that stigma. He never worked in television again.

John Nathan-Turner was just 54 when he died in 2002. The debate over his effect on *Doctor Who* continues. In 2013 he was the subject of a biography containing revelations about his private life that made the front page of the *Daily Mirror*. Later in the year he was a prominent part of a *Newsnight* feature that asked "Was *Doctor Who* rubbish in the 1980s?"

"I think history has judged JNT unfairly," says Russell. "Critics of the 1980s' episodes rarely understand or take into account the way the BBC was run in those days, the amount of money that was available and the stifling production methods that had become entrenched."

Graeme Harper takes a similarly positive view of his old producer. "He was delightful, cheeky and gossipy," he says. "He gave me such an important break when he asked me to direct *The Caves of Androzani* (1984). Once we'd finished I could feel that he was really excited about what we'd made. I know he enjoyed making *Doctor Who*, but I think he stayed on it for too long. *John* knew he stayed on it for too long."

As befits a man with such a conflicting reputation, it is JNT's grudging longevity that led to his greatest achievement. For without that commitment, *Doctor Who* would surely not have survived in its original form for as long as it did.

Top left: In 1980 Terrance Dicks' *Day of the Daleks* novelisation was published in Japan by Hayakawa as *The Dalek Race's Counterattack!* The cover artwork was by Michiaki Sato.

Above: August saw the publication of the latest *Doctor Who Annual*, with a cover photograph from *The Armageddon Factor*.

Below left: Peter Howell's new arrangement of the theme tune was issued in this 7" picture sleeve in October.

SEASON 18
Producer: John Nathan-Turner
Executive producer: Barry Letts
Script editor: Christopher H. Bidmead

5N *The Leisure Hive*
20 March – 21 April
written by David Fisher
directed by Lovett Bickford

Part One	30 August
Part Two	6 September
Part Three	13 September
Part Four	20 September

5Q *Meglos* 25 June – 12 July
written by John Flanagan and Andrew McCulloch
directed by Terence Dudley

Part One	27 September
Part Two	4 October
Part Three	11 October
Part Four	18 October

5R *Full Circle*
23 July – 23 August
written by Andrew Smith
directed by Peter Grimwade

Part One	25 October
Part Two	1 November
Part Three	8 November
Part Four	15 November

5P *State of Decay*
30 April – 31 May
written by Terrance Dicks
directed by Peter Moffatt

Part One	22 November
Part Two	29 November
Part Three	6 December
Part Four	13 December

1981

The E-Space trilogy came to an end with *Warriors' Gate*, written by Stephen Gallagher and realised with abstract flourishes by director Paul Joyce. Romana and K-9 stayed behind in the pocket universe, leaving Adric as the Doctor's only companion on his visit to *The Keeper of Traken*.

Johnny Byrne's four-part story helped to establish a new cast of characters for the next stage of John Nathan-Turner's reinvention of *Doctor Who*. Nyssa (Sarah Sutton) was not created as one of the Doctor's companions, though she would join the TARDIS in the next story. Nyssa's father, Consul Tremas (Anthony Ainley), fell prey to the Master (Geoffrey Beevers), who stole his body in the closing moments of the final episode.

The solemn atmosphere became positively funereal as the ringing of the TARDIS' Cloister Bell predicted the Doctor's demise in *Logopolis*. Tom Baker's final story began with yet another addition to the regular cast – Australian air hostess Tegan Jovanka (Janet Fielding).

Christopher H Bidmead's script saw the Doctor defeat the revitalised Master but make the ultimate sacrifice in his effort to prevent nothing less than the destruction of the universe. He falls from a radio telescope and his last moments are spent with his three young companions, two of whom – Nyssa and Tegan – he's only just met. The most sophisticated regeneration sequence yet seen culminated in the arrival of the fresh-faced Doctor Number Five (Peter Davison).

There would be a nine-month wait before any new episodes were broadcast. In an effort to bridge the gap, and partly to remind audiences that there had been other Doctors before Baker's lengthy ownership of the role, Nathan-Turner organised a season of repeat screenings for BBC2. The Five Faces of Doctor Who began in November and comprised *100,000 BC*, *The Krotons*, *Carnival of Monsters*, *The Three Doctors* and *Logopolis*. The two Jon Pertwee stories garnered average ratings of 5.2 million – equal to or higher than some of the first-run stories from earlier in the year.

A year pregnant with possibilities ended with the broadcast of the spin-off episode *K-9 and Company* on 28 December. Fans would have to wait until 2007 before they would see a full series featuring Sarah Jane Smith and her robot dog.

The first adventure with the new Doctor was, however, merely days away...

FACE VALUE

Episodes featuring all five Doctors were shown by BBC Television in 1981, and when Season 19 began in January 1982 the opening story, *Castrovalva*, devoted unprecedented time to the new Doctor's post-regenerative recovery. In a further reminder to audiences that such change was part of the show's heritage, the confused Fifth Doctor lapsed into uncanny impressions of his first three selves.

But how did the unusual concept of regeneration come about? Like so much of the Doctor's back

Above: June Hudson's final costume design for Romana was worn by Lalla Ward in *Warriors' Gate*.

Top right: John Nathan-Turner asked Hudson to prepare this montage of her *Warriors' Gate* costume designs.

Right: One of two Privateer spaceship models designed and built by Mat Irvine for *Warriors' Gate*. The other model was blown up during visual effects filming.

story, it was an ingenious response to an off-screen problem. The haphazard evolution of the idea means that the 'rules' relating to the turnover of leading men are complex and self-contradictory. "That's the trouble with regeneration," says the Doctor in Part One of *Castrovalva*. "You never quite know what you're going to get..."

On 16 July 1966 Heather Hartnell wrote in her diary: "Bill decides to give up *Dr Who* in October." Producer Innes Lloyd was able to reassure William Hartnell that the programme he loved would continue, albeit with a different actor in the lead role. Story editor Gerry Davis added a new sequence to Episode 4 of *The Tenth Planet*, explaining the change. Possibly affected by his energy-draining encounter with the planet Mondas, the Doctor states that "this old body of mine is wearing a bit thin" and collapses in the TARDIS control room. As scripted, he was supposed to prefix this with: "No, no, I can't go through with it – I can't. I can't. I will not give in," but in the event his face is bathed in a bright light and he quietly transforms into a much younger man.

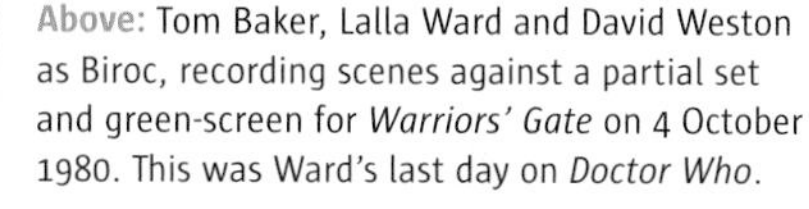

Above: Tom Baker, Lalla Ward and David Weston as Biroc, recording scenes against a partial set and green-screen for *Warriors' Gate* on 4 October 1980. This was Ward's last day on *Doctor Who*.

Left: June Hudson's design for the Tharils in *Warriors' Gate*.

Below: One of the Tharil belt buckles, made by the Costume Department from Hudson's design.

Far left: Volumes One and Two of *The Doctor Who Programme Guide* by Jean-Marc Lofficier were published in hardback on 21 May 1981. The cover artwork was by Bill Donohoe. Volume One was an episode guide, and Volume Two was an encyclopedia of characters and other fictional elements.

Above left: Target's novelisation of the first *Doctor Who* story was published in October, with cover artwork by Andrew Skilleter.

Above: The German-language edition of *Doctor Who and the Dalek Invasion of Earth* was published by Schneider-Buch, with a cover by David Hardy.

Below: Peter Logan's design for the servo shutdown device in *The Keeper of Traken* incorporated the data sphere prop from *Destiny of the Daleks*.

Below right: The Doctor, Adric (Matthew Waterhouse) and Tremas (Anthony Ainley) in Part Two of *The Keeper of Traken*.

"It's part of the TARDIS," says the 'rejuvenated' Doctor in *The Power of the Daleks*. "Without it, I couldn't survive." David Whitaker's draft script contained a more detailed description that was never recorded: "No one would ever submit to it voluntarily," the Doctor tells his companions. "I fight it every time – but I cannot resist..."

In 1969 a different explanation had to be found to explain the transition from the Second to Third Doctors. This time it was the Doctor's own people, the Time Lords, who forced his change of appearance as part of his criminal sentence. It wasn't until *Planet of the Spiders*, and the change from the Third to the Fourth Doctors, that the word 'regeneration' was used.

The first two regenerations had been the most matter-of-fact. From the mid-1970s onwards passing the baton became an accepted part of the programme's mythology and producers started exploring the idea.

A comedic sequence in Part One of *Robot* had the new Doctor try on a number of potential outfits in impossibly quick succession. In *Destiny of the Daleks* script editor Douglas Adams paid homage with a sequence where Time Lady Romana assumed three very different bodies before settling on her favourite. "Romana's regeneration was jokey," admitted producer Graham Williams, "but the whole concept of regeneration is jokey!"

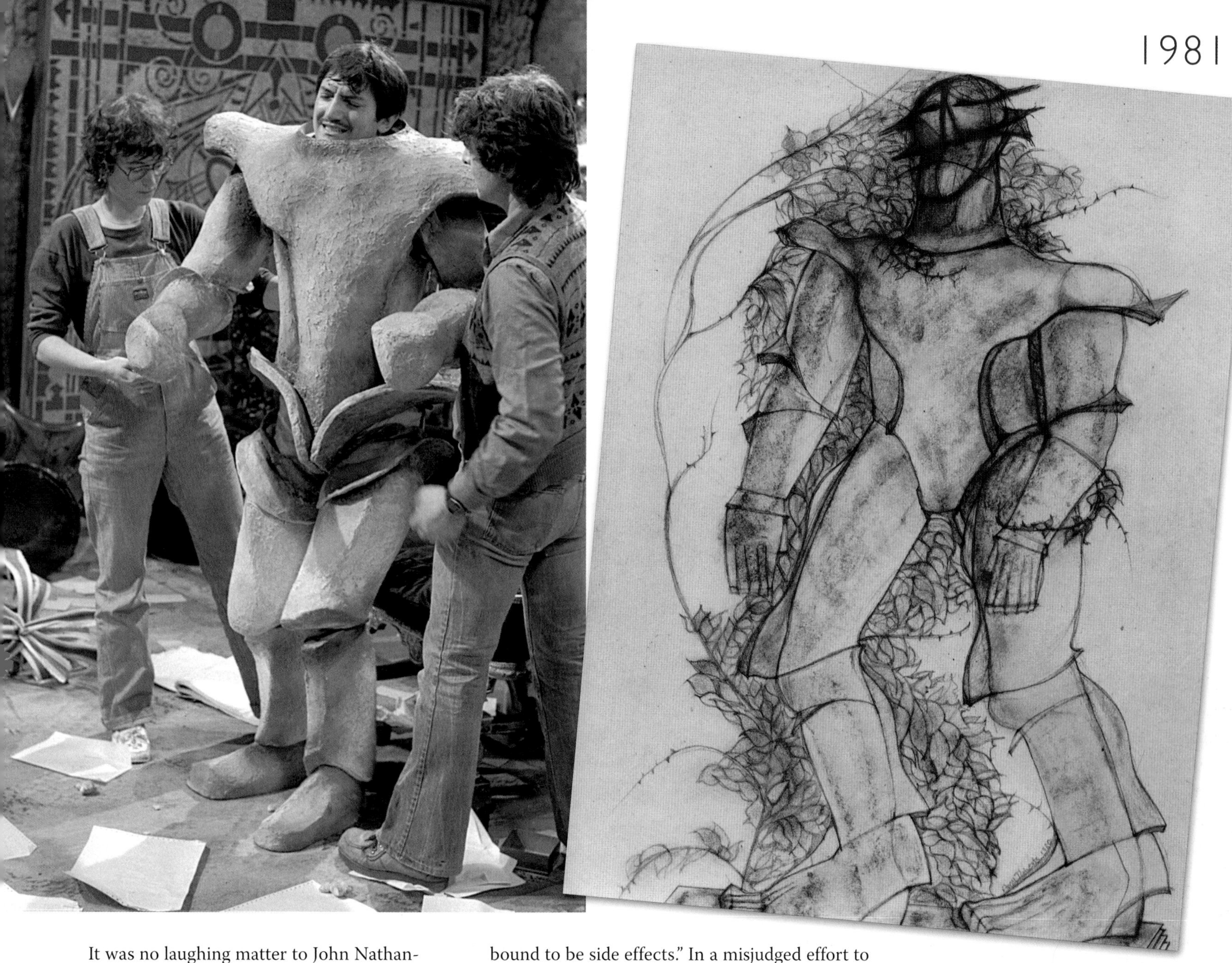

It was no laughing matter to John Nathan-Turner's original script editor Christopher H Bidmead. He wrote the Fourth Doctor's final story, *Logopolis*, and structured most of its follow-up, *Castrovalva* (1982), around the problems of regeneration. Eric Saward, one of Bidmead's successors, oversaw the transition from Fifth to Sixth Doctors. Like *Logopolis* before it, Robert Holmes' *The Caves of Androzani* (1984) was a sombre farewell to the incumbent Doctor. The next story, however, was quite unlike anything that had been seen before. In Anthony Steven's *The Twin Dilemma* (1984) the Sixth Doctor (Colin Baker) says "Regeneration in my case is a swift but volcanic experience. A kind of violent biological eruption in which the body cells are displaced, changed, renewed and rearranged. There are bound to be side effects." In a misjudged effort to create a contrast with the Fifth Doctor's enfeebled debut, *The Twin Dilemma* gave us the unwelcome sight of a disorientated Doctor attempting to strangle his companion Peri (Nicola Bryant).

"Our thinking was to focus more on the *process* of regeneration, so he became unstable," says Saward. "With hindsight, we were probably wrong. It's a device, really, to establish a new actor into the role, but we dwelled upon it for far too long and alienated the character from the audience, which was a mistake."

The Twin Dilemma was the first story to describe a Doctor's life as an 'incarnation', and this term has stuck. The introduction of the Doctor's seventh incarnation (Sylvester McCoy) in *Time and the Rani* (1987) was compromised by Colin

Above left: Graham Cole struggles inside the Melkur costume during studio recording for Part Four of *The Keeper of Traken* on 7 November 1980.

Above: Costume designer Amy Roberts created this sketch of the Melkur for *The Keeper of Traken*. While visually striking, the finished suit proved difficult to manipulate, so director John Black limited the character's movements as much as possible.

Left: These *Doctor Who* Viewmaster slides retold *Full Circle* in 21 stereoscopic images and were produced by the GAF Corporation. A second set presented the Fifth Doctor story *Castrovalva* in 1983.

Below: *The Doctor Who Quiz Book* was compiled by Nigel Robinson and published by Target in December.

Bottom: Janet Fielding (as Tegan), Sarah Sutton (as Nyssa), Matthew Waterhouse and Tom Baker outside the BBC Overseas Monitoring Station in Caversham, Berkshire, on 18 December 1980. The picture was taken during location filming for Baker's final episode, Part Four of *Logopolis*.

Baker's refusal to reprise his portrayal for the sake of a smooth handover. In contrast, the producers of the 1996 TV Movie must have left much of their American network audience baffled by dwelling on the death of Doctor Number Seven and the resurrection of Doctor Number Eight (Paul McGann).

Ninth Doctor Christopher Eccleston hit the ground running in *Rose* (2005), with only a brief mention of his new appearance. His self-sacrifice in *The Parting of the Ways* (2005) ushered in Tenth Doctor David Tennant, who was birthed in the golden glow of 'regenerative energy'. In *The Christmas Invasion* (2006) the Doctor still had enough of that energy to regrow a severed hand, 15 hours after his transformation. This short-term flexibility finally lent some logic to Romana's seemingly reckless window shopping all those years before.

Left and below: The latest *Doctor Who Annual* was edited by Brenda Apsley and published in August. This edition was unique in featuring two Doctors, although at this point artist Glenn Rix didn't know what the new Doctor would be wearing.

Bottom: The set for the main gantry of the Pharos Project radio telescope from *Logopolis*. The fatal struggle between the Doctor and the Master was recorded here on 9 January.

The Tenth Doctor was able to control his regenerative energy in *The End of Time* (2010), delaying the spectacular transition to his successor, played by Matt Smith. In common with all the regeneration scenes from 2005 onwards, this one occurred with its subjects standing up, leaving them better placed for a memorable quip in the immediate aftermath.

New Doctors no longer arrive with the sort of gimmicks that producers once attached to their newly regenerated predecessors (a recorder for the Second Doctor, a yo-yo for the Fourth, musical spoons for the Seventh) but the ordeal itself doesn't get any easier. And as the series continues beyond its 50th year, the concerns surrounding the Doctor's longevity become ever more pressing. Despite contradictory statements in *The War Games* (1969) and spin-off series *The Sarah Jane Adventures* in 2010, it was enshrined in *The Deadly Assassin* (1976) and several subsequent stories that Time Lords can only regenerate 12 times. On those terms, the Doctor is nearing the end of his life.

A way will surely be found to extend this lifespan. After all, *Doctor Who*'s unique talent for regeneration relies on the Doctor's ability to regenerate.

SEASON 18
(continued)
Producer: John Nathan-Turner
Executive producer: Barry Letts
Script editor: Christopher H. Bidmead

5S *Warriors' Gate*
24 September – 4 October 1980
written by Steve Gallagher
directed by Paul Joyce

Part One	3 January
Part Two	10 January
Part Three	17 January
Part Four	24 January

5T *The Keeper of Traken*
5 November – 17 December 1980
written by Johnny Byrne
directed by John Black

Part One	31 January
Part Two	7 February
Part Three	14 February
Part Four	21 February

5V *Logopolis*
16 December 1980 – 24 January 1981
written by Christopher H. Bidmead
directed by Peter Grimwade

Part One	28 February
Part Two	7 March
Part Three	14 March
Part Four	21 March

1982

With the casting of 29-year-old Peter Davison in the title role, John Nathan-Turner's revitalisation of *Doctor Who* was complete. Davison's first season was largely produced in 1981, with Eric Saward taking over from temporary script editor Antony Root in April.

Breaking with a tradition established in 1963, the new series was taken away from Saturday nights and broadcast on Mondays and Tuesdays from January 1982. This experiment proved successful, boosting the show's viewing figures to between nine and ten million.

Davison was a prominent television star at the time, and this – combined with the fact that new episodes were now on the air for just three months of the year – meant that he would never become as closely identified with the role as his predecessor.

Davison's low-key performance also made him distinct from Tom Baker. The Fifth Doctor's struggle to cope with his regeneration in *Castrovalva* set the tone for his vulnerability, with Davison adding occasional reminders that this Doctor was much older than his demeanour suggested. In relation to his youthful crew – Adric, Nyssa and Tegan – he assumed a paternal role in the crowded TARDIS.

Castrovalva was recorded fourth in the schedule to allow Davison the chance to make a confident on-screen debut. *Four to Doomsday*, the next serial to be transmitted, had actually been first on the studio floor and was more conventional in every respect. The disturbing surrealism in *Kinda* stretched perceptions of the show, and early evening drama, even if it over-stretched the BBC's ability to construct a convincing jungle scenario and the snake-like Mara.

The Visitation provided a more convincing alien threat in the Terileptils, the first *Doctor Who* monsters to benefit from the technology of animatronics. The 1920s backdrop to *Black Orchid* was impeccably realised, although the two-parter's lack of science fiction elements puzzled regular viewers.

The season's biggest hit with fans was *Earthshock*. Plotted as a crowd-pleaser by writer Eric Saward and vigorously paced by director Peter Grimwade, the serial reintroduced the Cybermen at the end of Part One and disintegrated Adric at the end of Part Four.

The season finale *Time-Flight* featured a disappointing rematch with the Master, and could have been nothing but anti-climactic in comparison.

TAKING OVER THE ASYLUM

Above: The Mark III Sonic Screwdriver was first seen in *Frontier in Space*, and used regularly from that point on until John Nathan-Turner decided it made things too easy for the Doctor. It was destroyed in Part Three of *The Visitation* and never returned during the original run of the series.

Above centre: June Hudson's idea for 'the look' of the Fifth Doctor.

Above right: Costume designer Colin Lavers submitted this proposal for *Four to Doomsday*, the first story recorded by Peter Davison.

In 1982 *Doctor Who* was responding more than ever to its increasingly savvy, and increasingly vocal, followers. The multi-faceted *Kinda* was one of the most sophisticated (detractors would say impenetrable) stories in the show's history, while the surprise return of the Cybermen in *Earthshock* could have been designed to reward long-time enthusiasts.

The way that writers were chronicling the show was also beginning to change. *Doctor Who Monthly*, as the old weekly had been retitled, looked at the series' history in greater depth than before. The magazine was squarely aimed at fans, whom publisher Marvel Comics now considered to be the only significant market.

John Nathan-Turner was similarly happy to facilitate books such as *Doctor Who: The Making of a Television Series*, which devoted most of its 60 pages to the production of *The Visitation*, explaining the exact roles played by key crew members and BBC departments.

With such informative material at their fingertips, it was perhaps inevitable that *Doctor Who* fans would find employment on their favourite show. Andrew Smith, the writer of *Full Circle*, had learned how to format script submissions to the production office by following the advice laid down by Malcolm Hulke and

Terrance Dicks in *The Making of Doctor Who*. Adric, one of the characters he created in that story, was played by Matthew Waterhouse – a fan who had submitted a letter to *Doctor Who Weekly* before the first issue had even been published.

Adric had proved unpopular with a large element of the Doctor Who Appreciation Society, and was killed off in Season 19. The end credits of *Earthshock*'s final episode rolled in silence, underlining the tragedy of Adric's self-sacrifice. Many fanzine editors saw the character's demise as a cause for celebration instead. "*Kinda* taught us that one thing evil cannot withstand its own reflection," says

Above left: *Doctor Who: The Making of a Television Series* was first published in hardback by André Deutsch in July. Writer Alan Road and photographer Richard Farley trailed the cast and crew of *The Visitation*, covering every aspect of the serial's production.

Above right: A *Doctor Who* Easter egg manufactured by Tobler Suchard.

Below: Inside each Easter egg box was a comic strip panorama entitled *Doctor Who's little book of Villains* and a set of rub-down transfers.

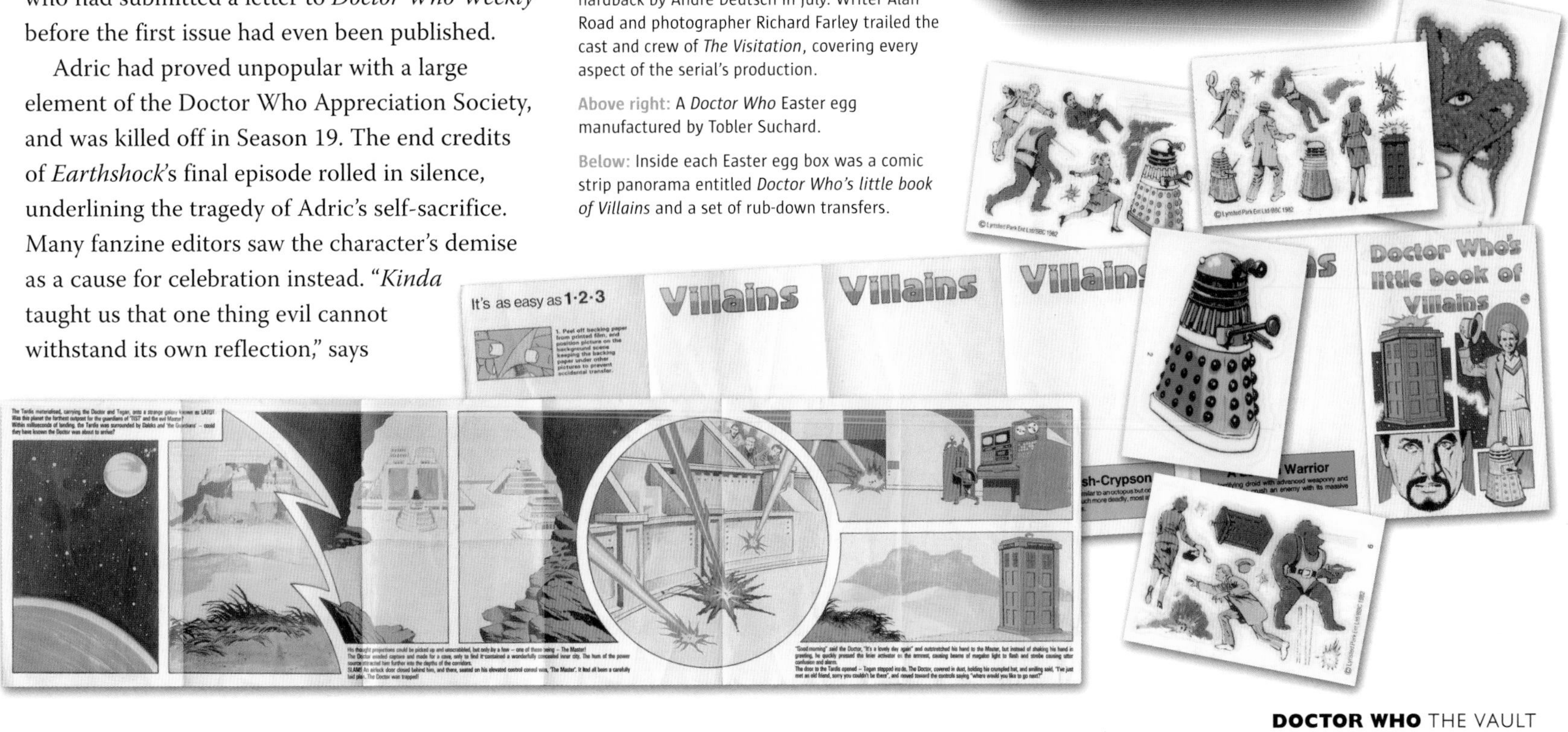

Gareth Roberts. "Adric proves as much about fans."

The response to Adric's death was an early example of dissent from the 'powerless elite' described by authors John Tulloch and Manuel Alvarado in their 1983 book *Doctor Who: The Unfolding Text*. Such fans were starting to resemble the cliché of football supporters, frustrated by their inability to influence or replace their faltering team's manager.

In 2004 Gary Downie, *Doctor Who*'s former production manager and the late John Nathan-Turner's partner, expressed his disdain for such carping. "What I hate about the fans is that they all think they can do it better," he said. "They're working at Tesco service tills or as warehousemen, but they all know how to produce the show better than John did. It annoys me, that. I read stupid things about him – never in the press, but in fan magazines. Yeah, it's fanzine oriented – ignoramuses who think they know everything."

Prior to the widespread adoption of the Internet in the late 1990s it was estimated that there had

Above: Three examples of the *Doctor Who* fanzines that were becoming commonplace by 1982.

Left: The Terileptil leader (Michael Melia) in *The Visitation*. Richard Gregory's company Imagineering provided three Terileptil costumes for this story, but the heads were built by Peter Litten. The leader's head was fitted with animatronics to move its mouth and gills. Litten would later attempt to make a *Doctor Who* feature film with his production company Coast to Coast.

Below: At Quainton Road railway station John Nathan-Turner (far right) watches as the TARDIS prop is put into position for location filming on *Black Orchid*.

been almost 600 *Doctor Who* fanzines over the previous two decades, although the true figure could well have been more. One of the most outspoken was launched in 1983; Downie implied that the views expressed by *Doctor Who Bulletin* may actually have hastened the show's cancellation.

"It's extraordinary with *Doctor Who*, how judged it is," says Sylvester McCoy. "And they go over every line and breath and every piece of logic. Well, there was no way anyone could have ever created anything to go through that scrutiny. It wasn't meant for that. There was no time for that... When the episodes came out, that's what it was for. Instant television. It wasn't designed to be pored over!"

Above and right: The original artwork used for the sleeves of Thorn EMI's VHS and Betamax tapes of the two Dalek films. These were the first commercially available *Doctor Who* videos.

Top right: The front cover of *Time Out* magazine Issue 593, dated 1-7 January.

Right: One of four jigsaws issued by Waddingtons in 1982. This one composites a picture of the Doctor with a scene from *Earthshock*.

Below: Scott (James Warwick), the Doctor and Nyssa attempt to open a mysterious underground hatch in Part Two of *Earthshock*.

Bottom: Imagineering produced these troopers' helmets for *Earthshock*, based on Dinah Collin's design. The lights on either side were operated by the actors.

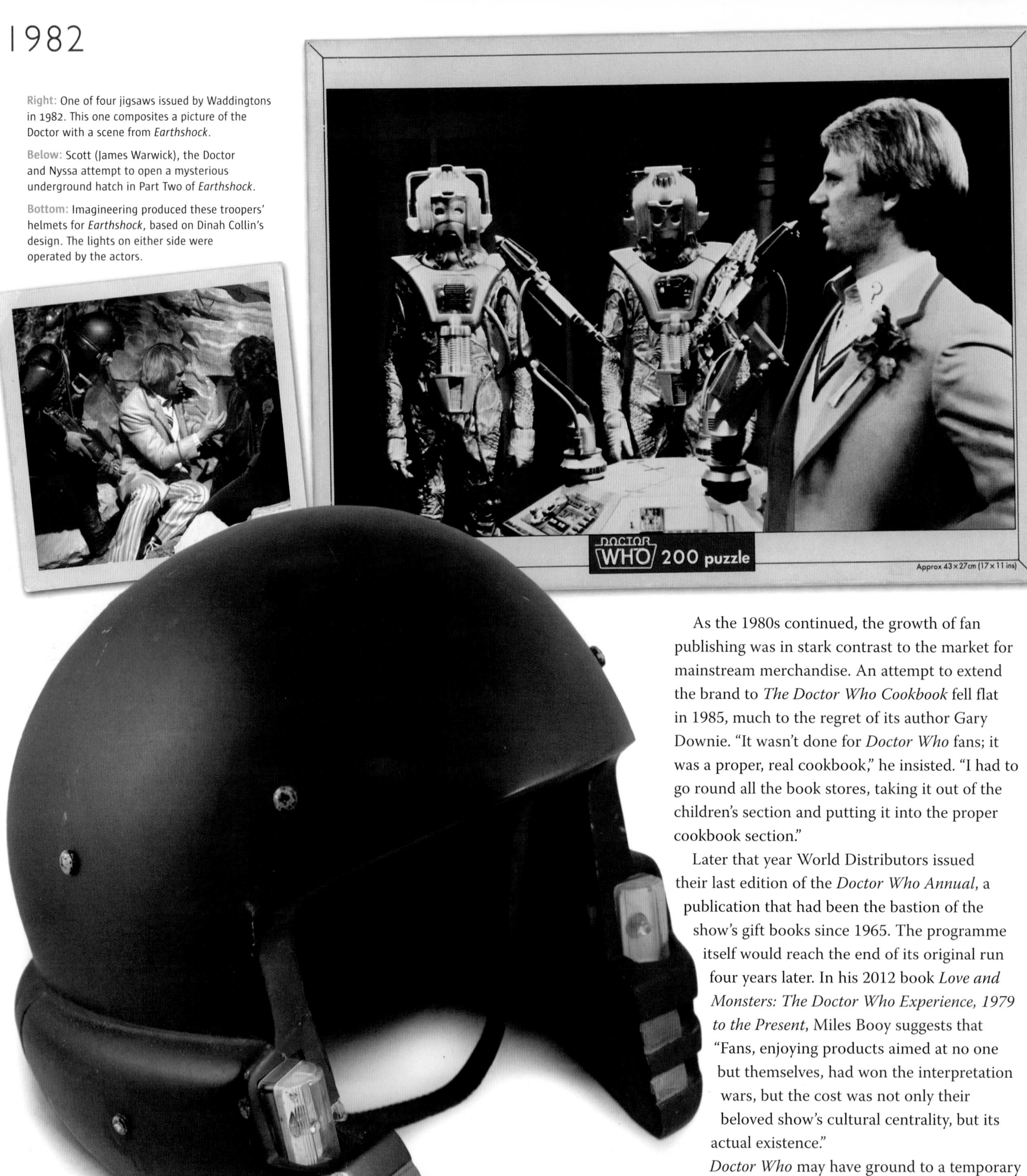

As the 1980s continued, the growth of fan publishing was in stark contrast to the market for mainstream merchandise. An attempt to extend the brand to *The Doctor Who Cookbook* fell flat in 1985, much to the regret of its author Gary Downie. "It wasn't done for *Doctor Who* fans; it was a proper, real cookbook," he insisted. "I had to go round all the book stores, taking it out of the children's section and putting it into the proper cookbook section."

Later that year World Distributors issued their last edition of the *Doctor Who Annual*, a publication that had been the bastion of the show's gift books since 1965. The programme itself would reach the end of its original run four years later. In his 2012 book *Love and Monsters: The Doctor Who Experience, 1979 to the Present*, Miles Booy suggests that "Fans, enjoying products aimed at no one but themselves, had won the interpretation wars, but the cost was not only their beloved show's cultural centrality, but its actual existence."

Doctor Who may have ground to a temporary halt, but the numerous industries that fans cultivated, and in some cases created, continued.

Virgin Publishing's range of *Doctor Who* fiction included the 1996 novel *Damaged Goods*, written by Russell T Davies. The same year Steven Moffat, another future showrunner, contributed to the short story collection *Decalog 3: Consequences*.

From 1999 *Doctor Who Magazine* was partly sustained by news and reviews of the *Doctor Who* audio dramas created by Big Finish. This provided another path for influential fans to come to prominence. In 2000 Big Finish producer Gary Russell was contacted by actor Toby Longworth, who explained that he had a friend who would very much like to appear in one of the company's *Doctor Who* stories. In 2000 Russell gave David Tennant his first Big Finish role, playing a German guard in a serial called *Colditz*. "I think the reason David asked Toby to put me in touch with him was that he was a *Doctor Who* fan," says Russell. "He really wanted to be in an episode of *Doctor Who*, and at that point this was the only way to do it."

Four years later *Doctor Who* was readied for its television comeback by a team that included co-executive producer Russell T Davies, producer Phil Collinson (a former DWAS member), Steven Moffat and several writers familiar from *Doctor Who* books, fanzines and audio productions. "Mark [Gatiss], Paul [Cornell] and Rob [Shearman] enter into rewrites with glee and laughter, never complaining about the time or the energy or the loss," Davies told *Doctor Who Magazine* in 2004. "Within the industry this is remarkable and rare, and d'you know what? I think they were taught by *Doctor Who*. Since they were kids, they've been reading the wise words of Mr Dicks and Holmes and Saward and Cartmel and Whitaker and all those glorious script editors of ages past. They know that rewriting is vital, because their own fandom taught them so... It's a bit neat – the programme's heritage, and our obsession with it, is rewarding the show itself."

Below: The latest *Doctor Who Annual* was published in August and edited by Brenda Apsley. This was the first edition to feature articles about the making of the series.

Below left: Peter Howell's arrangement of the *Doctor Who* theme was reissued in February.

Bottom: Anthony Ainley browses *Doctor Who Monthly* Issue 61 during a break in the studio recording of *Time-Flight*. Unfortunately the front cover included a mis-spelling of Peter Davison's surname as 'Davidson'.

SEASON 19

Producer: John Nathan-Turner

5Z *Castrovalva*
1 September – 1 October 1981
written by Christopher H. Bidmead
directed by Fiona Cumming
Script editor: Eric Saward

Part One	4 January
Part Two	5 January
Part Three	11 January
Part Four	12 January

5W *Four to Doomsday*
13 April – 30 April 1981
written by Terence Dudley
directed by John Black
Script editor: Antony Root

Part One	18 January
Part Two	19 January
Part Three	25 January
Part Four	26 January

5Y *Kinda*
29 July – 14 August, 11 November 1981
written by Christopher Bailey
directed by Peter Grimwade
Script editor: Eric Saward

Part One	1 February
Part Two	2 February
Part Three	8 February
Part Four	9 February

5X *The Visitation*
1 May – 5 June 1981
written by Eric Saward
directed by Peter Moffatt
Script editor: Antony Root

Part One	15 February
Part Two	16 February
Part Three	22 February
Part Four	23 February

6A *Black Orchid*
5 October – 21 October 1981
written by Terence Dudley
directed by Ron Jones
Script editor: Eric Saward

Part One	1 March
Part Two	2 March

6B *Earthshock*
29 October – 26 November 1981
written by Eric Saward
directed by Peter Grimwade
Script editor: Antony Root

Part One	8 March
Part Two	9 March
Part Three	15 March
Part Four	16 March

6C *Time-Flight*
6 January – 3 February
written by Peter Grimwade
directed by Ron Jones
Script editor: Eric Saward

Part One	22 March
Part Two	23 March
Part Three	29 March
Part Four	30 March

1983

Following the sweeping changes of recent years, John Nathan-Turner felt confident enough to preside over a 20th anniversary series where every serial contained an element from the show's past.

Largely shown on Tuesday and Wednesday evenings, *Doctor Who*'s 20th season opened with *Arc of Infinity*. The story had been scheduled as second in production, in an effort to coincide with better weather for the location filming in Amsterdam. The returning villain was Omega (Ian Collier), making another attempt to escape the universe of anti-matter where *The Three Doctors* had encountered him ten years earlier.

The title *Snakedance* made it clear that the Doctor (and the still-possessed Tegan) would be facing the Mara in the next story. There was, however, considerable fan debate over the true identity of its publicity-shy author. Many years later the bemused Christopher Bailey revealed that his name wasn't a pseudonym at all.

Mawdryn Undead had been written in the hope that William Russell would reprise his role as Ian Chesterton, now teaching at a boys' school. When Russell proved unavailable Nicholas Courtney stepped in to play Alastair Lethbridge-Stewart, with confusing results for keen followers of UNIT continuity.

At the end of the story the Doctor, Nyssa and Tegan were joined by Turlough (Mark Strickson), one of the pupils at the retired Brigadier's school. Over the course of the next two stories Turlough would continue his attempts to kill the Doctor on behalf of the vengeful Black Guardian.

Nyssa stayed behind to help the victims of Lazar's disease in *Terminus*, and in *Enlightenment* Turlough finally redeemed himself. The season came to an end with *The King's Demons*, in which the Master was prevented from rewriting medieval history and the Doctor inherited a shape-changing robot called Kamelion (voiced by Gerald Flood).

The Five Doctors was screened as part of BBC1's annual *Children in Need* telethon on 25 November. Terrance Dicks' script corralled a staggering array of friends and foes from the show's history. Peter Davison was joined by Jon Pertwee and Patrick Troughton, while the late William Hartnell's place was taken by Richard Hurndall. Tom Baker declined to appear so was represented by clips from the unscreened *Shada*. The 90-minute special ended the anniversary year on a high, reinforcing *Doctor Who*'s seemingly unassailable status as a television institution.

INSTANT REPLAY

Above and top right: *Revenge of the Cybermen* became the BBC's first *Doctor Who* video release in October. The anachronistic packaging was revised for subsequent releases of the story.

Above right: After being relaunched in March 1982, the *Eagle* comic ran a collectible series of pages to make up a calendar for the following year. Issue 41 featured Peter Davison as the 'TV Superstar' for January and February.

In *Doctor Who*'s 20th anniversary year most fans were still relying on their memories – and Target Books – as a record of the programme's legacy. And yet by this time there were already 4.5 million video recorders in the UK, with 200,000 new owners every month.

If BBC Video was late to the party then bureaucracy was to blame. The company's earliest releases were restricted to sporting events, music concerts, documentaries and films. In 1983 an agreement between the BBC and unions representing actors, writers and musicians finally allowed the Corporation to exploit other titles from its archive. The first to be released under the new deal were *The Best of the Two Ronnies*, sitcoms *Butterflies* and *The Fall and Rise of Reginald Perrin*, and the 1975 *Doctor Who* story *Revenge of the Cybermen*.

BBC Video justified this choice by claiming it was the serial most voted for in a poll conducted at the 20th anniversary celebration in April. "BBC Video expects to release a *Dr Who* videocassette later this year," began the questionnaire. "We would like to know which of the many programmes would most appeal to you in home-video form."

Conspiracy theorists have sought to rationalise the eccentric result of this survey by suggesting that another story was *actually* chosen – possibly *The Tomb of the Cybermen* – but on discovering

Left: The programme and badge (inset) sold at the 20th anniversary celebration held in the grounds of Longleat House over the bank holiday weekend of 3 and 4 April. BBC Enterprises seriously underestimated interest in the event, and many visitors without pre-booked tickets were turned away.

Above: Lord Bath with some of the event's special guests: Elisabeth Sladen, Peter Davison, Carole Ann Ford, Jon Pertwee and Anthony Ainley.

Below right: Costume designer Dee Robson created a new look for Omega (Ian Collier) in *Arc of Infinity*. The mask, chest plate and gloves were made by Imagineering.

Below left: The Crown of Rassilon, used to gain access to the Matrix on Gallifrey in *Arc of Infinity*.

that it had been wiped, BBC Video substituted a Cyberman tale in colour as an alternative.

According to BBC research, 85 per cent of domestic video cassettes were rented, and it was at this market that their new releases were aimed. Research also suggested that customers responded well to family entertainment feature films; thus *Revenge of the Cybermen* was presented in a compilation format that edited all four of its episodes together. The packaging was similarly revisionist, showing the current *Doctor Who* logo, a picture of Tom Baker in his Season 18 costume and a Cyberman from *Earthshock*. The price tag of £39.95 excluded all but the most dedicated fans and managers of rental shops.

The 17 October edition of trade magazine *Video Business* featured an advert from BBC Video, promoting *Revenge of the Cybermen* as "one of the definitive adventures in the BBC's mega saga of time travel, pneumatic assistants and long scarves." Despite such marketing, the title proved successful enough to warrant further releases.

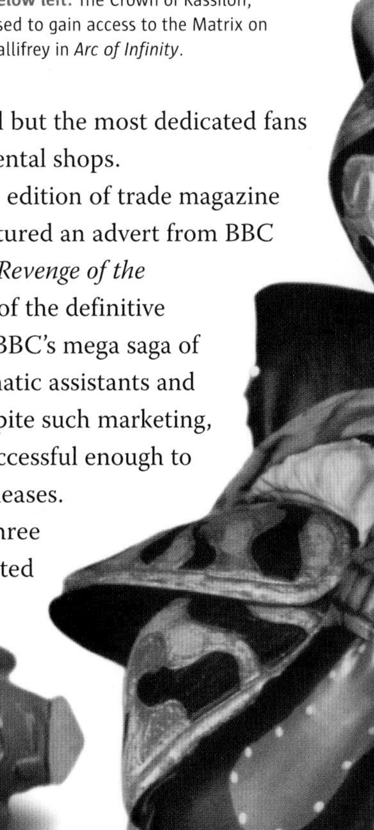

Over the next three years, similarly edited

Left: The Black Guardian (Valentine Dyall) returned in a trilogy of Season 20 stories: *Mawdryn Undead*, *Terminus* (pictured) and *Enlightenment*.

Below: Costume designer Dinah Collin used crow and dove motifs in the elaborate headdresses she created for the Black and White Guardians in Season 20.

VHS and Betamax tapes of *The Brain of Morbius, Pyramids of Mars, The Seeds of Death, The Five Doctors, Robots of Death* and *Day of the Daleks* were released for the rental market, generally priced at £24.95.

In 1987 BBC Video moved its focus onto the fast-growing 'sell-through' market, aiming titles directly at consumers. The price of new tapes was standardised at £9.99 with the first such release, a compilation version of *Death to the Daleks,* available from July. It became the most successful *Doctor Who* video ever, spending 11 weeks in the retail chart and selling more than 50,000 copies.

From here on, archive releases would become a regular, and affordable, addition to fans' collections. Once the novelty of being able to own vintage *Doctor Who* stories began to wear off, many started complaining about the way episodes were presented. They weren't the only ones. "What I don't like is when you view these programmes on video and they're all joined together," said director David Maloney, reflecting on the 1988 release of his six-part classic *The Talons of Weng-Chiang*. "It's very difficult to get the idea of the original pace... You have no idea of the height of the climax at which each episode closed."

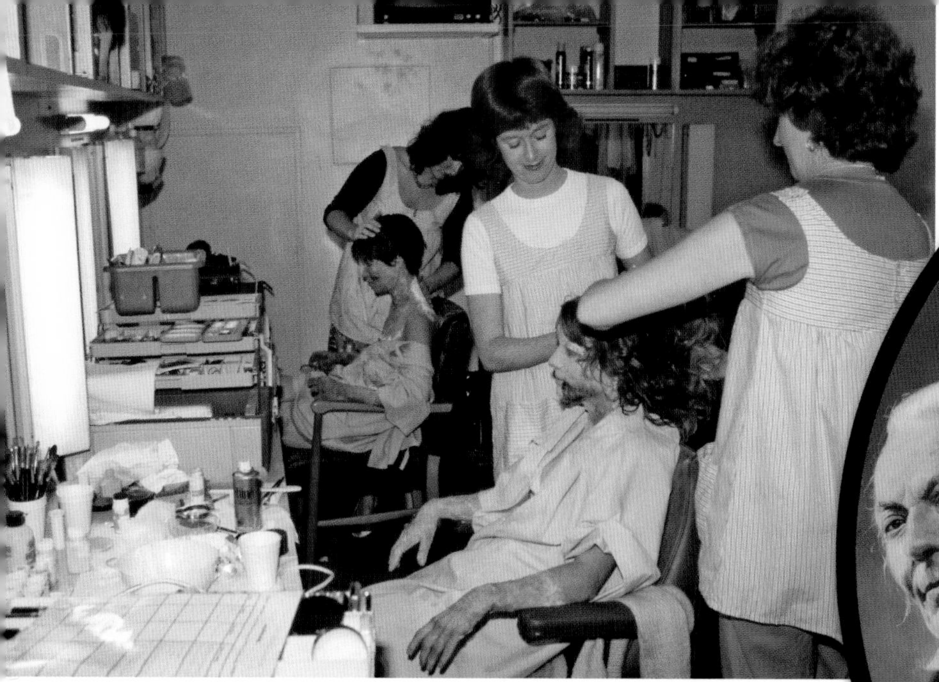

Stories finally started appearing in their original episodic format from 1989. Despite the flaws, their edited predecessors had already made an impact on fans. With the best of the surviving episodes being selected for home video release, it's possible that BBC Video were inadvertently compromising the perception of contemporary *Doctor Who*. The early videos certainly coincided with the most vociferous criticism of recent episodes, making it harder for John Nathan-Turner to ascribe these unflattering comparisons to faulty memories.

The *Doctor Who* videos also led to a democratisation of opinion about the past; with *Death to the Daleks* now available for less than £10 most fans could have an informed view of the story – not just a handful of journalists or members of the DWAS Reference Department.

Above: On 8 September 1982 Janet Fielding and Sarah Sutton get ready for the scenes in which Tegan and Nyssa are infected in Part Four of *Mawdryn Undead*.

Right: This 12" picture disc was released in the US in November. Side A featured a selection of tracks from *Doctor Who: The Music* (released in Britain earlier in the year), and Side B contained tracks from 1978's *Doctor Who Sound Effects*.

Below and below right: Target's parent company WH Allen published *Doctor Who: A Celebration* on 15 September. Compiled by Peter Haining, with a jacket illustration by Graham Potts, this was the most lavish *Doctor Who* book to date.

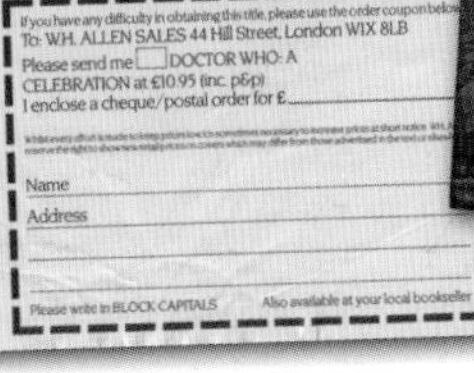

Above: *Radio Times'* first *Doctor Who* cover for ten years promoted *The Five Doctors* with an illustration by Andrew Skilleter.

Above right: Editor Brian Gearing oversaw the *Radio Times' 20th Anniversary Special*, which was published on 3 November. One of the highlights was a short story by Eric Saward entitled *Birth of a Renegade*.

Below: The Coronet and Ring of Rassilon – these original props from *The Five Doctors* were based on designs by Colin Lavers.

In the early 1980s an informal network of fans – including members of some DWAS local groups – had sprung up to facilitate video piracy. This network now started to fragment. The recovery of *The Tomb of the Cybermen* in 1991 made possible the first video whose sales hadn't been at least partly compromised by such bootleggers. Tellingly, it became the only *Doctor Who* title to get to number one in the retail charts when it was rush-released the following year.

In 1995 a 'Special Edition' of *The Five Doctors* integrated unused footage, new effects and a remixed score for a version that ran more than 17 minutes longer than the original. In 1996, the video of the *Doctor Who* TV movie was released a week before its UK transmission. It went on to sell more than 40,000 copies.

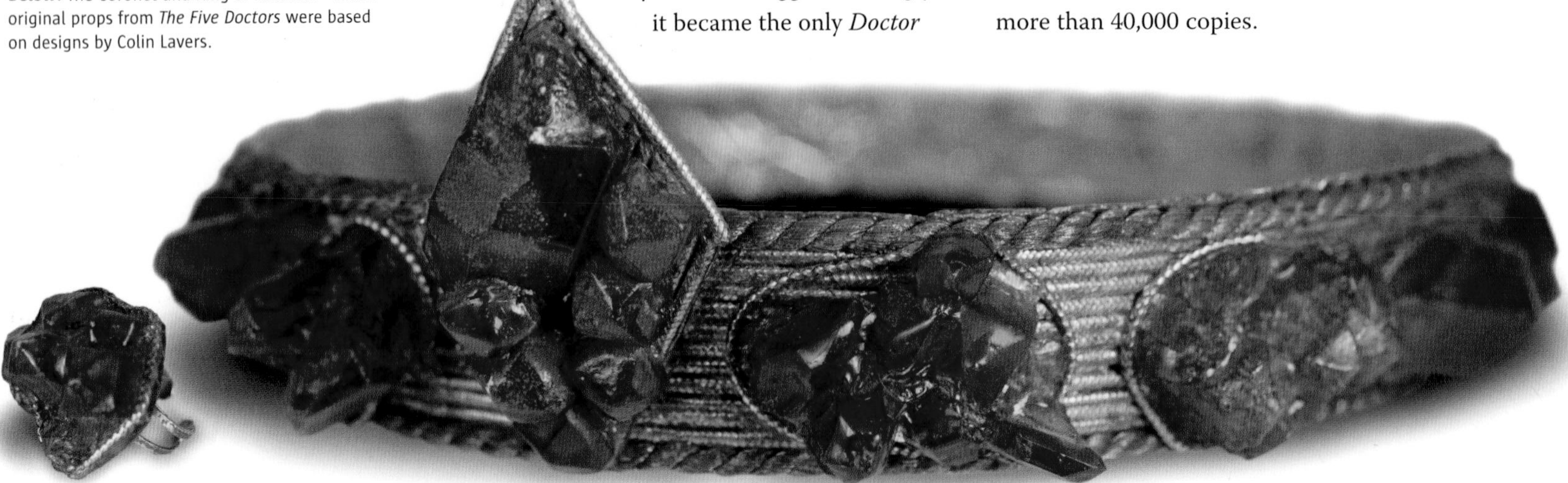

In November 1999 the *Five Doctors* Special Edition became the first story to appear on DVD. In 2003 the VHS releases came to an end with *Invasion of the Dinosaurs* – an appropriate place to say goodbye to a format that digital snobs now regarded as prehistoric.

Like the later VHS releases, the *Doctor Who* DVDs are co-curated by the Restoration Team. This collective of industry professionals use their knowledge of the show to return old episodes to their original glory, or the best possible approximation. Alongside these restorations, each disc includes a wealth of special features that puts the releases of other archive television – and most feature films – to shame.

The 2010 DVD of *Revenge of the Cybermen* includes a documentary examining the underground culture of copyright infringement that existed before BBC Video's sell-through releases. One interviewee admits to being part of a syndicate that paid £350 for a colour copy of *Doctor Who and the Silurians* in 1984. At the time the BBC only held the story in black and white, and were showing no sign of repeating or releasing it anyway.

For fans old enough to remember the early days of video, the dream of owning every existing *Doctor Who* episode is now a very real possibility. Some consolation, perhaps, for sacrificing the irreplaceable thrill of scarcity.

Above left: Mike Kelt supervised filming of the racing ships for *Enlightenment* at the BBC Visual Effects Workshop in November 1982.

Above: John Nathan-Turner with *Doctor Who*'s Saturn Award and a copy of *Doctor Who: The First Adventure*, a computer game for the BBC Micro. Unlike the television Doctor, the game afforded players a maximum of 15 regenerations.

Left: This year's *Doctor Who Annual* was published in August. It included a quiz on the history of the programme, an article about costume designers and a story featuring the Brigadier.

SEASON 20

Producer: John Nathan-Turner
Script editor: Eric Saward

6E *Arc of Infinity* 3 May – 2 June 1982
written by Johnny Byrne
directed by Ron Jones

Part One	3 January
Part Two	5 January
Part Three	11 January
Part Four	12 January

6D *Snakedance*
31 March – 28 April 1982
written by Christopher Bailey
directed by Fiona Cumming

Part One	18 January
Part Two	19 January
Part Three	25 January
Part Four	26 January

6F *Mawdryn Undead*
24 August – 24 September 1982
written by Peter Grimwade
directed by Peter Moffatt

Part One	1 February
Part Two	2 February
Part Three	8 February
Part Four	9 February

6G *Terminus*
28 September – 18 December 1982
written by Steve Gallagher
directed by Mary Ridge

Part One	15 February
Part Two	16 February
Part Three	22 February
Part Four	23 February

6H *Enlightenment*
3 November 1982 – 1 February 1983
written by Barbara Clegg
directed by Fiona Cumming

Part One	1 March
Part Two	2 March
Part Three	8 March
Part Four	9 March

6J *The King's Demons*
5 December 1982 – 16 January 1983
written by Terence Dudley
directed by Tony Virgo

Part One	15 March
Part Two	16 March

6K *The Five Doctors*
5 March – 31 March
written by Terrance Dicks
directed by Peter Moffatt
25 November

1984

Peter Davison stuck to his original intention to play the Doctor for three years, and newspapers announced his departure on 29 July 1983. His replacement, Colin Baker, was named several weeks later. Baker was already familiar to many fans – as Commander Maxil he had seemingly executed the Fifth Doctor in *Arc of Infinity*.

Season 21 began with *Warriors of the Deep*, which established *Doctor Who*'s latest broadcast slot of Thursdays and Fridays. It was a faltering start for Davison's final series – disruption of the planned studio recording at Television Centre meant that the visual effects team in particular were ill-prepared. In many scenes, the combined return of the Sea Devils and Silurians failed to make the intended impact.

The Awakening was a two-part story that evoked the spirit of *The Daemons* in its depiction of an alien horror discovered within a rural church. This was the last *Doctor Who* assignment for designer Barry Newbery, who had first contributed to the show in 1963.

Christopher H Bidmead's bleak portrayal of *Frontios* featured the subterranean Tractators and an alarming cliffhanger where the TARDIS appeared to have been destroyed, leaving only the control room's hat stand in its stead.

Ratings improved for Eric Saward's *Resurrection of the Daleks*, which was broadcast in two 45-minute instalments because of a scheduling conflict with the BBC's coverage of the Winter Olympics. Davros was now played by Terry Molloy, and a disenchanted Tegan left amidst the carnage of the story's conclusion.

Planet of Fire was partly filmed on location in Lanzarote. The story saw the departure of both Turlough and the technically problematic Kamelion. American botany student Peri Brown (Nicola Bryant) joined the Doctor, her character's nationality the latest attempt by John Nathan-Turner to broaden the international appeal of the show.

The Fifth Doctor met his demise, poisoned by Spectrox Toxaemia, in *The Caves of Androzani*. Robert Holmes' script was equal to his finest work from the 1970s, and Graeme Harper's direction offered a tantalising glimpse of a *Doctor Who* freed from its archaic multi-camera production methods.

The garish costume, arrogant demeanour and bizarre post-regenerative behaviour of the Sixth Doctor made *The Twin Dilemma* an uncomfortable viewing experience. The programme was about to test the loyalty of its audience as never before...

BENEATH THE SURFACE

Bottom: Visual effects designer Mat Irvine created this model of Sea Base 4, the location for the Silurian and Sea Devil attack in *Warriors of the Deep*.

Below right: The model was filmed at the BBC Visual Effects Workshop in July 1983.

Below: Terrance Dicks' novelisation of Johnny Byrne's script made its hardback debut in May. Andrew Skilleter's cover showed one of the samurai-style Sea Devils.

In 2009 *Doctor Who Magazine* conducted a poll to commemorate the series' 200th story. The aim was to rank every story by popularity, and more than 6,700 readers responded with their views.

The number one adventure beat several previous poll-winners, as well as recent favourites starring Christopher Eccleston and David Tennant. According to *Doctor Who Magazine* readers in 2009, the greatest story ever made was *The Caves of Androzani*.

Broadcast in March 1984, *The Caves of Androzani* was the final regular appearance of Peter Davison's Doctor, but something of a comeback for its author, Robert Holmes, who hadn't written for the series since he delivered *The Power of Kroll* in 1978. "I wanted to write a story with an underlying moral about the evils of gun-running, armaments supply etc, because I didn't remember *Doctor Who* ever having touched on that particular subject," he said. "The fact that the script went into production unchanged was a source of satisfaction. But, I suppose, even more satisfying was the fact that after an absence of six or seven years I still seemed to have the knack of writing for the programme."

Analysing that knack reveals much about how to write successful scripts for the series. Holmes was the quintessential *Doctor Who* writer, yet the best of his work was made both accessible and unusual by the juxtaposition of familiar elements from other stories and genres.

Nine years before *Star Wars, The Space Pirates* was science fiction partly configured as a Western; *Pyramids of Mars* owed its iconography to *The Mummy* and the faux Victoriana of *The Talons of Weng-Chiang*

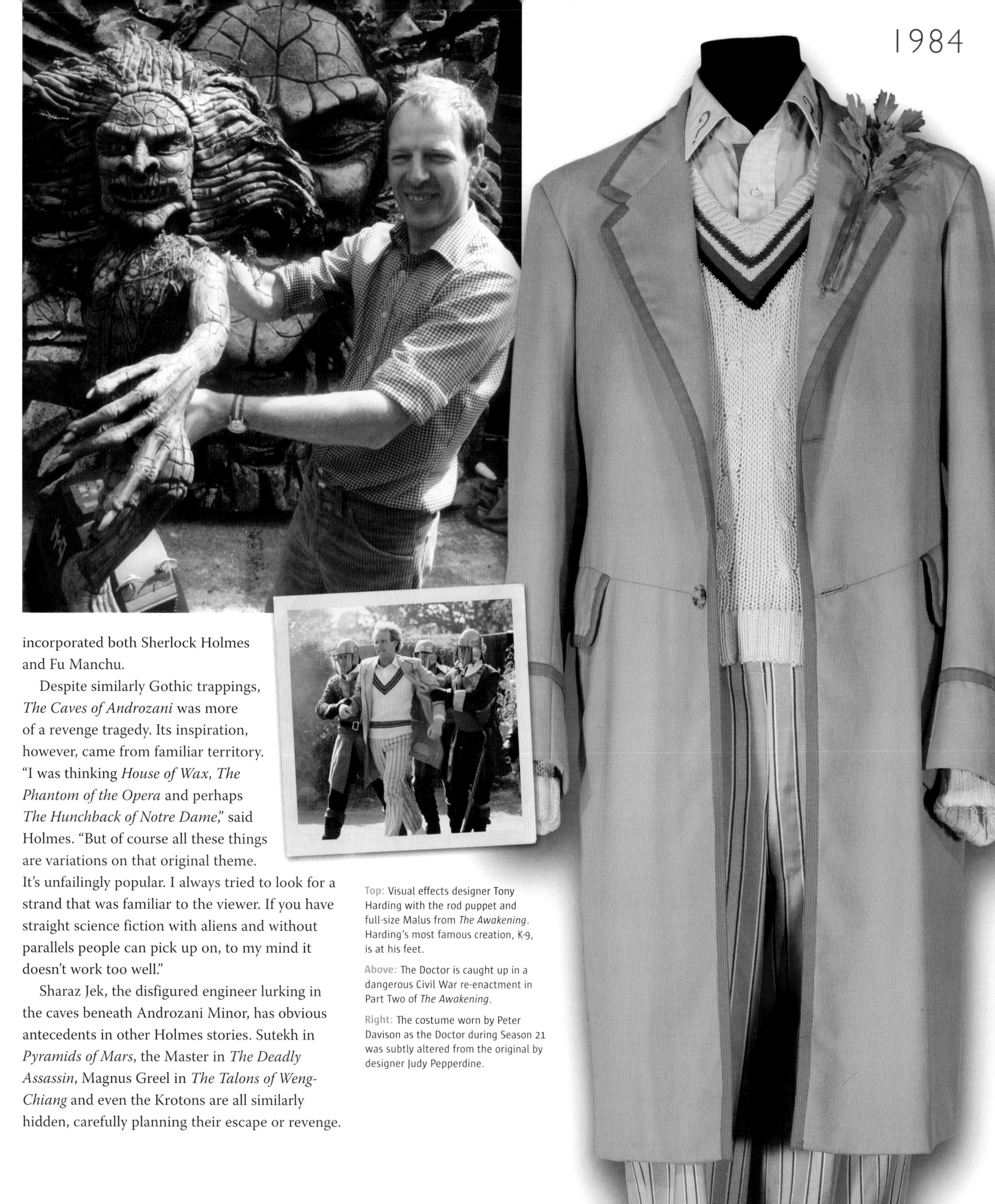

incorporated both Sherlock Holmes and Fu Manchu.

Despite similarly Gothic trappings, *The Caves of Androzani* was more of a revenge tragedy. Its inspiration, however, came from familiar territory. "I was thinking *House of Wax, The Phantom of the Opera* and perhaps *The Hunchback of Notre Dame*," said Holmes. "But of course all these things are variations on that original theme. It's unfailingly popular. I always tried to look for a strand that was familiar to the viewer. If you have straight science fiction with aliens and without parallels people can pick up on, to my mind it doesn't work too well."

Sharaz Jek, the disfigured engineer lurking in the caves beneath Androzani Minor, has obvious antecedents in other Holmes stories. Sutekh in *Pyramids of Mars*, the Master in *The Deadly Assassin*, Magnus Greel in *The Talons of Weng-Chiang* and even the Krotons are all similarly hidden, carefully planning their escape or revenge.

Top: Visual effects designer Tony Harding with the rod puppet and full-size Malus from *The Awakening*. Harding's most famous creation, K-9, is at his feet.

Above: The Doctor is caught up in a dangerous Civil War re-enactment in Part Two of *The Awakening*.

Right: The costume worn by Peter Davison as the Doctor during Season 21 was subtly altered from the original by designer Judy Pepperdine.

Left: One of the phosphor lamps used by the colonists on *Frontios*.

Above: The Doctor takes one of the lamps as he explores the area beneath the research room in Part Two.

Right: *Frontios* writer Christopher H Bidmead novelised his story for Target. The book was first published in hardback on 20 September, with a cover illustration of the Gravis by Andrew Skilleter.

But there was light as well as dark in Holmes' scripts. He delighted in absurdist humour, pricking the Third Doctor's pomposity in Episode One of *Carnival of Monsters*. "These creatures may look like chickens," he tells Jo, "but for all we know they're the intelligent life form on this planet."

Holmes' love of language provided numerous wry exchanges – Vira's confusion at the Doctor and Harry's colloquialisms in *The Ark in Space*, the obsequious Hade's mispronunciation of 'mahogany' in *The Sun Makers* and Garron's florid patter in *The Ribos Operation* are just a few examples. *The Sun Makers* was also a vehicle for his thinly veiled attack on the Inland Revenue; other satirical touches included his mockery of the broadcast media, as personified by Commentator Runcible in *The Deadly Assassin*.

His typically detailed characters often worked together, beginning with the vaudeville double act of Vorg and Shirna in *Carnival of Monsters* and continuing through the roguish Dibber and Glitz in *The Trial of a Time Lord* (1986). *The Talons of Weng-Chiang*'s amateur detectives Jago and Litefoot even went on to their own series of audio adventures.

Doctor Who was Holmes' favourite of the many television shows he worked on, but he was under no illusion that it represented anything other than formulaic entertainment. "If anyone decides that *Doctor Who* is an art form its death knell will be sounded," he said with typical candour. "It is good, clean, escapist hokum, which is no small thing to be."

As one of *Doctor Who*'s most successful script editors, Holmes was well aware of its limited budget and format restrictions. His best scripts showed how much could be achieved while respecting the boundaries of both.

Within these constraints he created some of the most memorable moments in the show's history. He named the Doctor's home planet, wrote the first scripts to feature Sarah Jane Smith and the Master, created the Autons and Sontarans, and defined the perception of Time Lord society that still holds sway today. Given its hostile reaction from the Doctor Who Appreciation Society, *The Deadly Assassin* was Holmes at his most transgressive. "I had to decide what sort of people the Time Lords were," he said. "I noticed that over the years they had produced quite a few galactic lunatics – the Meddlesome Monk, the Master, Omega, Morbius... How did this square with

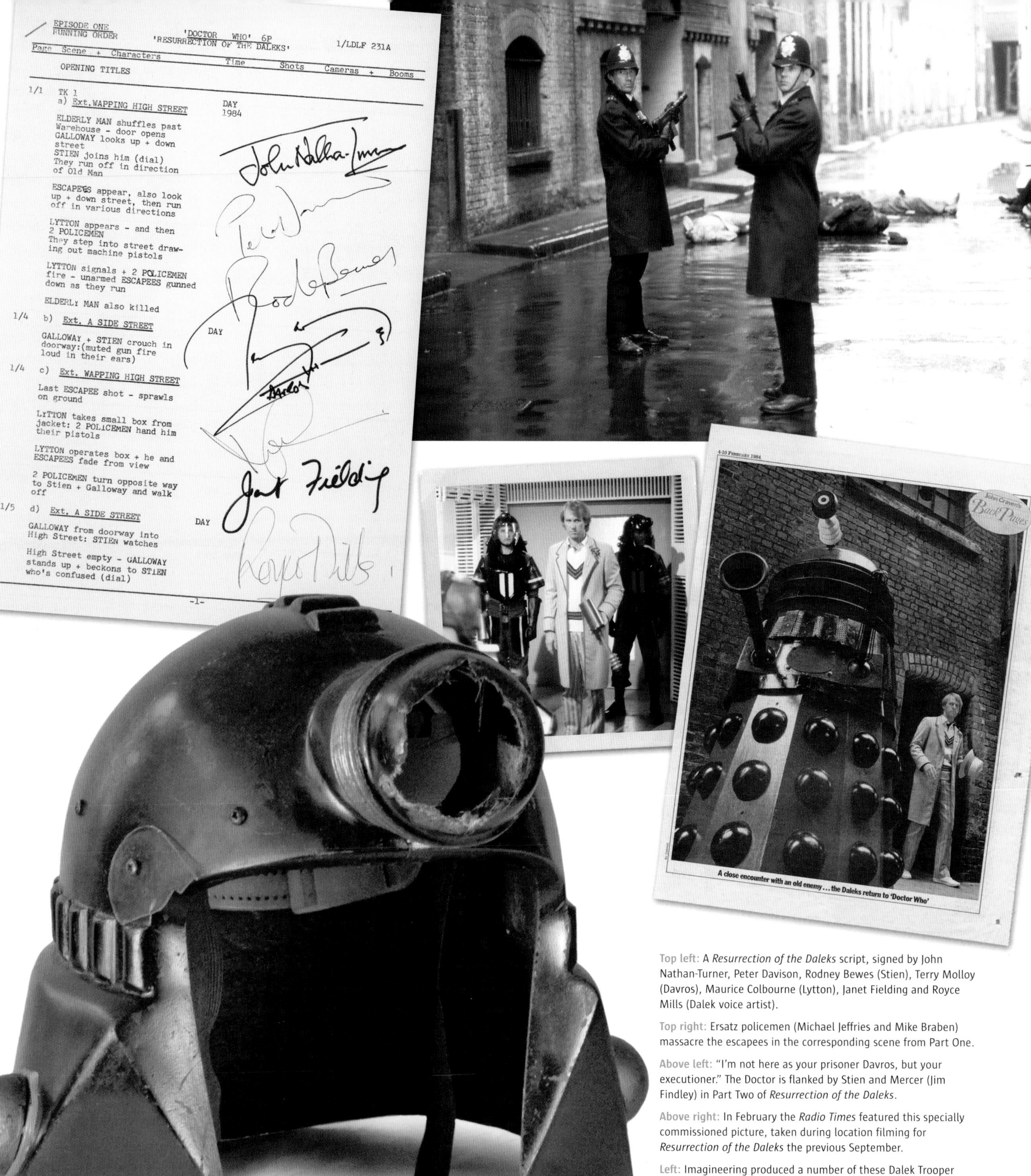

EPISODE ONE
RUNNING ORDER

'DOCTOR WHO' 6P
'RESURRECTION OF THE DALEKS'

1/LDLF 231A

Page	Scene + Characters	Time	Shots	Cameras + Booms
	OPENING TITLES			
1/1	TK 1 a) Ext.WAPPING HIGH STREET	DAY 1984		
	ELDERLY MAN shuffles past Warehouse - door opens GALLOWAY looks up + down street STIEN joins him (dial) They run off in direction of Old Man			
	ESCAPEES appear, also look up + down street, then run off in various directions			
	LYTTON appears - and then 2 POLICEMEN They step into street drawing out machine pistols			
	LYTTON signals + 2 POLICEMEN fire - unarmed ESCAPEES gunned down as they run			
	ELDERLY MAN also killed			
1/4	b) Ext. A SIDE STREET	DAY		
	GALLOWAY + STIEN crouch in doorway:(muted gun fire loud in their ears)			
1/4	c) Ext. WAPPING HIGH STREET			
	Last ESCAPEE shot - sprawls on ground			
	LYTTON takes small box from jacket: 2 POLICEMEN hand him their pistols			
	LYTTON operates box + he and ESCAPEES fade from view			
	2 POLICEMEN turn opposite way to Stien + Galloway and walk off			
1/5	d) Ext. A SIDE STREET	DAY		
	GALLOWAY from doorway into High Street: STIEN watches			
	High Street empty - GALLOWAY stands up + beckons to STIEN who's confused (dial)			

-1-

Top left: A *Resurrection of the Daleks* script, signed by John Nathan-Turner, Peter Davison, Rodney Bewes (Stien), Terry Molloy (Davros), Maurice Colbourne (Lytton), Janet Fielding and Royce Mills (Dalek voice artist).

Top right: Ersatz policemen (Michael Jeffries and Mike Braben) massacre the escapees in the corresponding scene from Part One.

Above left: "I'm not here as your prisoner Davros, but your executioner." The Doctor is flanked by Stien and Mercer (Jim Findley) in Part Two of *Resurrection of the Daleks*.

Above right: In February the *Radio Times* featured this specially commissioned picture, taken during location filming for *Resurrection of the Daleks* the previous September.

Left: Imagineering produced a number of these Dalek Trooper helmets for *Resurrection of the Daleks*.

Top: The Doctor asks Sharaz Jek (Christopher Gable) to help him treat Peri's Spectrox Toxaemia in *The Caves of Androzani* Part Four.

Above: *The Offical Doctor Who Magazine* Issue 90 was edited by Alan McKenzie and reviewed Davison's tenure as the Doctor. The cover picture was taken at Stokeford Heath in Dorset, during the location shoot for *The Caves of Androzani*.

the perceived notion that the Time Lords were a bunch of omnipotent do-gooders? Could it be that this notion had been put about by the Time Lords themselves? Heresy!"

Shortly after *The Deadly Assassin* was broadcast, Holmes was subjected to a profile by Jean Rook of the *Daily Express*. "Mr Holmes is tall, grey-haired and bloodless, in a cape-shaped fawn mac," she wrote on 11 February 1977. "He looks like Sherlock Holmes playing Dracula. He reads Poe, Arlen and Bradbury in bed."

In 1998 Barry Letts painted a rather kinder picture. "I was very fond of Bob, who was a very humble man in a way. After one of his *Doctor Who* stories aired I remember there was a review which said, 'This is very fine writing indeed. I can't think why Robert Holmes isn't writing *The Wednesday Play*.' The script editor of *The Wednesday Play* got in touch with him and said, 'Well, how about it?' and Bob said he'd love to, but it never happened. It was down to self esteem. If Bob had more, he would have been much better known."

The Caves of Androzani initiated Holmes' final stint on *Doctor Who*, brought back into the fold by admiring script editor Eric Saward. By the time Holmes died in 1986 he had become the programme's longest serving writer, and the most prolific. His closest rival for the latter distinction is Russell T Davies, who cites *The Ark in Space* as his favourite story. "I must have watched this a hundred times," he admits. "It's not enough."

In 2007 Davies advised *The Daily Telegraph*'s Richard Johnson to take a look at *The Talons of Weng-Chiang*. "Watch episode one. It's the best dialogue ever written. It's up there with Dennis Potter. By a man called Robert Holmes. When the history of television drama comes to be written, Robert Holmes won't be remembered at all because he only wrote genre stuff. And that, I reckon, is a real tragedy."

Left: "I feel almost as though this part was made for me, or I was made for this part," Colin Baker told the *Radio Times*' Katie Griffiths.

Below: John Nathan-Turner's production of *Cinderella* opened at Southampton's Gaumont Theatre on 26 December. The pantomime was still running when location filming for *Revelation of the Daleks* began in nearby Portsmouth two weeks later.

Bottom: The *Doctor Who Annual* published in August featured illustrations by Mel Powell and articles by Brenda Apsley on set design and visual effects.

17-23 March 1984

John Craven's Back Pages

A dream come true for Doctor Who

When he appears in green shoes, orange spats, yellow-and-black-striped trousers and jacket with a hotchpotch of patterns on Thursday (BBC1), Colin Baker will be making his first fully-fledged appearance as the sixth Doctor Who. But he has already been accepted by some of his most demanding fans, those who attended the Doctor Who Convention in America last month. 'I was introduced as Peter Davison's replacement, and after the natural reserve of the first five minutes they were very warm to me.'

The new Doctor's first, eventful adventure will be 'The Twin Dilemma', a story containing a nasty slug-like monster called the Gastropod. Colin will emerge as a spiky, articulate, sometimes arrogant but very witty character. It was, in fact, Colin's own ability to be highly entertaining which originally impressed the show's producer, John Nathan-Turner.

'We were at a wedding of someone working on the series,' says Colin, 'and it was one of those days in a million when you are really on form.' He emphasises, however, that he is dependent on the script and cannot go around

as Paul Merroney, 'a prototype J. R. Ewing'. When travelling in Holland, Sweden and Israel, where *The Brothers* attracted a cult following, he had a chance to sample the sort of fame that he can now expect as the Doctor. Recently his work has included theatre appearances in Sweden. During a break from **Doctor Who** Colin will be in Norway for five weeks, producing, directing and playing in Agatha Christie's *The Mousetrap*. And he is playing another doctor, Dr Dudgeon, in BBC2's current serial *Swallows and Amazons For Ever!*

Perhaps the most difficult thing he has had to adjust to on the series has been the tight schedule. 'It is important to muster reserves of energy as there is a lot of standing around to do.'

He has already been filming in the chalk pits and clay pits that Terry Wogan jokes about. 'When it is raining it is all right for the crew – they can put wellies on. I have to walk around with plastic bags on my legs and keep wiping the mud from my shoes.'

But he is almost embarrassed about enjoying himself so much. 'It is everybody's dream to play their hero, whether it is Lancelot or Biggles or Doctor Who, because they are characters in modern mythology. I always suspected it would be good fun. I feel almost as though this part was made for me, or I was made for this part.'

KATIE GRIFFITHS

SEASON 21

Producer: John Nathan-Turner
Script editor: Eric Saward

6L *Warriors of the Deep*
23 June – 15 July 1983
written by Johnny Byrne
directed by Pennant Roberts

Part One	5 January
Part Two	6 January
Part Three	12 January
Part Four	13 January

6M *The Awakening*
19 July – 6 August 1983
written by Eric Pringle
directed by Michael Owen Morris

Part One	19 January
Part Two	20 January

6N *Frontios*
24 August – 9 September 1983
written by Christopher H. Bidmead
directed by Ron Jones

Part One	26 January
Part Two	27 January
Part Three	2 February
Part Four	3 February

6P *Resurrection of the Daleks*
11 September – 7 October 1983
written by Eric Saward
directed by Matthew Robinson

Part One	8 February
Part Two	15 February

6Q *Planet of Fire*
14 October – 11 November 1983
written by Peter Grimwade
directed by Fiona Cumming

Part One	23 February
Part Two	24 February
Part Three	1 March
Part Four	2 March

6R *The Caves of Androzani*
15 November 1983 – 12 January 1984
written by Robert Holmes
directed by Graeme Harper

Part One	8 March
Part Two	9 March
Part Three	15 March
Part Four	16 March

6S *The Twin Dilemma*
24 January – 16 February
written by Anthony Steven
directed by Peter Moffatt

Part One	22 March
Part Two	23 March
Part Three	29 March
Part Four	30 March

1985

Season 22 opened with *Attack of the Cybermen* and returned *Doctor Who* to Saturday nights, albeit in the 45-minute slot familiar from the previous year's *Resurrection of the Daleks*. Some viewers struggled with *Attack*'s references to old storylines, while others expressed reservations over the Doctor turning a gun on the Cybermen at the climax.

There was more graphic imagery, this time involving an acid bath, in *Vengeance on Varos*. Nabil Shaban proved memorable as the avaricious Sil, a slug-like alien who revelled in the televised torture of criminals.

The methods of despatch in *The Mark of the Rani* were rather more novel, including a land mine that turned its victim into a tree. For this story the Master was joined by the Rani (Kate O'Mara), a renegade Time Lady with a talent for bio-chemistry.

The Two Doctors brought the Second Doctor and Jamie back to the programme, joining the Sixth Doctor and Peri for a skirmish with the Sontarans. Robert Holmes' light-hearted script was partly filmed in Santa Cruz and Gerena in Spain.

The Two Doctors was another story that led critics to question the levels of violence in the show. More seriously, around this time John Nathan-Turner learned that Michael Grade, the Controller of BBC1, and Jonathan Powell, the BBC's Head of Drama, were also unhappy about the direction the series was taking. It wasn't just the violence that concerned Grade – he later told journalists that he was also worried about its cost and declining viewing figures. Plans for the next season were scrapped while the programme was prescribed an enforced rest and rethink.

Fans perceived the decision as a near-cancellation. They were further dismayed by the next story *Timelash*, even though it boasted one of the series' most visually arresting villains in the Borad (Robert Ashby).

The season closed on a return to form, as Graeme Harper directed Eric Saward's *Revelation of the Daleks*. With Davros' macabre scheme confounded, Part Two was due to end with the Doctor offering to take Peri to Blackpool. The comment was designed to lead into *The Nightmare Fair*, a Season 23 story written by former producer Graham Williams.

Revelation of the Daleks was instead curtailed on a freeze frame, as the future of *Doctor Who* hung in the balance.

SIX OF ONE

Above: In 1985 the former *Doctor Who Weekly* underwent its latest name change, becoming *The Doctor Who Magazine*. Issue 100 was edited by Cefn Ridout and published in April.

Right: The version of the Sixth Doctor's costume worn by Colin Baker in Season 22. Baker had hoped for a black velvet suit, but that was considered too similar to the costume worn by Anthony Ainley as the Master.

Below: The Doctor (Colin Baker) and Peri (Nicola Bryant) arrive at a familiar junkyard in *Attack of the Cybermen* Part One.

"I can't understand why they brought you back looking like *that*," said Selena Scott, surveying the latest Doctor's costume on *Breakfast Time* on 22 March 1984.

"Would you suggest an alternative?" Colin Baker asked.

"Well, someone suave and charismatic..." she began.

The persona of the Sixth Doctor was certainly a bold departure from the Fifth Doctor's more introspective character, but it was also a reflection of the programme's supreme confidence in the wake of the 20th anniversary celebrations. *Doctor Who*'s profile and viewing figures were high, and even the fans seemed satisfied with the latest storylines. It was only from such a position of strength that John Nathan-Turner could consider unleashing a Doctor seemingly designed to provoke both the audience and the critics.

"He has become a nasty fellow, manic and violent, with a conceited, vacuous smile," wrote Judith Simmons in the *Daily Express*, the day after the *Breakfast Time* interview. "This ploy gives us an interval to accept that Peter Davison has vacated the role, and Baker will have a different personality. Baker's first new costume is a multi-patterned masterpiece of mixed-up tailoring. I congratulate the designer on a style suiting both the Doctor's current fractured psyche and his habit of hamming up his lines like school Shakespeare."

While Nathan-Turner had very clear ideas about the costume from the outset, the details of the new Doctor's character were still being worked out by Colin Baker during production of his first story, *The Twin Dilemma*. "He was very, very unsure what sort of Doctor to be," said director Peter Moffatt in 1997. "Every time we got a new scene he'd say, 'How do you think I should play this? Authoritatively? Jokey?' We thought up the basis of the character very much between us, working off each other. Colin had ideas, but he wanted to be different; he didn't want to steal mannerisms from any of the other doctors."

Baker still defends the Sixth Doctor's prickly demeanour. "I thought *The Twin Dilemma* was

DOCTOR WHO
THE CAVES OF ANDROZANI
TERRANCE DICKS

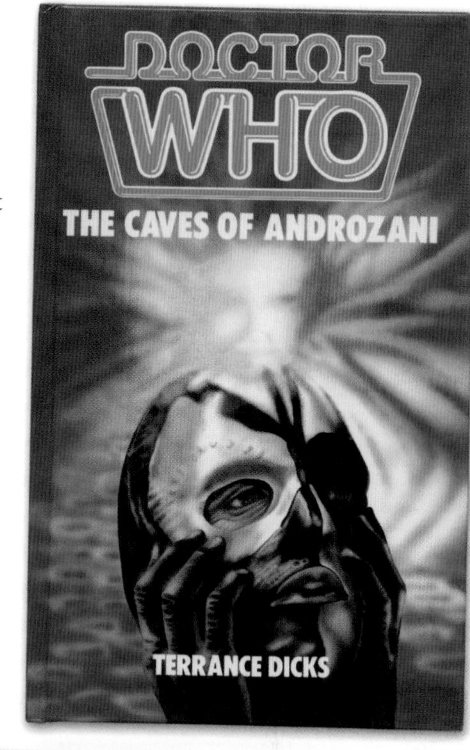

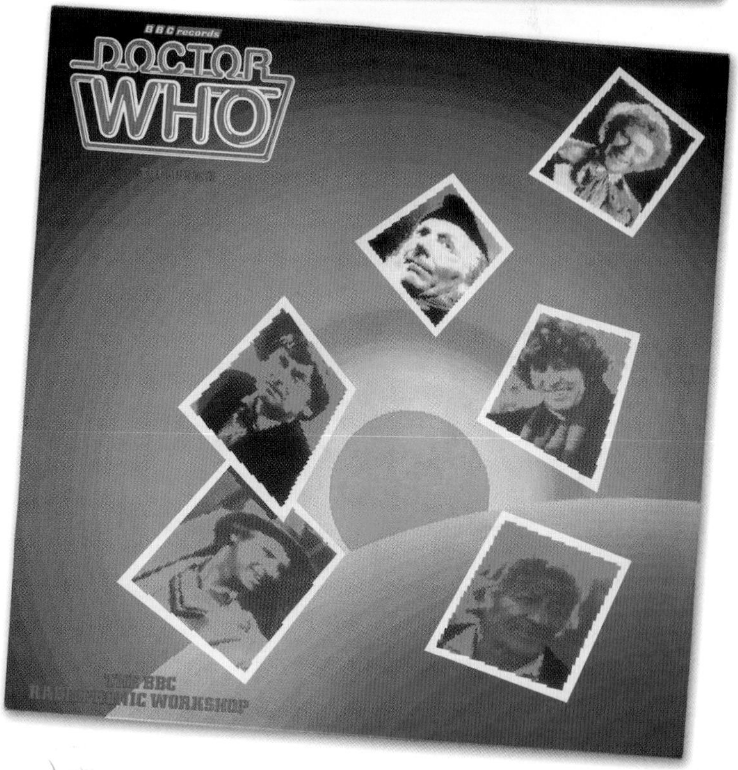
BBC records
DOCTOR WHO

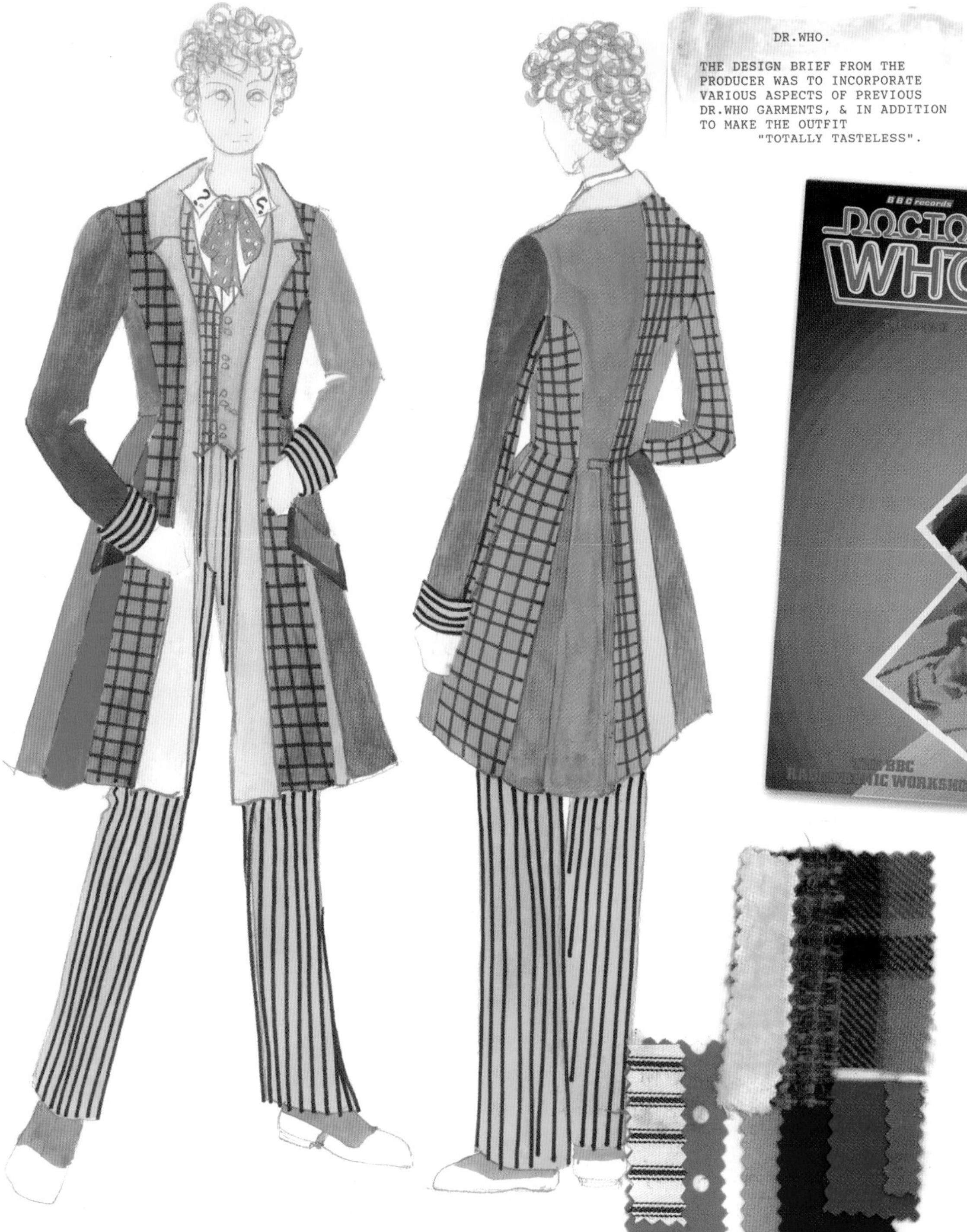
DR.WHO.
THE DESIGN BRIEF FROM THE PRODUCER WAS TO INCORPORATE VARIOUS ASPECTS OF PREVIOUS DR.WHO GARMENTS, & IN ADDITION TO MAKE THE OUTFIT "TOTALLY TASTELESS".

Top: Target Books were unable to secure the rights to use Colin Baker's likeness, but Andrew Skilleter's cover art for *The Caves of Androzani* seemed to suggest the Sixth Doctor emerging from his regeneration.

Above: A second volume of *Doctor Who: The Music* was released by BBC Records in February. It included incidental music by Malcolm Clarke, Jonathan Gibbs, Peter Howell and Roger Limb.

Left: Pat Godfrey's design for the Sixth Doctor's costume was completed in 1984.

Left: Lytton (Maurice Colbourne) undergoes the conversion process in *Attack of the Cybermen* Part Two.

Below: *The Third Doctor Who Quiz Book* was published in October. This was another Target Book that avoided showing the Sixth Doctor's face.

Bottom: Visual effects designer Chris Lawson adapted the guns seen in *Attack of the Cybermen* (inset) from a design previously used in *Earthshock* and *The Five Doctors*.

brave," he says, "but not a mistake. I thought the programme was big enough – and I still do, despite the uneasy reaction – to withstand that kind of shock, and I think it possibly needed it. We needed to wrong-foot the viewer."

It was a portrayal that proved unpopular in some quarters, not least with certain executives in BBC management, but after his debut story did the Sixth Doctor actually break any of the unwritten rules governing the way the character should behave?

The Doctor is sketchily defined, but there are a number of characteristics that have remained consistent throughout all his different incarnations. He is empathic, essentially liberal and likes travelling with pretty Earth girls, although his insatiable curiosity takes him to many other planets too.

The Sixth Doctor's belligerence was nothing new – the First Doctor could be similarly unsociable. A frequently childish, possibly slightly senile character, Doctor Number One plotted a haphazard course around the universe, never quite in control of his ship. He does, however, firmly establish the character as a pacifist when he tells Tyler (Bernard Kay) in *The Dalek Invasion of Earth*: "I never take life. Only when my own is immediately threatened."

Left: The new *Doctor Who Annual* was published in August. Falling sales and the long gap between seasons 22 and 23 prompted World Distributors to make this their last.

Above: Sil (Nabil Shaban) debates with the Governor (Martin Jarvis) in Part Two of *Vengeance on Varos*.

Right: The BBC's Programme Correspondence Section responded to a multitude of complaints following the announcement that *Doctor Who* was to be 'rested'.

Below: 'Doctor in Distress', a charity record protesting at the series' hiatus, was recorded on 7 and 8 March. Among the singers were Anthony Ainley, Colin Baker, Nicola Bryant and Nicholas Courtney.

BBC

BRITISH BROADCASTING CORPORATION
BROADCASTING HOUSE LONDON W1A 1AA
TELEX: 265781 CABLES: BROADCASTS LONDON TELEX
TELEPHONE 01-580 4468
DIRECT TELEPHONE LINE: 01-927

21st March 1985

Reference: 28/JB

Dear Mr. Ware,

Thank you for your recent letter, which I have been asked to acknowledge.

We are sorry that you should be concerned about the decision to 'rest' "Doctor Who". In broadcasting, no programme can, of course, be regarded as an entirely fixed institution, but time for reflection on a programme's future is seldom time wasted. What will, in fact, happen, is that in 1986 "Doctor Who" will be on the air, as it has been for each of the past twenty-two years. It will, however, return to what was until recently its traditional placing in the Autumn, and each episode will be produced at the traditional twenty-five minute duration.

The producer and his team welcome the return to the old form, which will enable them to concentrate on the essential themes which have proved so popular over the years. Shorter programmes will mean that it will be possible to present a longer series, extending over a greater number of weeks.

Thank you again for your letter. We appreciate your interest in writing, and shall bear all your comments in mind.

Yours sincerely,

Jane Barrow (Miss)
Programme Correspondence Section

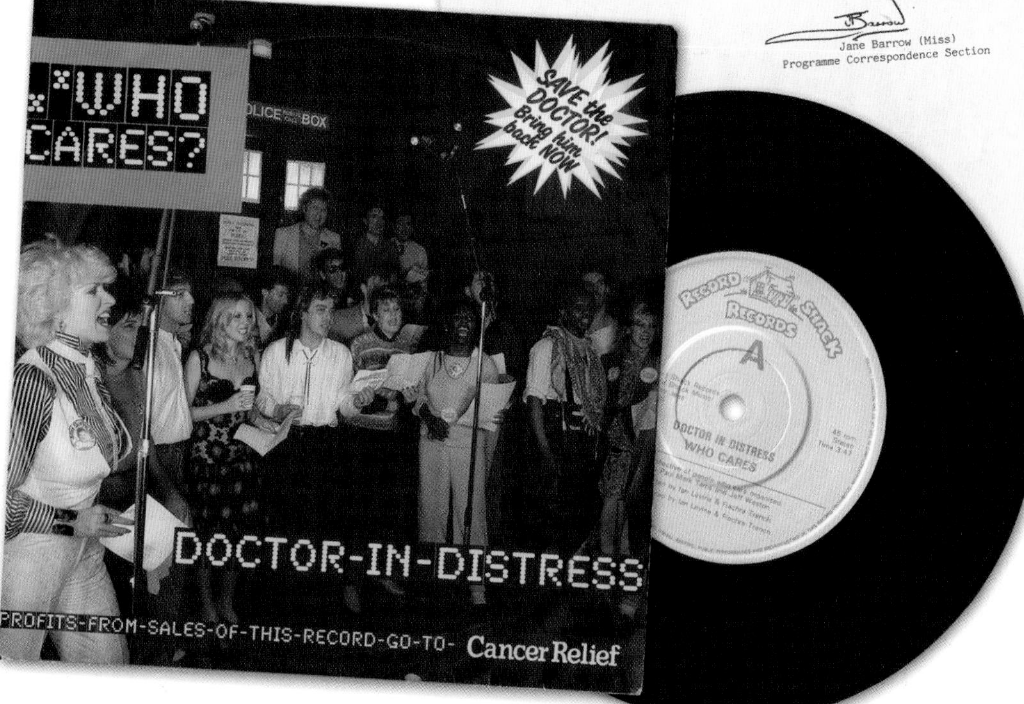

The Sixth Doctor also echoed some of the pomposity of his third incarnation. Especially abrasive during his exile on Earth, this Doctor was a bon viveur whose velvet smoking jacket and allegiance to UNIT created the impression that he was something of an establishment figure. This was entirely superficial, as the Brigadier and jobsworth bureaucrats soon discovered.

The parameters of the role were succinctly summarised by Tom Baker in the 1977 documentary *Whose Doctor Who*: "One of the problems about playing the Doctor, which makes it interesting for the actor who plays it – and everybody's been successful therefore it would seem to be actor-proof – is that it's not an acting part, in the sense that the character is very, very severely limited. There are boundaries over which the Doctor can't go. He can't suddenly become interested in romance – he doesn't have those kind of emotions. He's not at all acquisitive, he couldn't be suddenly gratuitously violent. So therefore, in the ordinary sense of acting, the character can't develop. So the problem for the actor is to surprise the audience constantly."

Tom Baker achieved this element of surprise by allowing his own personality to inform his portrayal. "To play someone from outer space, a

benevolent alien who still looks like a human being and who has secrets – how do you suggest he is alien? So I felt the best way to suggest I was an alien – and had dark thoughts, wonderful thoughts – was to be Tom Baker. And so that's what I did."

In comparison with the manic Fourth Doctor, Number Five came across as someone swept along by the tide of events. Doctor Number Six was predictably disparaging about his predecessor. "I was never happy with that one," he tells Peri in Part One of *The Twin Dilemma*. "It had a sort of feckless charm which simply wasn't me."

So the Sixth Doctor's personality was not such an innovation – more an exaggeration of traits previously displayed by his first, third and fourth incarnations. Given that the character has remained essentially the same, Colin Baker believes that "just about any actor in Equity could play the Doctor." But there is one portrayal that has become a benchmark for the role.

Patrick Troughton's depiction of the Second Doctor is hard to define – his Doctor can be secretive (*The Evil of the Daleks*), manipulative (*The Tomb of the Cybermen*) and prone to childish panic, but this unassuming vagabond conceals a fierce intelligence.

Matt Smith, Doctor Number 11, has been particularly fulsome in his praise of the scruffy Second Doctor. "Just as every actor since Patrick Troughton has done, he fell in love with Troughton's version," says Steven Moffat. "It seems to be the bedrock, really. I was looking at *The Invasion* recently, and thinking that all the dialogue

Above: *The Two Doctors* (with Patrick Troughton in full Androgum make-up) during location filming at the olive grove at Dehera Boyer, near Genera in Spain, during early August 1984.

Left: One of the original Sontaran costumes from *The Two Doctors*, as displayed at the Doctor Who Experience exhibition in Cardiff.

Below right: Stike (Clinton Greyn) meets a grisly end in Part Three of *The Two Doctors*.

would fit for David Tennant's Doctor. Completely different actor, different performance, but he is the Doctor as we know him now... He was the very first person to play the Doctor like that which is, broadly speaking, how he's been portrayed ever since. So he was inventing the idea that he could be the hero and the boffin. He could be a bit mad and a bit funny, a bit exotic and a bit common."

Troughton revived that endearing humility in 1985, when he appeared alonsgside Colin Baker in *The Two Doctors.* Here were two vastly different portrayals of recognisably the same character, side by side. Fortunately, as the Second Doctor surmises in the final episode, "the time continuum should be big enough for the both of us."

Above: Davros (Terry Molloy) in *Revelation of the Daleks* Part One. A washing machine motor was used to spin the prop head for the death scene.

Above right: The cover of *The Doctor Who Magazine* Issue 102 featured the mutant incubator from *Revelation of the Daleks* Part Two.

Right: In *Revelation of the Daleks* Part One the Doctor and Peri wore the colour of mourning on Necros.

Left: The original cloak designed by Pat Godfrey for the episode.

SEASON 22

Producer: John Nathan-Turner
Script editor: Eric Saward

6T *Attack of the Cybermen*
29 May – 6 July 1984
written by Paula Moore
directed by Matthew Robinson

Part One	5 January
Part Two	12 January

6V *Vengeance on Varos*
18 July – 2 August 1984
written by Philip Martin
directed by Ron Jones

Part One	19 January
Part Two	26 January

6X *The Mark of the Rani*
22 October – 20 November 1984
written by Pip and Jane Baker
directed by Sarah Hellings

Part One	2 February
Part Two	9 February

6W *The Two Doctors*
9 August – 28 September 1984
written by Robert Holmes
directed by Peter Moffatt

Part One	16 February
Part Two	23 February
Part Three	2 March

6Y *Timelash*
4 December 1984 – 30 January 1985
written by Glen McCoy
directed by Pennant Roberts

Part One	9 March
Part Two	16 March

6Z *Revelation of the Daleks*
7 January – 1 February
written by Eric Saward
directed by Graeme Harper

Part One	23 March
Part Two	30 March

1986

The only new *Doctor Who* produced during the long break between seasons was a Radio 4 serial called *Slipback*. Colin Baker and Nicola Bryant recorded Eric Saward's script on 10 June 1985.

The state of television *Doctor Who* remained precarious. The show had been granted a 14-episode probation, but the episodes themselves would revert to the old 25-minute format. The next series would run to exactly half the length of John Nathan-Turner's debut, Season 18.

Saward proposed a single 14-episode story, in which the Doctor was again put on trial by the Time Lords. Pre-production would be tortuous, with numerous script rejections and rewrites leading to a terminal breakdown in the already strained relationship between producer and script editor.

Part One of *The Trial of a Time Lord* was broadcast on Saturday 6 September 1986. The courtroom drama was presided over by the Inquisitor (Lynda Bellingham), with the mysterious Valeyard (Michael Jayston) acting as prosecutor. Evidence was relayed to the court as projections from the Matrix, the repository of all Time Lord knowledge.

The historical evidence that comprised the first four-episode segment was written by Robert Holmes and unofficially dubbed *The Mysterious Planet*. The second segment – *Mindwarp* by Philip Martin – showed the return of Sil and the recent death of Peri on Thoros Beta. The third segment – written by Pip and Jane Baker and known as *Terror of the Vervoids* – took place in the Doctor's future and saw him travelling with computer programmer and fitness fanatic Melanie Bush (Bonnie Langford).

The two-part conclusion to the season was unfinished when Robert Holmes died. The final episode was written by Saward, according to Holmes' plot, but Saward withdrew the script following his final argument with Nathan-Turner. The producer then commissioned Pip and Jane Baker to hastily complete the story, which is now widely known as *The Ultimate Foe*.

Perhaps unsurprisingly, the ending proved baffling. The Master gains access to the Matrix and the Valeyard is revealed as an evil future version of the Doctor. Declining the presidency of Gallifrey, the reprieved Doctor leaves with Mel – someone he is technically yet to meet.

The verdict of the programme's real-life trial proved academic to Colin Baker. Before the series finished Nathan-Turner was forced to inform him that his contract would not be renewed.

WHAT THE PAPERS SAY

Above: Bonnie Langford and Colin Baker defy gravity at the Aldwych Theatre in January.

Left: This model of the TARDIS was built by visual effects assistant Mike Tucker for the motion control sequence that opened *The Trial of a Time Lord*.

During the first half of the 1980s John Nathan-Turner excelled at generating publicity for *Doctor Who*. It was his belief that maintaining the show's high profile would be one of the best safeguards against cancellation, and he worked closely with journalists to give them what they wanted. It was a relationship that initially benefited the programme, and Nathan-Turner, but in 1986 the tabloid press predictably turned on its benefactor.

The casting of the effervescent Bonnie Langford was just the tonic for a series that had earned a reputation – at least within the BBC – for gratuitous violence. But the *Doctor Who* fan press was alarmed by Langford's imminent arrival, and Nathan-Turner's latest publicity stunt appeared to confirm their worst fears.

Langford was playing Peter Pan at London's Aldwych Theatre when it was confirmed she had been cast as the new *Doctor Who* companion. Colin Baker joined her at the theatre for her photocall, and both were rigged with Kirby wires to create the impression they were flying. "Peter Pan to rescue of Dr Who," announced the London *Evening Standard* on 23 January 1986. "Television executives are hoping that some of Peter Pan's immortality will rub off on the programme, which has the threat of BBC1 Controller Michael Grade's axe hanging over it."

Baker cringed when he saw the pictures, and some of the cruel comments that accompanied them. "I have rarely said no to anything, because I don't like to be unhelpful," he says, "but I had to wear a thick leather harness in between my costume so they could hoist me up... I had put on weight at that point, and so I've never felt so silly!"

The press proved rather more sympathetic when, almost a year later, Baker broke his silence about the non-renewal of his contract. On 6 January 1987 *The Sun* began a two-part interview entitled "Why I'll Never Forgive Gutless Grade, By Axed Dr Who".

"What I couldn't accept was that Grade didn't have the guts to tell me man-to-man," Baker

Above: From November, Golden Wonder gave away six abridged and colourised reprints of *Doctor Who Magazine* comic strips in multipacks of their crisps. The full-page advert ran in the *Eagle* and other titles.

Left: The underground dwellers of UK Habitat gloat over the unconscious Doctor in *The Trial of a Time Lord* Part One.

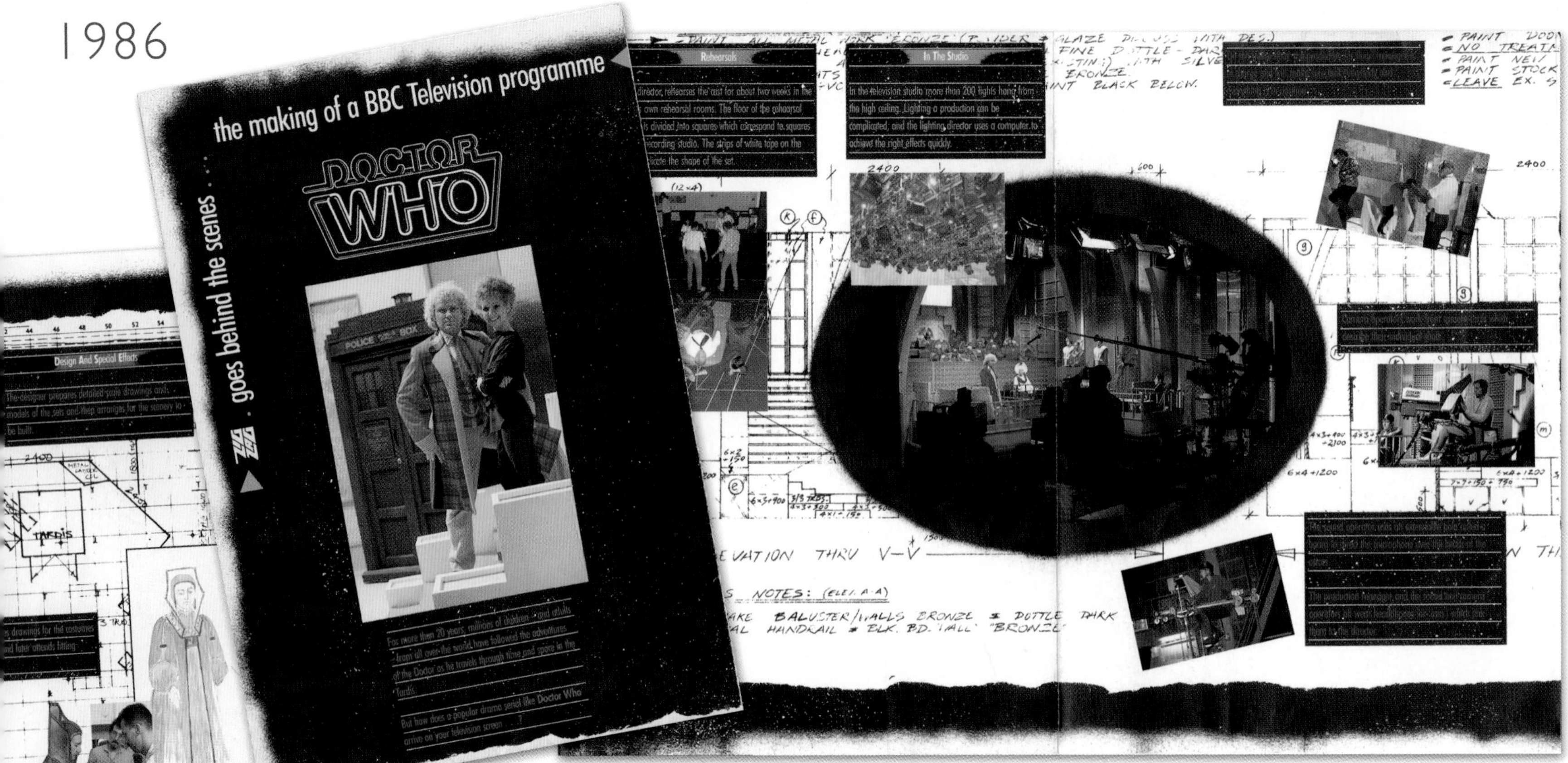

told Sue Carroll. "All I wanted was a proper explanation, but he was too much of a coward. Many people believe, as I do, that I have been treated shabbily."

The article, which exposed the inner workings of the show as never before, also revealed that Nathan-Turner no longer wanted to produce *Doctor Who*. "After seven years he had had enough," said Baker, "but they made him do it."

There was more bad publicity following the debut of Seventh Doctor Sylvester McCoy in *Time and the Rani* later that year. On 12 September the *Daily Mail* published the most notorious review in *Doctor Who*'s history. Negative comments about the new series were not uncommon at this time, but this one astonished fans because it was credited to Andrew Beech, the co-ordinator of the Doctor Who Appreciation Society.

"There are those (usually the under-20s) who enjoy the bright colours, the starry cast and the glitzy 'production values' which the show now embraces," he wrote. "But others believe that there is something radically wrong with a show which 24 years ago had something indefinable, but sufficiently attractive to capture the hearts and imaginations of the British public. Whether through the short-sighted ineptitude of the planners or the excesses of the production team, *Doctor Who* (as a *popular* television show) is slowly but surely being killed."

Ironically, this scathing attack had been partly facilitated by Nathan-Turner. The *Daily Mail* had contacted the producer, asking to speak to a fan who might want to review the new series. Nathan-Turner recommended Beech, but when the *Daily Mail* rang him they didn't mention that recommendation – only that they wanted a negative review of the first episode.

"Some of my colleagues on the committee of the fan club were extremely unhappy about the fact that it had been written," says Beech. "I explained to them that the *Daily Mail* were looking for

Top: Children's magazine *Zig Zag*, published by BBC Education, covered the production design and studio recording of *The Trial of a Time Lord*'s later episodes.

Left: *Doctor Who: The Early Years* was written by Jeremy Bentham and published by WH Allen on 15 May. This detailed analysis of the show's first three years was illustrated with pictures from the collection of designer Raymond Cusick.

Far left: Visual effects designer Mike Kelt with the refurbished TARDIS console used in Season 23.

Left: A Vervoid from *The Trial of a Time Lord* Part Twelve.

Below left: Visual effects designer Peter Wragg checks on Kiv (Christopher Ryan) during the recording of *The Trial of a Time Lord* Part Eight.

Below: One of two Vervoid masks made by visual effects designer Kevin Molloy for close-ups and dialogue sequences. More basic masks were made for background Vervoids.

someone to criticise the show, and I put myself in that position so I could control that criticism, rather than giving the opportunity to a fanzine editor who might be vindictive, destructively critical, and make fans look like obsessive maniacs. What I wrote was meant to be constructive."

Beech says the piece was "substantially what I'd written, but one or two of the points got made rather more aggressively than I'd intended. My only regret is that I didn't have the control to make it appear exactly as I wrote it. And it didn't give me any pleasure at all that it obviously upset John."

Nathan-Turner had rallied considerable support for the programme during the hiatus that followed

Below: "This is an illusion – I deny it!" Colin Baker rehearses the climax to *The Trial of a Time Lord* Part Thirteen at Camber Sands in June.

Right: The *Radio Times* covered the arrival of Bonnie Langford as Mel in the issue dated 1-7 November.

Bottom right: Dominic Glynn's arrangement of the theme was released by BBC Records and Tapes in September. The front cover of the 12" featured a hologram, and the reverse included a cassette.

1-7 November 1986

BACK PAGES

Bonnie's just what the Doctor ordered!

DOCTOR WHO doing press-ups in the Tardis! It could happen, for joining Colin Baker in **The Trial of a Time Lord** is Bonnie Langford, who plays the Doctor's energetic 26th companion Melanie.

Physical fitness is an important part of Bonnie's role, and in this week's **Doctor Who** (Saturday 5.45 BBC1) Melanie will put the Doctor on a diet and make him do keep-fit exercises because she thinks he's out of shape.

Melanie, or Mel as the Doctor calls her, is a computer programmer who comes from the village of Pease Pottage (yes, its a real name) in West Sussex.

She's already part of the Tardis crew at the start of this week's episode. 'It's set in the Doctor's future,' says Bonnie. 'You never learn how I got to travel with him.

'Melanie is a bit headstrong and nosey, so although she recognises the Doctor is the leader, she helps out by running off and investigating things for herself. She ends up in awful trouble and the Doctor has to save her.'

Bonnie says *Doctor Who* is a very happy series to work on, 'and we had lots of wonderful guest stars like Honor Blackman, Lynda Bellingham and Michael Jayston'.

Bonnie shot to fame at the age of six after an appearance on *Opportunity Knocks*.

She is currently on tour in Britain with the musical *Peter Pan*, which for a month over Christmas will be at Eastbourne.

'I'll be just along the road from Colin Baker who'll be doing pantomime at Brighton [written and directed by *Doctor Who* producer John Nathan-Turner]. No doubt we'll bump into one another.'

For the future she says she'd like 'to break away from the kind of work people expect me to do' . . . starting this week with *Doctor Who*.

PATRICK MULKERN

Come in number 26! Colin Baker with his new companion Bonnie Langford

Season 22, but over the following months much of that support turned to contempt. As the tide of opinion turned, the beleaguered producer struggled to find friends in Fleet Street. Worse still, during the Sylvester McCoy years it became increasingly difficult for him to generate *any* mainstream publicity. Unsympathetic scheduling of the show depleted its viewing figures, and *Doctor Who* all but disappeared from the pages of national newspapers.

When the news of *Doctor Who*'s belated return was announced in September 2003, tabloid cynicism was more vitriolic than ever. *The Sun* greeted the appointment of showrunner Russell T Davies with the headline "Duckie Who: Time Lord has gay show writer". This was soon followed by glib speculation over casting of the lead roles. In February 2004 the *Sunday People* even suggested that TV magician Paul Daniels might play the Doctor. "Paul Daniels!" exclaimed Davies in a later interview with *The Guardian*. "How low was this programme in people's minds?"

The casting of Christopher Eccleston, and ratings of 10.8 million for his first episode *Rose*, swiftly re-established *Doctor Who*'s credibility in 2005. On 31 March, only five days after that episode was transmitted, *The Sun* led with the leaked news that Eccleston had already decided to quit – thereby confounding the producers' plan to make the Doctor's imminent regeneration a surprise.

The departure and the casting of each successive Doctor, and even the companions, has been front page news in the British press ever since. In the era of social networking, traditional print publications remain some of the most difficult media to control. Information is still leaked to newspapers, and certain journalists continue to interpret the slightest fluctuation in viewing figures as the death knell for the programme. Generally speaking, however, *Doctor Who*'s current profile is a reassuring reflection of the show's popularity. And, if John Nathan-Turner was in fact correct, a useful insurance policy for the future.

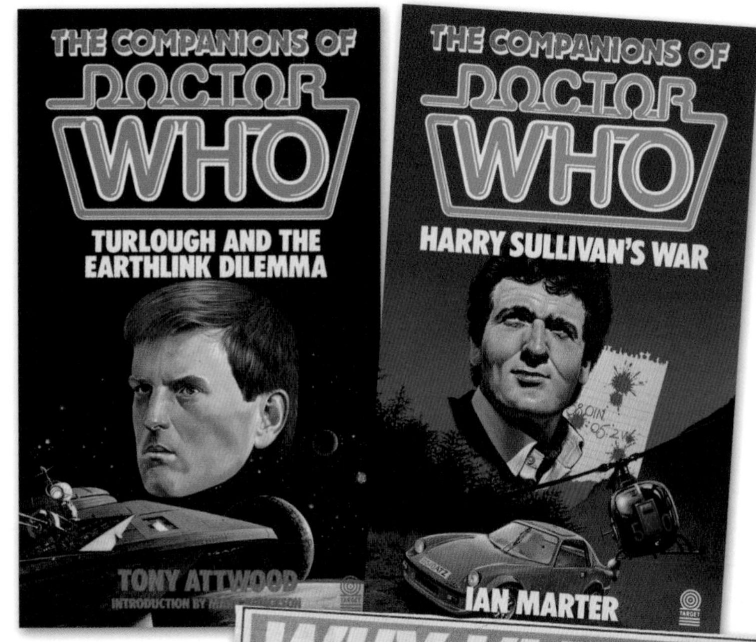

Left: The first two novels in Target's short-lived *The Companions of Doctor Who* series. *Turlough and the Earthlink Dilemma* was published in May, and *Harry Sullivan's War* followed in September. Disappointing sales meant that proposed titles featuring both Tegan and Brigadier Lethbridge-Stewart were never published.

Below: The first part of this controversial interview, reflecting on Colin Baker's departure from the programme, was printed in *The Sun* on 6 January 1987.

WHY I'LL NEVER FORGIVE GUTLESS GRADE, BY AXED Dr WHO

DR WHO star Colin Baker today blasts off against the man who sacked him, BBC Controller Michael Grade.

In this sizzling Sun series, he says he knew he would be exterminated, because Grade (left) gunned for him from day one.

But he was still furious when the axe finally fell, because the high-powered telly boss:

- SLAGGED off Dr Who in public after deciding to dump the show.
- BACKED out of facing Baker man-to-man to tell him why he was being fired.
- INSULTED millions of loyal Dr Who viewers by calling them "loonies."
- ACTED like a stern headmaster and treated grown men like schoolboys.
- ORDERED Baker's boss to persuade him to say that he had not been fired, but had resigned
- TRIED to make him look foolish by asking him to come back for four episodes—just to be killed off.

Coward who can't face me

EXCLUSIVE
By Sue Carroll

COLIN BAKER will never forget the day the bombshell dropped—by phone.

"Do you want the good news or the bad news?" asked the Dr Who producer on the line.

"The good news is that Grade wants the show to go on.

"The bad is that he wants to do it without you."

Colin, 43, and under the impression that he was contracted to the series for at least one more year, felt the blood drain from his veins.

He says: "I couldn't quite take it in, it was such a shock. I'd fought so hard for the show, I was stunned.

"What I couldn't accept was that Grade didn't have the guts to tell me man-to-man.

Excuses

"If I knew *why* I was sacked then I would feel better about it all.

"But I got fobbed off with excuses about Grade thinking that three years as Dr Who was long enough.

"The fact is, I only made 28 episodes before he cancelled the show.

"When it started again last year there were only 14 episodes. Hardly a long run, is it?

"All I wanted was a proper explanation, but he was too much of a coward. Many people believe, as I do, that I have been treated shabbily.

"Of course, I should have seen it coming when I first got the part, back in 1983.

"Michael Grade landed the job as controller just after I joined, and made his feelings about Dr Who very clear.

"He confirmed this one Christmas, when he saw me in the lobby at the BBC. I was in full costume, and there was no way he didn't know me.

"But there was not a flicker of recognition on his face. He just stomped straight past, without so much as a Happy Christmas.

"That's the sort of attitude Grade has introduced to the BBC.

"Instead of being a friendly organisation, it has turned into a cold and businesslike institution.

"In the old days controllers like Bill Cotton would have a word for everyone.

"Not Grade—nowadays the atmosphere at the BBC is very sinister. Everyone is very guarded and there is a lot of dark whispering.

"I'm sure Grade would want me to keep quiet about all of this, but the fact is that there are a lot of dissatisfied actors around.

"He didn't even want me say I had been fired.

"My boss Jonathan Powell, the head of series and serials, said that the BBC would stand by any statement I made.

"He strongly suggested to me that I should claim to be leaving for personal reasons.

"But the worst thing of all is, they actually wanted me to come back and do four more episodes, just so that I could be killed off and fit in with their plans!

Tragic

"I told them what they could do with their offer."

Until his bombshell sacking, life was looking sweet for Baker.

His prize role as Dr Who had helped him mortgage himself to the hilt on a £150,000 home in the country.

And the birth of daughter Lucy, had brought comfort to Colin and wife Marion after the tragic cot death of their son Jack.

Colin says: "I was by no means a rich man from Dr Who, because they never repeated any of the shows I did.

"And of course repeat fees bump up your earnings.

"I earned almost £1,000 an episode, and I was paid by Australian and American television companies who bought the show.

"But all the promises of extra money from spin-offs didn't really materialise.

"The deal was that I made 20 per cent of the earnings from any product the BBC sold with a likeness of me on it, like Dr Who mugs and T-shirts.

"Only small amounts of money dribbled in.

"But I was happy in my job and I was convinced that I was a good Dr Who—certainly on an equal footing with my predecessors.

"I would have liked to carry on being Dr Who for a good few years, and I believe that's what should have happened.

"Instead the show was moved about, cancelled, cut down, and criticised by Grade, the very man who should have been backing it."

Grade grounded Dr Who and his Tardis for a year, then brought it back—cut from 26 episodes a year to just 14.

Slump

The team, says Colin, were pleased with the series, titled Trial Of A Time Lord, but despondent that it was shown immediately after Roland Rat and before dusk, when people are not actually glued to the box.

"There was no encouragement from Grade," says Colin. "He never visited the set once.

"It doesn't take a genius to see that they brought the show back, just to kill it off.

"How could they expect viewing figures to rise, when it was slotted in at such a bad time?"

By the end of the last series, the audience had slumped to five million.

But Colin says: "Even producer John Nathan-Turner, a great friend and one of my allies, was fed up. He didn't want to do another series. After seven years he had had enough, but they made him do it.

"Even so, five million viewers isn't so bad. The Wogan show doesn't do much better than that, but you won't find Grade moaning about a show that's his brainchild.

"It's terrible to hear the way he talks about Dr Who. On one radio show he dismissed the fans of the show as 'loonies.'

Slammed

"And when a former Dr Who, Patrick Troughton, slammed the decision to cancel the show, Grade had one answer, 'Actors should stick to acting.'

"Well in that case, controllers should stick to controlling and not pop up on the box all the time. Maybe if they did, they'd be better at their jobs."

Colin Baker in his Dr Who costume . . . when Grade saw him in it he walked past without a word

'Boost Beeb' shock

TWO WEEKS after Colin Baker learned he had been axed, he was invited to boost the BBC by going to America for a week.

Colin was convinced there must be some mistake, and rang the BBC publicity office to check.

He says: "I was amazed when they told me it was not an error. They wanted me, as the existing Dr Who, to go off to the States with some other actors as the guest of BBC bigwigs Michael Grade, Bill Cotton and Alasdair Milne.

"I wasn't paid a penny as a fee, it was supposed to be a great privilege just to be invited along.

"I went mainly because I hoped I would be able to discuss my future with Grade.

"But he ducked and dived, and although he was pleasant to me, and terribly polite, we never had a chat.

"Yet publicly, at prestige dinners, he would introduce me along with the others as one of the BBC's most important actors. What an irony!

"I really resented the way, as a grown man of 43, I was made to feel like a frightened schoolboy, with Grade the stern headmaster."

TOMORROW: Losing my job could hit Cot Death cash

SEASON 23
Producer: John Nathan-Turner
Script editor: Eric Saward (except Part Fourteen)

7A *The Trial of a Time Lord*
(aka ***The Mysterious Planet***)
8 April – 12 May, 13 June
written by Robert Holmes
directed by Nicholas Mallett

Part One	6 September
Part Two	13 September
Part Three	20 September
Part Four	27 September

7B *The Trial of a Time Lord* (aka ***Mindwarp***)
27 May – 16 June
written by Philip Martin
directed by Ron Jones

Part Five	4 October
Part Six	11 October
Part Seven	18 October
Part Eight	25 October

7C *The Trial of a Time Lord*
(aka ***Terror of the Vervoids***)
23 June – 14 August
written by Pip and Jane Baker
directed by Chris Clough

Part Nine	1 November
Part Ten	8 November
Part Eleven	15 November
Part Twelve	22 November

7C *The Trial of a Time Lord*
(aka ***The Ultimate Foe***)
written by Robert Holmes, Pip and Jane Baker
directed by Chris Clough

Part Thirteen	29 November
Part Fourteen	6 December

1987

John Nathan-Turner believed he would finally be allowed to move on to another programme once he told Colin Baker his time as the Doctor was over. However, Jonathan Powell was unable to find anyone else willing to produce *Doctor Who*, so at the beginning of 1987 a reluctant Nathan-Turner resumed his duties. He faced the preparations for Season 24 with no Doctor, no script editor and no scripts.

Colin Baker had declined Powell's suggestion that he continue long enough to complete a regeneration handover, so Nathan-Turner asked the dependable Pip and Jane Baker to introduce the seventh incarnation at the beginning of the next episode. *Time and the Rani* saw the return of Kate O'Mara and a hasty regeneration apparently induced by severe buffeting within the TARDIS control room. The unscathed Mel regains consciousness to find the confused Doctor isn't the only victim of mistaken identity.

Any praise for *Time and the Rani* was largely reserved for the serial's visual effects, but Pip and Jane had simply played the difficult hand they were dealt. If the story veered towards broad comedy then this could partly be attributed to Nathan-Turner interpreting the wishes of BBC management and the show's new leading man falling back on his traditional talents.

Nathan-Turner had decided to audition Sylvester McCoy for the role of the Doctor when he saw him play *The Pied Piper* at the National Theatre in early January. McCoy's casting was announced in late February, leaving little time to devise a characterisation before recording started in April.

Script editor Andrew Cartmel had now also joined the programme, and over the coming months he asserted an increasing influence over its direction. For the remainder of 1987, however, the series continued on a naïve and freewheeling path as the Doctor and Mel liberated the colourful occupants of *Paradise Towers*, won a trip to a 1950s holiday camp in *Delta and the Bannermen* and discovered hidden treasure in *Dragonfire*. At the end of the series Mel chose to keep an eye on travelling rogue Sabalom Glitz (Tony Selby), leaving space in the TARDIS for teenage tomboy Ace (Sophie Aldred).

Season 24 would do little to restore *Doctor Who*'s battered reputation, but at the end of another difficult year the new team retreated to plot a more robust renaissance.

THE AMERICAN DREAM

Agent: Brian Wheeler 995 3934

BBC PRESS OFFICES
12 Cavendish Place, London W1A 1AA. Direct dial 01-927 4709 (4 lines)
Television Centre, London W12 7RJ. Direct dial 01-576 1865 (5 lines)

PRESS SERVICE

SYLVESTER McCOY IS THE NEW DOCTOR

Sylvester McCoy is to be the new 'Doctor Who'. He will join Bonnie Langford as Melanie when the long-running science fiction series returns to BBC-1 in the autumn.

Sylvester has just finished starring in the title role in the National Theatre's production of "The Pied Piper". Before that he played three Shakespearean roles, Feste in 'Twelfth Night' at the Leicester Haymarket and Tranio in 'The Taming of the Shrew' and Pompey in "Anthony and Cleopatra" in the Vanessa Redgrave/Timothy Dalton season at the Haymarket Theatre, London. Two plays have been specially written for him 'Satie's Faction' by Adrian Mitchell in which he played the composer Erik Satie and 'Buster's Last Stand' by Peter Fieldson, in which he played Buster Keaton. He worked with Bonnie Langford in 'The Pirates of Penzance', playing the role of Samuel and has made many appearances with 'The Ken Campbell Road Show'.

His numerous television and film work includes Birdie Bowers in 'The Last Place on Earth', 'Eureka', 'Tiswas', 'Big Jim and the Figaro Club', 'Jigsaw' and a leading role in the forthcoming feature film 'Fireworks'.

Producer John Nathan-Turner says: "I am delighted to have signed Sylvester to be the seventh Doctor Who. I saw him as 'The Pied Piper' and realised at once that he possessed all the qualities necessary to play the Doctor."

Sylvester's predecessors as Doctor Who were William Hartnell, Patrick Troughton, Jon Pertwee, Tom Baker, Peter Davison and Colin Baker.

For further information: Kevin O'Shea
BBC Drama Publicity (Series/Serials)
Tel: 01 576 1861

2nd March 1987

Left: The BBC press release announcing Sylvester McCoy as the Seventh Doctor.

Above: John Nathan-Turner attends McCoy's photocall at Cavendish Place on 2 March.

Below: Bonnie Langford and McCoy pose for photographers in front of the TARDIS.

On 28 February 1987 *The Sun* announced that "A zany Scot who used to make a living by stuffing ferrets up his trousers is the new Doctor Who." On the same day, and on the other side of the Atlantic, Sylvester McCoy made his first television appearance to promote the series when he joined Jon Pertwee and John Nathan-Turner at Mercer University, Atlanta for a live show hosted by Public Broadcast System (PBS) television.

"We're entering a new era," promised the producer, "and to back that up we're having a new logo and a new arrangement of the old music, so the whole thing will have a fresh look."

Interviewer Eric Luskin asked McCoy what was going through his mind as recording of that new season approached. "Panic," he replied.

The fact that a clearly bewildered McCoy had been whisked to America was a measure of how important the territory had become to Nathan-Turner.

McCoy's interview was fitted in between stops on the Doctor Who Celebration & Tour '87-'88, a mobile exhibition that was described by its

Above: Ken Trew's original design for the Seventh Doctor's costume incorporated the question mark motif at Nathan-Turner's request.

Right: The costume worn by the Seventh Doctor in seasons 24 and 25. The hat and umbrella lent a Chaplinesque quality.

Left: Langford, Kate O'Mara (returning as the Rani) and McCoy, pictured during location recording for *Time and the Rani* at Cloford Quarry, Somerset on 6 April.

Right: The May edition of the Doctor Who Appreciation Society's fanzine *Celestial Toyroom* paid tribute to Patrick Troughton, who died while attending a US convention in Columbus, Georgia on 28 March.

Below: In response to queries from fans, the *Doctor Who* production office produced regular newsletters. This one detailed the story titles and transmission order for Season 24.

Bottom: Urak (Richard Gauntlett) awaits instructions from his mistress in Part Two of *Time and the Rani*.

Bottom right: Two of the surviving Tetrap costumes. Susan Moore and Stephen Mansfield designed animatronic mechanisms for the heads.

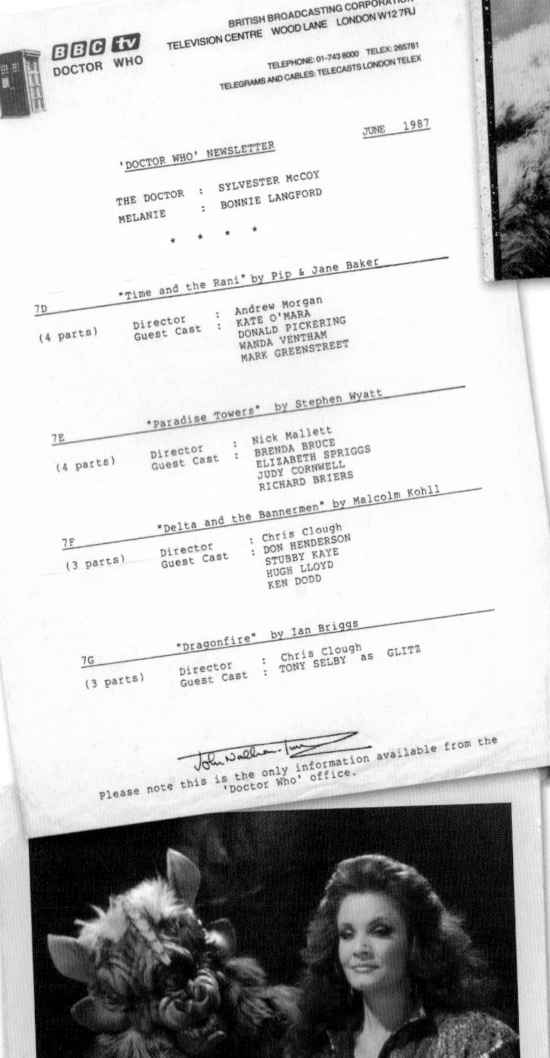

BBC tv
DOCTOR WHO

BRITISH BROADCASTING CORPORATION
TELEVISION CENTRE WOOD LANE LONDON W12 7RJ
TELEPHONE: 01-743 8000 TELEX: 265781
TELEGRAMS AND CABLES: TELECASTS LONDON TELEX

JUNE 1987

'DOCTOR WHO' NEWSLETTER

THE DOCTOR : SYLVESTER McCOY
MELANIE : BONNIE LANGFORD

* * * *

7D "Time and the Rani" by Pip & Jane Baker
(4 parts) Director : Andrew Morgan
Guest Cast : KATE O'MARA, DONALD PICKERING, WANDA VENTHAM, MARK GREENSTREET

7E "Paradise Towers" by Stephen Wyatt
(4 parts) Director : Nick Mallett
Guest Cast : BRENDA BRUCE, ELIZABETH SPRIGGS, JUDY CORNWELL, RICHARD BRIERS

7F "Delta and the Bannermen" by Malcolm Kohll
(3 parts) Director : Chris Clough
Guest Cast : DON HENDERSON, STUBBY KAYE, HUGH LLOYD, KEN DODD

7G "Dragonfire" by Ian Briggs
(3 parts) Director : Chris Clough
Guest Cast : TONY SELBY as GLITZ

Please note this is the only information available from the 'Doctor Who' office.

sponsors, Lionheart Television, as "the world's only experience theater on wheels". The travelling exhibition was devised by the BBC's Brian Sloman and housed inside a 48-foot trailer. Sloman accompanied the exhibition as it aimed to visit all 185 of the cities where Lionheart distributed *Doctor Who*. As well as McCoy and Jon Pertwee, luminaries such as Peter Davison, Anthony Ainley and Janet Fielding made guest appearances along the way.

Such efforts were necessary to maintain interest in a programme that was only kept on the air by donations from viewers to their local PBS stations. *Doctor Who*'s relationship with the American audience had long been fickle: the two Peter Cushing feature films made little impact in the 1960s, and the television series didn't arrive

BACKPAGESBACKPAGESBACKPAGESBACKPAGESBA

Who's taken to the cleaners!

JUST WHEN Melanie (Bonnie Langford) thought it was safe to dip her toes into the swimming pool, she got a nasty shock ... from a robot crab! It's just one of the many monsters – mechanical, humanoid and just indescribable – that rear their unpleasant heads in the new series of **Doctor Who** (Monday BBC1).

The creepy crabs are supposed to keep the pool clean, but they suddenly become human seeking – and Melanie is the nearest human!

But you'll have to wait a few weeks before they and other metal menaces like, of all things, robot cleaners, threaten the Doctor (now played by Sylvester McCoy – see Upfront, page 3) and Melanie when they arrive at Paradise Towers.

All the baddies are new for this series – not a Dalek in sight! – and the only returning arch-enemy is the Rani (Kate O'Mara), the renegade Time Lady who is as evil as the Doctor is good.

New monsters to watch out for in the weeks ahead include the fearful Bannermen, alien humanoids with red eyes and mouths, led by actor Don Henderson, and the Bio-mechanoid, an enormous, very bony, flat-headed thing that inhabits the Iceworld.

Other features of the series include new versions of the theme tune and opening titles, and appearances by stars like Ken Dodd and Richard Briers, who plays his first TV baddy.

In this week's opening episode, there's the first glimpse of the latest generation of monsters – the Tetraps, bat-like people from the planet Lakertya, who team up with the Rani.

All you will see of them is a strange-looking hand – very similar to the one that's creeping into the bottom of this page. Aaagh!!!

Dirty deeds! The Doctor helps Melanie fight off a crabby cleaner, but there are more monsters to face

until Time Life distributed a package of Third Doctor stories to a limited number of stations in 1972. Time Life had more success in 1978, when 98 Fourth Doctor episodes (from *Robot* through to *The Invasion of Time*) were sold to PBS television (see page 132). This package of stories rotated for several years, sometimes in production order and sometimes in a frustratingly random pattern. American fans struggled with the series' continuity and the largely pointless narration imposed by Time Life, who felt that viewers needed their memories refreshed at regular intervals. Howard Da Silva, the actor they hired for the voice-overs, later admitted he hadn't even seen the episodes. His work on *The Brain of Morbius* is typical of his uninformed but strangely endearing style: "Thanks to the Time Lords, Doctor Who gets re-routed to a planet where his welcome is less than hospitable..."

These early Fourth Doctor episodes were still doing the rounds when Tom Baker, Graham Williams, Barry Letts and Terrance Dicks went to Los Angeles on 1 December 1979 for the first American *Doctor Who* convention. Following Williams' departure from the series, Nathan-Turner became a regular guest at such events, often bringing stars from the show and tapes of stories that had not yet screened in America.

By 1983, the Time Life presentations had been superseded by more recent episodes. The 20th anniversary of *Doctor Who* was a catalyst for American enthusiasts: membership of the Doctor Who Fan Club of America hit 30,000, and in Chicago on 26-27 November the Ultimate

Top left: The 5-11 September issue of the *Radio Times* previewed the *Paradise Towers* cleaning robots and just a claw from one of *Time and the Rani*'s Tetraps.

Above: Young representatives of The Children's Society scrub the TARDIS clean during studio recording of *Paradise Towers* at Television Centre in June.

Right: Signed BBC postcards of Sophie Aldred and Sylvester McCoy were sent to fans who wrote to the production office requesting signatures.

Below: Ace's Nitro-9 was loaded inside old deodorants, but the Visual Effects Department created the props from empty cans of aerosol cream.

Bottom right: Mel carries the home-made Nitro-9 in Part One of *Dragonfire*. "It's just like ordinary nitroglycerin," explains Ace, "except it's got a bit more wallop."

Celebration convention lived up to its name with 12,000 attendees and a guest list headed by the four surviving Doctors. This fever pitch of activity engendered a quiet resentment from some British fans, who learned of the gathering via an advertisement in *Doctor Who Monthly* ("You are cordially invited to join John Nathan-Turner, your host and master of ceremonies, in salute of this prestigious event") or by watching the slightly bemused coverage on the BBC's national news bulletins.

For a while, some British conventions were compromised or even cancelled because they found it impossible to secure guests already lured to the United States by the promise of higher fees and a working holiday. Nathan-Turner's frequent presence at such US events probably had as much to do with the unconditional adoration he received, at a point when British fans were beginning to turn against him.

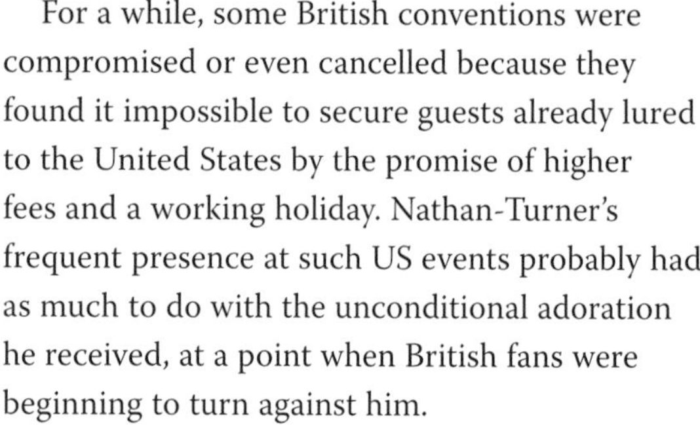

In 1984 Nathan-Turner attempted to further stoke US interest in the show by introducing American companion Peri, played by English actress Nicola Bryant. Robert Holmes' commission to write *The Two Doctors* had already included the instruction to set the story in New Orleans. "That's why I created the Androgums," he recalled. "I couldn't think of any reason why aliens should visit New Orleans and I recalled it was a jazz place – but not even I could envisage a race of aliens obsessed with jazz. Then I remembered it is the culinary centre of America with lots of restaurants so I invented the Androgums (an anagram of 'gourmand') who are obsessed with food. So they went to New Orleans for the food." When Nathan-Turner's ambitions for American filming fell through the setting was switched to the more affordable Seville, but the Androgums remained.

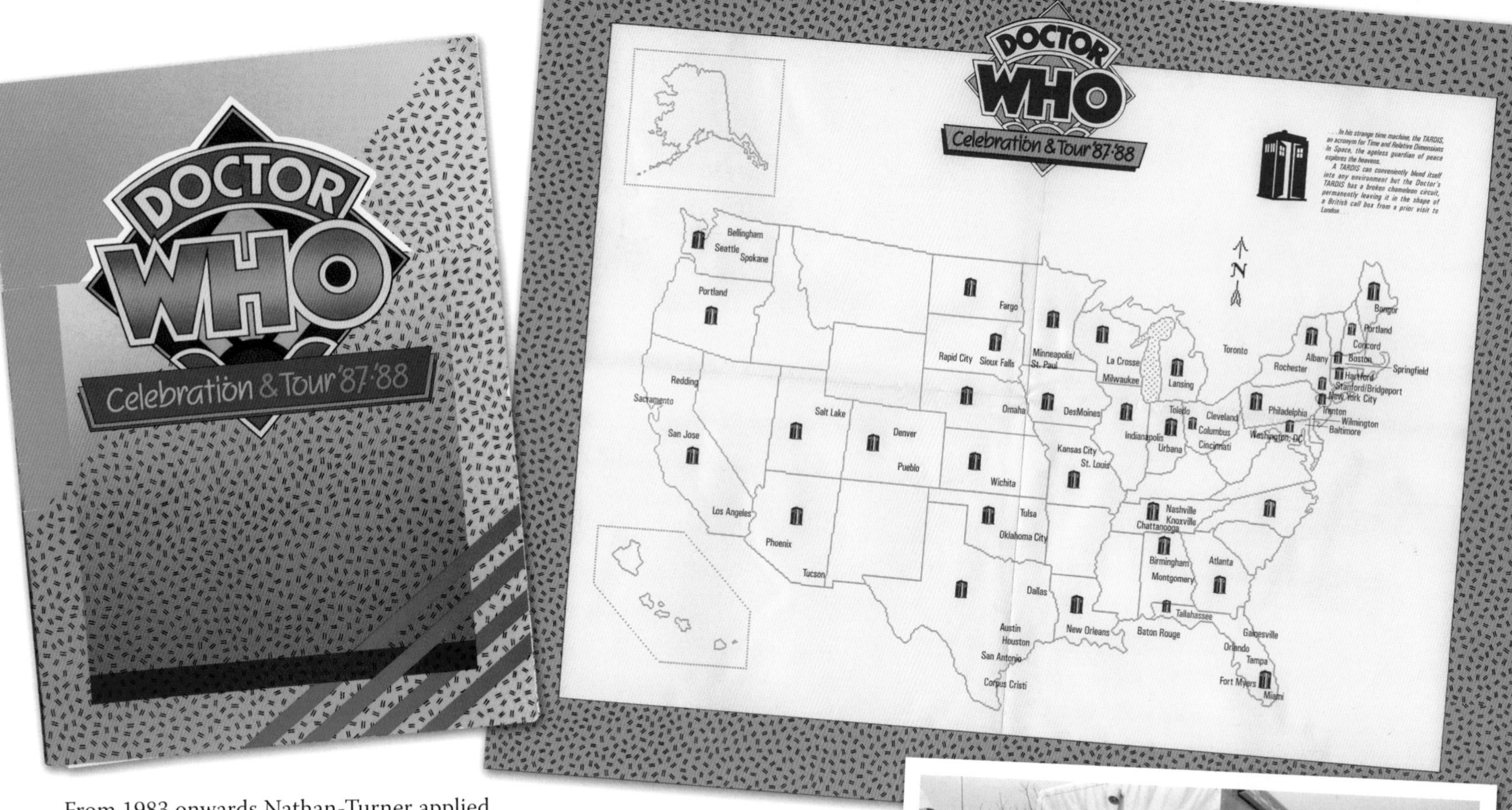

From 1983 onwards Nathan-Turner applied pressure on BBC Enterprises to market more memorabilia in the United States. One of the modest results of these efforts was the merchandise stand included in the Celebration & Tour trailer when it was first launched as the Doctor Who USA Tour in May 1986.

The trailer made its final stop in 1988, before being shipped back to England and ending its days in a scrap yard in Derbyshire. *Doctor Who*'s popularity in America was waning; the reduced number of episodes resulted in only four new stories a year and the launch of *Star Trek: The Next Generation* in September 1987 diverted large parts of fandom. The *Doctor Who* TV movie was an American co-production but failed to find a US audience when it was networked in 1996. By 1998 there were just five PBS stations still showing old episodes.

When *Doctor Who* returned in 2005 the new episodes were initially shown on the Sci Fi Channel, but from November 2006 the series has been a mainstay of the cable and satellite network BBC America. *Doctor Who* attracts relatively high ratings and is now a more prominent part of American popular culture than at any time since the 20th anniversary. Like James Bond and *Monty Python*, *Doctor Who* possesses a quintessential British quality that some US viewers continue to find fascinating.

The BBC are once again keen to build on that popularity. In 2006 second unit location recording for *Daleks in Manhattan/Evolution of the Daleks* was conducted in New York, and in 2010 the main cast went to Utah for *The Impossible Astronaut/Day of the Moon*. The stars of the series attend the annual Comic-Con event in San Diego, and BBC America produces its own behind the scenes content for broadcast on television and the internet. "It's endless, it's timeless," says Matt Smith, in a video message designed to entice American viewers. "It's the best. Watch *Doctor Who*!"

Top left: The folder containing press information for the Doctor Who Celebration & Tour '87-'88.

Top: The press pack included this map, showing some of the cities the tour would be visiting.

Above: The touring exhibition was housed inside this 48-foot trailer, the largest vehicle allowed on American highways. The artwork on the sides was designed by Andrew Skilleter.

SEASON 24
Producer: John Nathan-Turner
Script editor: Andrew Cartmel

7D *Time and the Rani*
4 April – 5 May
written by Pip and Jane Baker
directed by Andrew Morgan

Part One	7 September
Part Two	14 September
Part Three	21 September
Part Four	28 September

7E *Paradise Towers*
21 May – 19 June
written by Stephen Wyatt
directed by Nicholas Mallett

Part One	5 October
Part Two	12 October
Part Three	19 October
Part Four	26 October

7F *Delta and the Bannermen*
24 June – 12 August
written by Malcolm Kohll
directed by Chris Clough

Part One	2 November
Part Two	9 November
Part Three	16 November

7G *Dragonfire*
28 July – 13 August
written by Ian Briggs
directed by Chris Clough

Part One	23 November
Part Two	30 November
Part Three	7 December

1988

In 1987 *Doctor Who* had been transmitted at 7.35 on Monday evenings, opposite Britain's most popular soap opera *Coronation Street*. When Season 25 began it remained at 7.35 but moved to Wednesdays – now opposite *Coronation Street*'s second episode of the week. *Doctor Who* would perform admirably under the circumstances, improving on the previous year's ratings by consistently scoring between five and five-and-a-half million viewers.

The new season began with a pre-credits sequence of an ominous-looking spaceship approaching Earth. *Remembrance of the Daleks* would usher in a more sophisticated style, with Sylvester McCoy adding a brooding, contemplative quality to his portrayal of the Time Lord. Critics admired the explosive visual effects and the conviction of an impressive guest cast. Fans were similarly won over, intrigued by the first signs of a Machiavellian streak in the Seventh Doctor.

The second story to be broadcast was produced fourth. *The Happiness Patrol* drew derision in some quarters because of the (intentional) similarity between its robotic executioner the Kandy Man and the Liquorice Allsorts emblem Bertie Bassett. In 2010 *Newsnight* uncovered a deeper allegory, citing the dictatorial Helen A (Sheila Hancock) as evidence that *Doctor Who* may have been harbouring "a nest of anti-Thatcherite subversives".

As with the Dalek story, the Doctor was taking care of "unfinished business" in *Silver Nemesis*, the serial that saw redesigned Cybermen, neo-Nazis and the time-travelling Lady Peinforte (Fiona Walker) fight for possession of a statue made of Gallifreyan living metal. The first episode was broadcast on 23 November, *Doctor Who*'s 25th anniversary.

The second story in production order was broadcast at the end of the season. *The Greatest Show in the Galaxy* created a disquieting world of surreal characters and situations – no mean feat considering its studio recording dates were cancelled following an asbestos scare at Television Centre. John Nathan-Turner saved the production by reconstructing the Psychic Circus scenes inside a tent erected in the car park of the BBC's Elstree Studios.

Earlier in the year The Timelords had topped the charts with 'Doctorin' the Tardis', the most successful *Doctor Who*-inspired record ever released, and rumours circulated of a *Doctor Who* feature film.

The series was gaining popularity, and even a degree of credibility. For many, there was finally cause to be optimistic for the future.

SCRIPT DOCTORS

Right: The Doctor and Ace lure a hidden menace in *Remembrance of the Daleks* Part One. Kew Bridge Steam Museum near Brentford was used as the location for I.M. Foreman's junkyard, first seen in *An Unearthly Child*. Unfortunately the name was mis-spelled as 'I.M. Forman' when painted on the gates in *Remembrance*.

Below: In June, model railway manufacturer Dapol launched a range of *Doctor Who* toys, beginning with these action figures of the Doctor and Mel. From 1994 until 2003 the company's factory in Llangollen hosted The Doctor Who Experience, an exhibition of original props and costumes.

Generally speaking, the original series of *Doctor Who* was run by only a handful of permanent staff – the producer, his or her secretary, and the script editor. It was the script editor's responsibility to commission writers and develop their work until it was ready to be recorded or filmed. These duties naturally gave him a huge influence on the creative path of the show.

Many consider that the quality of *Doctor Who* improved markedly during the late 1980s, and credit this turnaround to script editor Andrew Cartmel. While the programme was struggling to reconnect with a mainstream audience, fans were encouraged by what they described as 'The Cartmel Master Plan' – a character arc that would transform the Doctor from the clownish buffoon of *Time and the Rani* to an occasionally sinister game-player who, it was suggested in stories such as *Silver Nemesis* and *Battlefield* (1989), may have been more than just a Time Lord.

Cartmel claims to have entirely initiated *Doctor Who*'s direction in the final few seasons of the original show, although he is quick to point out that producer John Nathan-Turner had ultimate approval on every decision. "If you're a parent in a large family I suspect you're restrictive and prescriptive with the first kid, you're slightly more loose with the second kid and third kids, and by the time you've got five or six you just let them get on with it," he says. "Perhaps there was an element of that with John; he'd been doing the show for so long that any desire to put his absolute stamp on it had gone. I think the show was starting to be very good when we left it, and was heading for even better. Anything he strongly objected to just

THE GRIEF

KILLJOY UNDERGROUND NEWS

PATROLS ON THE INCREASE

Beware Bluesy Street Zone

INDULGE YOUR DEPRESSIONS

Above: Tony Masero's artwork for the 1988 novelisation of *Time and the Rani* was rejected by Target because he had mistakenly painted the sleeping Tetraps in an upright position. The book ultimately appeared with a photograph by Chris Capstick on the cover (left) and the McCoy version of the *Doctor Who* logo.

Above right: A prop newspaper from Helen A's desk in *The Happiness Patrol*.

Far right: The Doctor, Susan Q (Leslie Dunlop) and Ace with the Pipe People in a publicity still from *The Happiness Patrol*.

Below: Dapol's *Doctor Who* toys sometimes differed from their screen counterparts. These were two of the earliest: this version of K-9 was green, and the TARDIS console only had five sides.

wasn't going to happen – but then he virtually never strongly objected to anything. For the most part we got along admirably smoothly."

When *Doctor Who* began, the script editor's job had the title 'story editor' but entailed exactly the same duties. The first producer, Verity Lambert, attributed much of *Doctor Who*'s groundwork to her original story editor, David Whitaker. However, Whitaker, like all of his successors on the original series, also had to contend with the more mundane challenge of polishing inadequate scripts.

One of the most notorious last-minute overhauls occurred in 1965 while story editor Donald Tosh was working with writer Terry Nation on *The Daleks' Master Plan*. "I spoke to Terry on the phone," remembers Tosh. "He said, 'I've got to catch a plane to

Above: The Timelords' 'Doctorin' the Tardis' was released on 28 May and went to number one in the charts.

Above right: The cast and crew of *Silver Nemesis* on location at Black Jack's Mill Restaurant in Harefield on 5 July.

Below: The Ace jacket worn by Sophie Aldred's stunt double Tracey Ebdon in *Silver Nemesis* and other stories.

Below right: The Cybermen step out for the location recording of *Silver Nemesis* at the Arundel Castle Estate in late June.

America tonight. I will deliver the scripts to your flat, and then I'm afraid I will be incommunicado for a few weeks.' At about half past eight the doorbell rang, so I went downstairs and opened the door. It was Terry. He said, 'There you are Donald. Don't worry – it'll work.' Then he dashed straight into a taxi and vanished in the direction of Heathrow. I was left holding a large brown envelope, in which were 26 pages of foolscap. He had typed on only one side of each page, and only half of each page at that. This was supposed to be six episodes. I was shattered! I rang [producer] Johnny Wiles and said, 'Terry has delivered.' He said, 'Oh wonderful, wonderful.' Then I said, 'No, no, stop there dear fellow – I have to tell you it's only 26 pages long.' There was a long pause on the other end of the line, before he said, 'Donald, I'm afraid you're going to be very, very busy...'"

Tosh ended up co-writing the first six episodes himself, accepting that this was part of the job. "I never minded who was being paid for what or anything else as long as I received my salary, which I felt was perfectly adequate. What always mattered most to me was what was up on the screen."

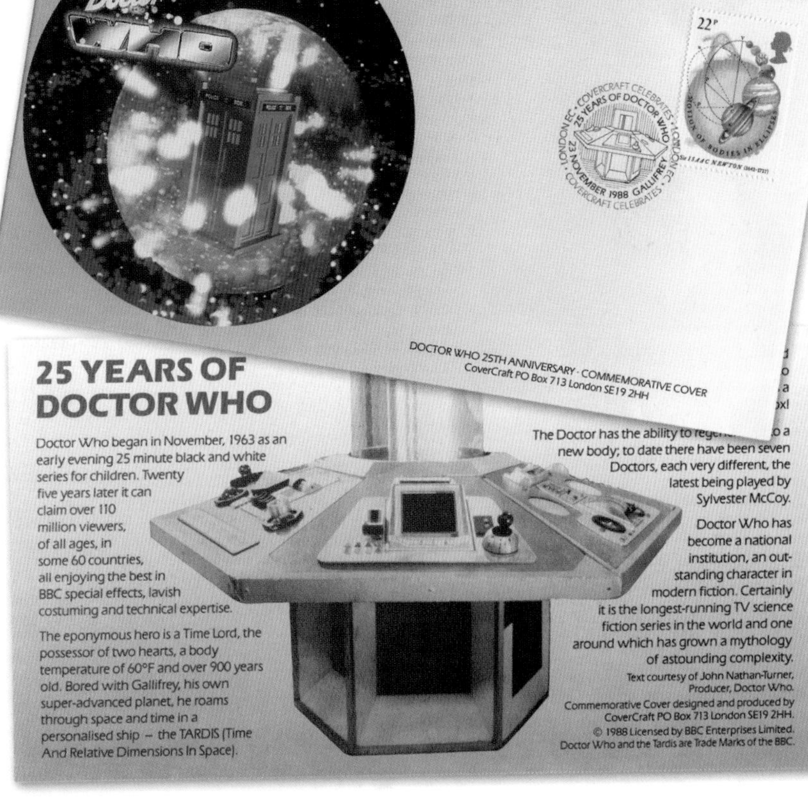

Doctor Who
22p
25 YEARS OF DOCTOR WHO
23 NOVEMBER 1988 GALLIFREY
COVERCRAFT CELEBRATES
DOCTOR WHO 25TH ANNIVERSARY · COMMEMORATIVE COVER
CoverCraft PO Box 713 London SE19 2HH
25 YEARS OF DOCTOR WHO
Doctor Who began in November, 1963 as an early evening 25 minute black and white series for children. Twenty five years later it can claim over 110 million viewers, of all ages, in some 60 countries, all enjoying the best in BBC special effects, lavish costuming and technical expertise.
The eponymous hero is a Time Lord, the possessor of two hearts, a body temperature of 60°F and over 900 years old. Bored with Gallifrey, his own super-advanced planet, he roams through space and time in a personalised ship – the TARDIS (Time And Relative Dimensions In Space).
The Doctor has the ability to regen… to a new body; to date there have been seven Doctors, each very different, the latest being played by Sylvester McCoy.
Doctor Who has become a national institution, an outstanding character in modern fiction. Certainly it is the longest-running TV science fiction series in the world and one around which has grown a mythology of astounding complexity.
Text courtesy of John Nathan-Turner, Producer, Doctor Who.
Commemorative Cover designed and produced by CoverCraft PO Box 713 London SE19 2HH.
© 1988 Licensed by BBC Enterprises Limited.
Doctor Who and the Tardis are Trade Marks of the BBC.

THE
FRAME
NOV 1988
TREVOR BAXENDALE

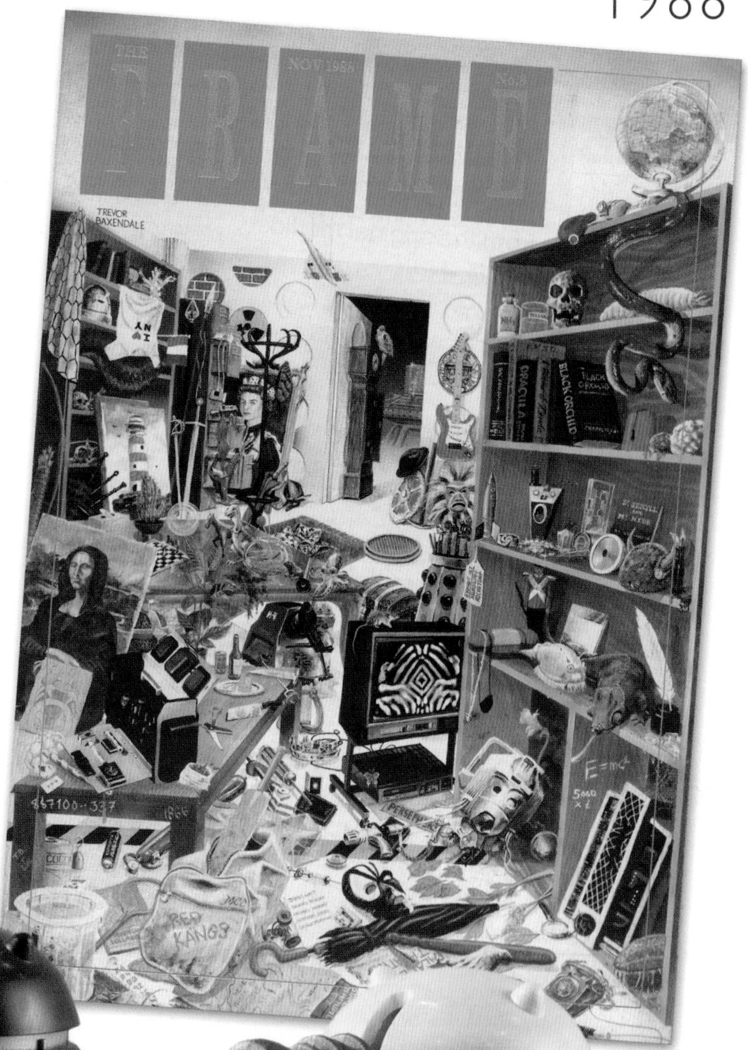

Many script editors took it upon themselves to maintain the show's continuity – in 1965 Tosh had to remind writers that the hero's name was not actually 'Doctor Who'. In a guide compiled in 1980 Christopher H Bidmead was still driving the point. "The protagonist is never referred to as 'Dr Who' either in speech headings or by other characters," he insisted. "His name is the 'Doctor'; 'Who?' is the mystery!"

Restoring that inscrutability was high on Cartmel's agenda when he joined the programme in 1987. "I think the problem began when they introduced the Time Lords," he says. "The Doctor is supposed to be an enigma; the more of a back story you give him the more you diminish him. This reached its nadir in *The Trial of a Time Lord*, where you've got Colin

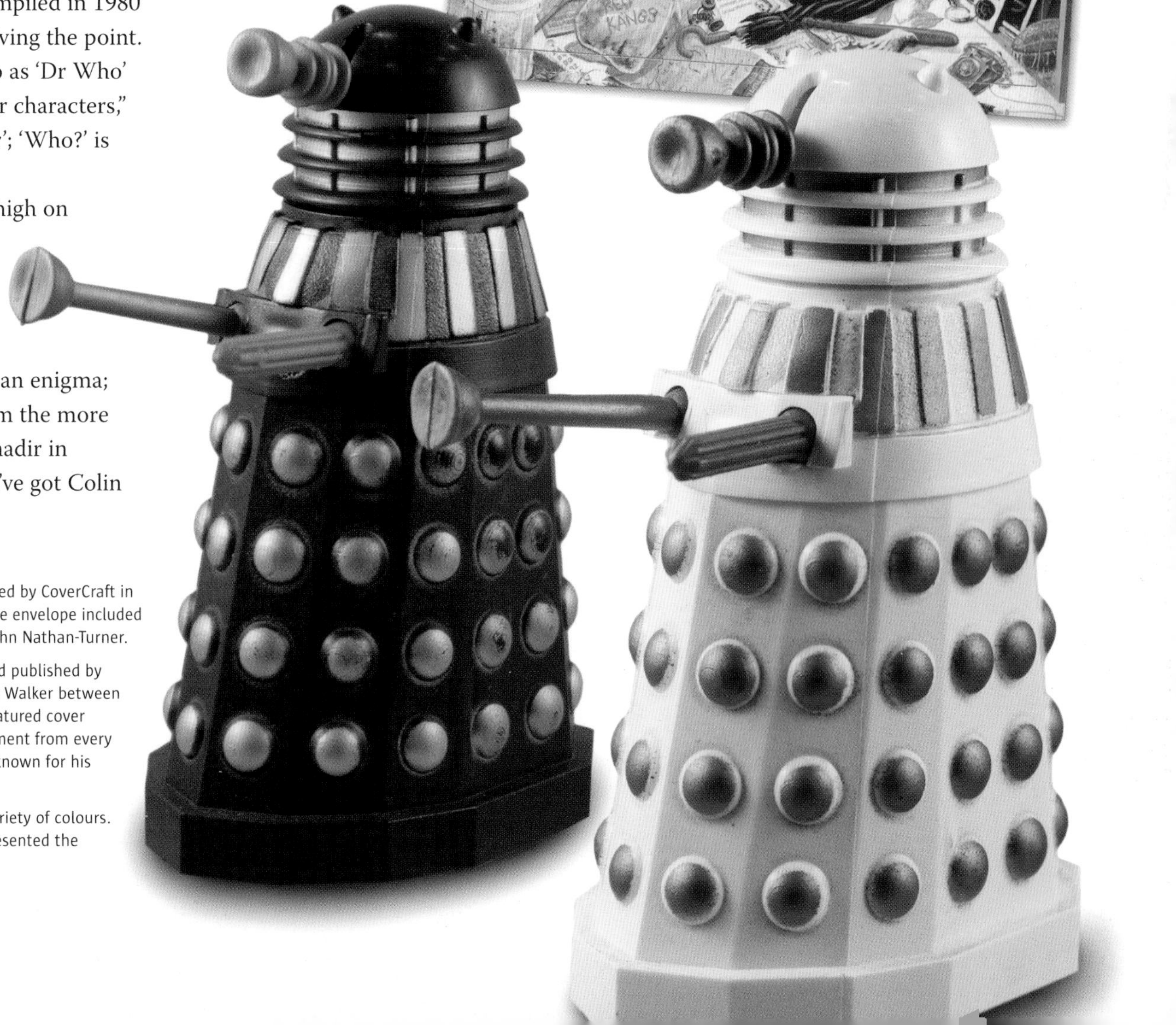

Top: This 25th anniversary first day cover was issued by CoverCraft in November, complete with a Gallifrey postmark. The envelope included a card with a brief history of the programme by John Nathan-Turner.

Top right: Fan magazine *The Frame* was edited and published by David J Howe, Mark Stammers and Stephen James Walker between 1987 and 1993. Issue 8, dated November 1988, featured cover artwork by Trevor Baxendale that included an element from every broadcast story to date. Baxendale is now better known for his *Doctor Who* prose fiction.

Right: Dapol produced Dalek action figures in a variety of colours. These versions were issued in November and represented the opposing factions in *Remembrance of the Daleks*.

1988

Right: The Doctor faces the Gods of Ragnarok in Part Four of *The Greatest Show in the Galaxy*.

Below: One of the robot Bus Conductor heads designed by Steve Bowman and constructed by Mike Tucker for *The Greatest Show in the Galaxy*.

Below right: Bowman's designs for the Bus Conductor's head and the robot excavated by Captain Cook.

Bottom right: The Bus Conductor (Dean Hollingsworth) from Part One of *The Greatest Show in the Galaxy*, and a sketch showing how an actor would wear the prop head.

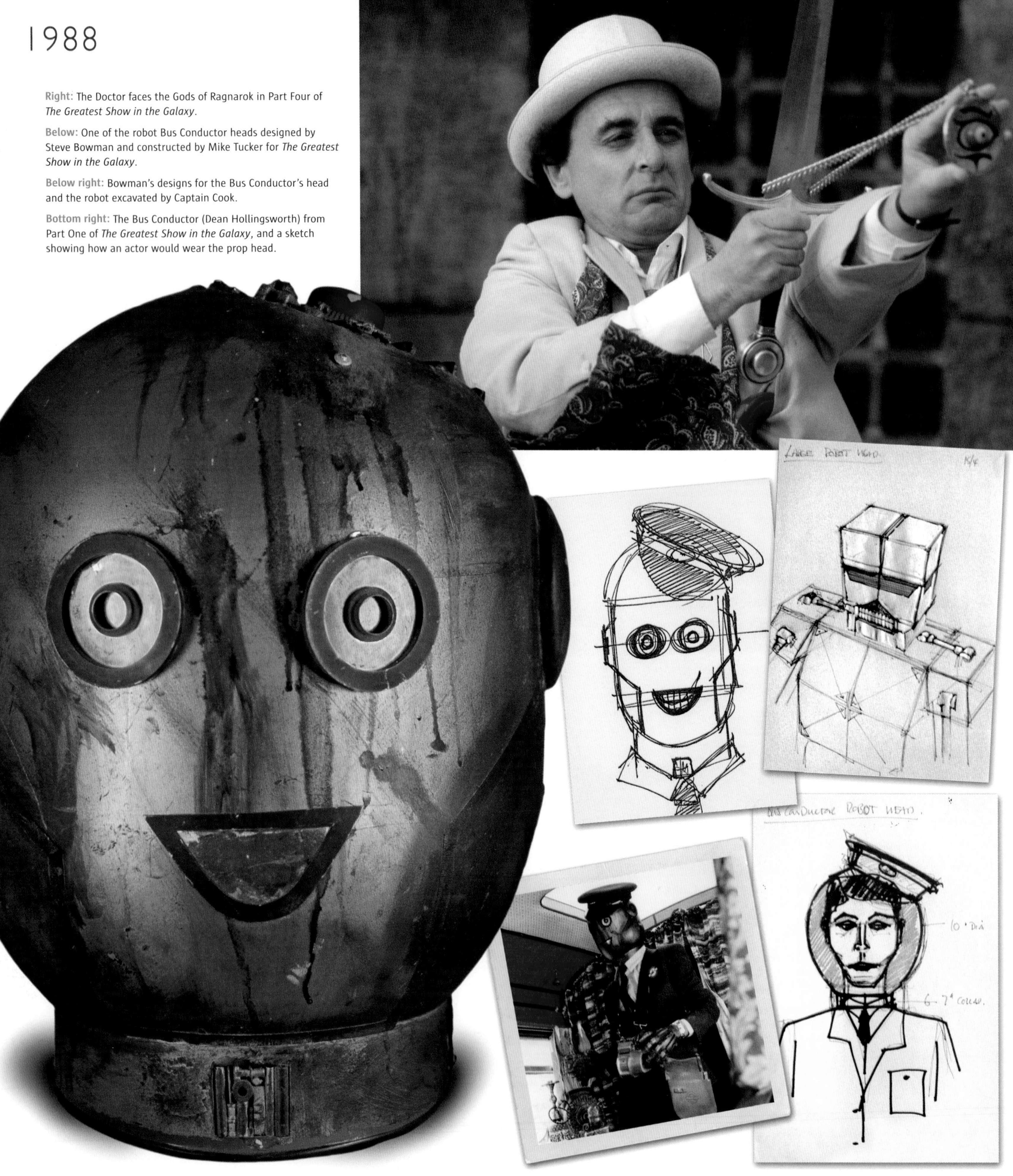

Above left: One of the prop posters used to advertise *The Greatest Show in the Galaxy*'s Psychic Circus.

Above centre: Jessica Martin, who played Mags in *The Greatest Show in the Galaxy*, signed this postcard for Sophie Aldred.

Above: Script editor Andrew Cartmel and Jessica Martin on location at Warmwell Quarry, Dorset in May.

Baker, who's a fine actor and a great guy, lumbered with a risible costume and not the best persona of the various Doctors. And now he's on trial, rendered powerless by his peers. I thought this was the ultimate example of what not to do with *Doctor Who*.

"By the end of my first season I'd seen what Sylvester was capable of and I had a clear idea of what was lacking. At the show's worst, and in *The Trial of a Time Lord*, there had been a tendency for the Doctor to be acted upon rather than acting. More recently he had been something of a clown – and not in a good Troughtonesque way either. I wanted to restore the Doctor's aura of mystery and potency. And in Sylvester I saw the ideal vehicle for those ideas."

Doctor Who still employs script editors, but they now have a very different purview. "We don't commission anything and we don't rewrite anything," says Gary Russell, who script edited the 2009 stories *The Waters of Mars* and *The End of Time*. "Those aspects are now pretty much taken care of by the showrunner. Once the executive producers decide on a story, and the person they'd like to write it, the script editor's job is to liaise with that writer with the aim of getting him or her to produce a synopsis and then a first draft for the execs. There are other things involved, but this is basically how it continues for the rest of the writing process – you're between the execs, the producer, the director, and the writer."

Andrew Cartmel worked on *Doctor Who* for three years, before leaving in 1989 to script edit another BBC series, *Casualty*. While his 'master plan' couldn't save the programme, an important part of his ethos was present when its 2005 comeback placed an increased emphasis on the companion and gave the Doctor a secret history.

"I wanted the Doctor to be a dark, mysterious, forceful figure," says Cartmel, summing up his contribution to the character. "I wanted him to once again be the powerful hero of his own show."

SEASON 25
Producer: John Nathan-Turner
Script editor: Andrew Cartmel

7H *Remembrance of the Daleks*
4 April – 29 April
written by Ben Aaronovitch
directed by Andrew Morgan

Part One	5 October
Part Two	12 October
Part Three	19 October
Part Four	26 October

7L *The Happiness Patrol*
26 July – 11 August
written by Graeme Curry
directed by Chris Clough

Part One	2 November
Part Two	9 November
Part Three	16 November

7K *Silver Nemesis*
22 June – 5 July
written by Kevin Clarke
directed by Chris Clough

Part One	23 November
Part Two	30 November
Part Three	7 December

7J *The Greatest Show in the Galaxy*
14 May – 18 June
written by Stephen Wyatt
directed by Alan Wareing

Part One	14 December
Part Two	21 December
Part Three	28 December
Part Four	4 January 1989

1989

In Season 26 Andrew Cartmel continued to develop the Doctor as a more mysterious, manipulative figure while shifting much of the emphasis onto Ace and her own troubled background.

The season began with *Battlefield*, which had been the second story in production. Brigadier Lethbridge-Stewart and Bessie were brought out of retirement to help the Doctor, Ace, and the new-look UNIT combat Arthurian knights from another dimension.

Recording of the story was interrupted by strike action and an industrial accident that placed Sophie Aldred and nearby crew members in jeopardy. She was rapidly hauled from a water-filled tank just before its glass front shattered.

While *Battlefield* was being produced Jon Pertwee portrayed the Third Doctor in the stage play *Doctor Who: The Ultimate Adventure*. Sixth Doctor Colin Baker took over as the tour continued through the summer.

The second story to be transmitted was the last on the studio floor. *Ghost Light* was an atmospheric mystery with an ingenious narrative by scriptwriter and long-time fan Marc Platt.

The Curse of Fenric was set in a wartime code-breaking facility and further explored Ace's history. Highlights included the vampiric Haemovores and an impressive guest performance by Nicholas Parsons. Initially dismissed as another example of John Nathan-Turner's publicity-minded casting, Parsons gave a moving portrayal of the tragic Reverend Wainwright.

Survival, the concluding story of the season, featured Anthony Ainley's final television performance. It was also one of his most accomplished, as the Master desperately sought to escape from the planet of the Cheetah People and its pernicious influence. Just as importantly, the plot of *Survival* exemplified an increased emotional literacy and was partially staged against the recognisable backdrop of suburban Perivale. In common with the themes of *Ghost Light* and *Survival*, *Doctor Who* was evolving.

The series remained in the same time slot, but this year fared less well against *Coronation Street*. Jonathan Powell, now the Controller of BBC1, had already decided not to renew *Doctor Who* for a 27th season.

On 23 November, Sylvester McCoy returned to Television Centre to record a poignant narration for the closing moments of *Survival*. The final episode was broadcast 13 days later.

It seemed the Doctor's travels in time and space were over.

THE ULTIMATE FOE

Above: A promotional flyer for the opening week of *Doctor Who: The Ultimate Adventure*. Jon Pertwee played the Doctor for the first part of the tour, except when his understudy David Banks took over for two performances on 29 April. Banks co-starred in the play as Karl, but was best known for playing the Cyber Leader in numerous *Doctor Who* stories during the 1980s.

Above right: Jon Pertwee and Colin Baker pose on the TARDIS set of *The Ultimate Adventure*. Baker took over from Pertwee as the Doctor from 5 June until the final performance on 19 August.

The axe fell quietly. On 11 September 1989 John Nathan-Turner wrote to Sylvester McCoy and Sophie Aldred, formally explaining what they already knew: the BBC wouldn't be taking up the options on their contracts for another series of *Doctor Who*.

The programme had effectively come to an end when Head of Series Peter Cregeen informed Nathan-Turner of the BBC's decision in the summer. He pointed out that *Doctor Who* would possibly be continuing as an independent production, but that the chances of that happening as early as 1990 were remote.

Both *Doctor Who* and Nathan-Turner had long been unpopular with some of the most influential figures in BBC management. By 1989 Michael Grade had left to become the chief executive of Channel 4, but the animosity he felt towards the programme lingered in the BBC's corridors of power.

Doctor Who's episodic format and multi-camera techniques were relics of the 1960s, and its status as an in-house production was similarly archaic. When Nathan-Turner eventually took redundancy from the Corporation in August 1990 he was the last salaried drama producer on the books.

In the late 1980s the BBC considered science fiction to be a problematic genre, largely because it was expensive to produce and tended to attract niche audiences. Over the last few years *Doctor Who* had exactly fulfilled that perception.

Of all the mitigating factors, the most conspicuous was its scheduling; since 1987 every episode of *Doctor Who* had begun five minutes after the beginning of *Coronation Street*. But this argument cut no ice with the executives who had put the programme there in the first place. "I don't expect *Doctor Who* under any circumstances to beat *Coronation Street*," said Grade in 1987, "but I do expect it to pick up a decent audience."

Over the life of the original series the BBC used two principal means of gauging a programme's appeal. The Reaction Index (RI) was a measure of approval, calculated by canvassing a panel of viewers the day after broadcast. For many episodes of *Doctor Who* in the 1960s and '70s these figures were accompanied by select comments from those viewers in the form of an Audience Research Report. Some of these now make for unintentionally hilarious reading. The document on Episode 4 of *The Moonbase*

Above: Jean Marsh returned to *Doctor Who*, 24 years after her last appearance in *The Daleks' Master Plan*, when she guest starred in *Battlefield*. She joined Sylvester McCoy for this publicity picture taken during location recording at Rutland Water in May.

Right and below: Ken Trew modified his design for the Doctor's costume during pre-production of *The Curse of Fenric*, the first story to be recorded for Season 26. The new jacket was intended to reflect the Doctor's darker character during this run of stories.

1989

Right: The Haemovores swarm through the graveyard in *The Curse of Fenric*. This sequence was recorded at St Lawrence's Church in Hawkhurst, Kent in April.

Below right: *Ghost Light* was the last story to be produced in *Doctor Who*'s original run. This picture shows rehearsals on the Upper Observatory set at Television Centre, and features (left to right) Sylvia Syms, Katharine Schlesinger (kneeling), Ian Hogg (seated), Sophie Aldred and Frank Windsor.

Below: Dapol's Cyberman figure was released in August. It was based on the *Earthshock* design, rather than the more recent *Silver Nemesis* version.

found it had impressed "a customs officer" ("this is what we love – makes us shiver"), but was rather less popular with "a consulting scientist" who was dubious about the way Polly melted the Cybermen's chest units: "the solvents mentioned would NOT attack plastics!"

Up until 1981 the RI was calculated by the BBC's Audience Research Department and based on samples of varying sizes. *An Unearthly Child* was seen by a panel of just 124 people ("what claptrap!" exclaimed one), while *The Moonbase* was watched by 323. In 1981 the RI was renamed the Appreciation Index and became the responsibility of the Broadcasters' Audience Research Board (BARB), a new organisation jointly established by the BBC and ITV. BARB regularly sampled the opinions of between 2,000 and 3,000 viewers.

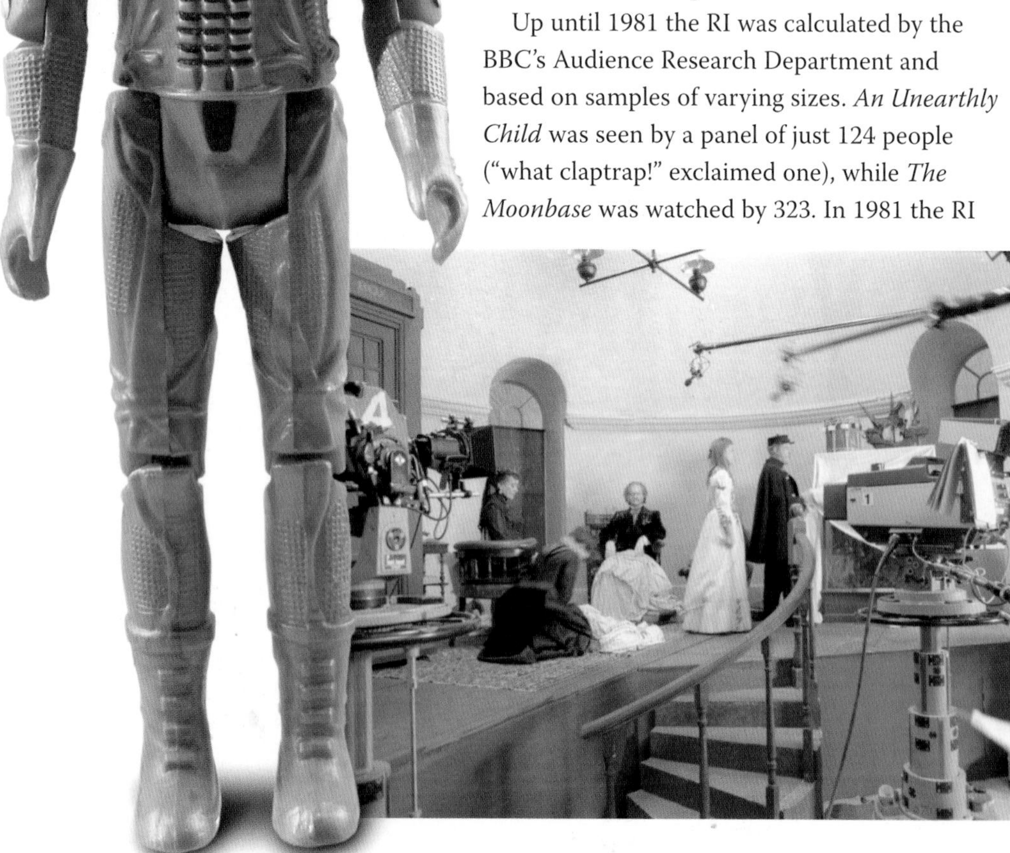

The RI/AI was entirely unrelated to audience size, which remains the most common measure of a show's popularity. The BBC usually includes a combination of both appraisals in any analysis, as divorcing viewing figures from their context can lead to misleading results. On paper, the 16.1 million that watched *City of Death* Part Four make it the most popular episode of *Doctor Who* ever made, until one considers that the only thing ITV was broadcasting in opposition was an apology for services being disrupted due to industrial action.

Appreciation Indexes, viewing figures and other factors including demographics were all included in a survey compiled by Samantha Beere of BBC Broadcasting Research in 1990. *Doctor Who Audience Data 1963-1989* depressingly refers to the programme in the past tense, and is the closest it got to a post mortem. Here was the entire history of the original show laid bare in statistics, from the RI peak of 75 (*Image of the Fendahl* Part Two) to the trough of 30 (*The OK Corral*).

Far left: Sylvester McCoy watches Nicholas Parsons, Joann Kenny and Joanne Bell record a scene for *The Curse of Fenric* Part Two on 12 April.

Left: The Doctor and Ace inside the vestry of St Jude's church in *The Curse of Fenric* Part Three.

Below: The Ancient One was designed by Ken Trew and sculpted by Susan Moore and Stephen Mansfield. The creature was first seen emerging from the sea in *The Curse of Fenric* Part Four.

Below left: Fenric was imprisoned inside this flask, which was designed by Graham Brown and made by Andy Frazer.

In the early 1980s *Doctor Who* had embarked on a nomadic journey through the television schedules, trying to catch the loyalty of a new audience. Beere came to the surprising conclusion that "the changes to the transmission times did not inconvenience viewers too much, as a similar size of audience continued to watch after a change as had before. However, from the time of the first transmission time change onwards audiences gradually declined and it could be that the change of transmission times in the later

Anthony Ainley –
The Master
"Survival" (cat flap)

Shirt with
Stand Collar

Reveres in Pale Silk
Snake Silk Shirt
(Needs 2. crash Sequence)

Double Ended
Cravat with
Built on Bow
Jewel Stud

→ Belt also in Snake Silk

Snake Buckle

Black leather
Gloves + Shoes
from Original
Costume

Above: Anthony Ainley (as the Master) and two Cheetah People on location for *Survival* at Warmwell Quarry in Dorset. Six days of location work began there on 18 June.

Above right: For the Master's first story in three years, costume designer Ken Trew gave the character a new look. Ainley was spared the discomfort of wearing the Master's usual black velvet suit for the summer location work.

Right: *The Nightmare Fair* was the first 'missing episodes' novel published by Target Books, based on unused scripts and storylines for the abandoned version of Season 23. Graham Williams' *The Nightmare Fair* was published on 18 May, with cover artwork by Alister Pearson and Graeme Wey. This was followed by Wally K Daly's *The Ultimate Evil* and Philip Martin's *Mission to Magnus*.

Far right: The Master and the Doctor fight for *Survival* on the planet of the Cheetah People. This scene was in the last episode of the original series, broadcast on 6 December 1989.

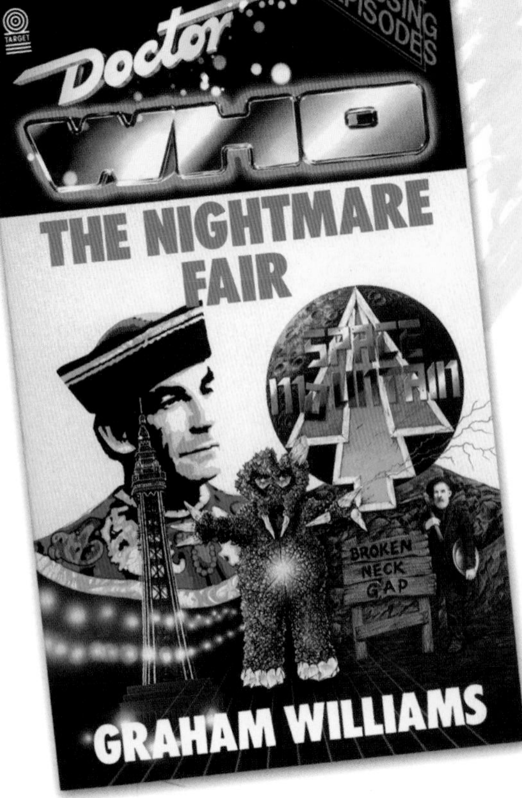

years did have some small and cumulative effect on audience size. It is obvious though that this was by no means the only or main cause of the decline in viewing figures."

Beere's summary also highlighted the rejuvenating effect that a change of lead actor had on the programme, noting that after the peak of 1979 "the following season saw those figures nearly halve in a pattern often repeated before a change of Doctor. Peter Davison's advent in the title role immediately pushed the figures up despite the change in transmission times. These dropped in his second and final season[s] but only rose marginally with the advent of Colin Baker in a break from the pattern of previous years. However the audience figures remained steady at around 7.2 million until the series was given a year-long rest in 1985. After its return the audience size struggled at under 6 million for the next 3 years. However the drop in Colin Baker's last season from an average 7.2 million to 4.8 million was consistent with the previous Doctors in their last season, although in Sylvester McCoy's first season the figures were only pulled up by 200,000 and the programme never again reached the heights of the early William Hartnell or Tom Baker years."

Beere's analysis of the show's changing demographic is amongst the most illuminating data she presented. The earliest such survey was conducted in 1976, when it was revealed that *Doctor Who*'s audience comprised 44.1 per cent children and 55.9 per cent adults. By 1986, this had become 22 per cent children and 78 per cent adults.

For Jonathan Powell, who firmly believed that *Doctor Who* was a programme for younger viewers, this was possibly the most alarming decline of all.

Samantha Beere's report is now filed in the BBC archive. This lengthy dossier remains one of the most meticulous surveys of *Doctor Who* ever compiled, yet it has nothing to say about the indomitable spirit of a programme that refused to die.

Top left: The latest version of the TARDIS key was designed and made by visual effects assistant Mike Tucker. It featured in *Ghost Light* and *Survival*.

Top centre: Issue 71 of fanzine *DWB* (originally *Doctor Who Bulletin*) was edited by Gary Leigh and dated November 1989. Its cover broke the news that the programme would not return in 1990.

Above: Sylvester McCoy relaxes between takes on 11 June, the second day of location work on *Survival*.

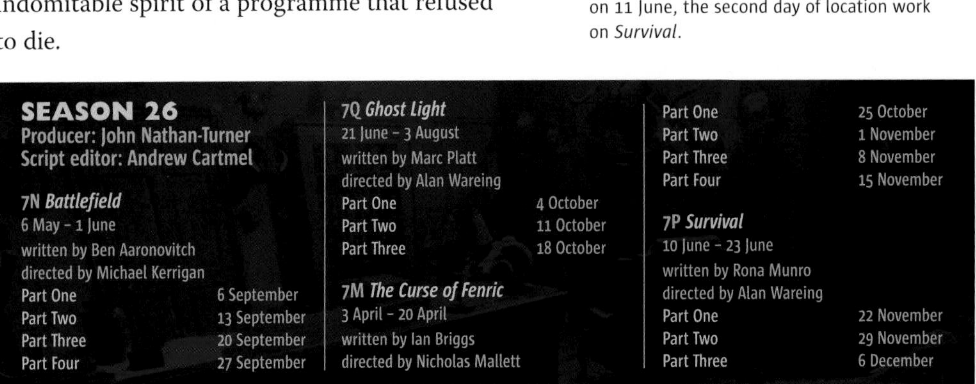

SEASON 26
Producer: John Nathan-Turner
Script editor: Andrew Cartmel

7N *Battlefield*
6 May – 1 June
written by Ben Aaronovitch
directed by Michael Kerrigan

Part One	6 September
Part Two	13 September
Part Three	20 September
Part Four	27 September

7Q *Ghost Light*
21 June – 3 August
written by Marc Platt
directed by Alan Wareing

Part One	4 October
Part Two	11 October
Part Three	18 October

7M *The Curse of Fenric*
3 April – 20 April
written by Ian Briggs
directed by Nicholas Mallett

Part One	25 October
Part Two	1 November
Part Three	8 November
Part Four	15 November

7P *Survival*
10 June – 23 June
written by Rona Munro
directed by Alan Wareing

Part One	22 November
Part Two	29 November
Part Three	6 December

1990–1995

The future of *Doctor Who* seemed to lie in its past. The staff and writers of *Doctor Who Magazine* turned their attention to retrospective articles, often synchronising issues with the increasingly busy schedule of releases from BBC Video. On 22 and 23 September 1990 John Nathan-Turner co-presented a weekend of archive episodes, documentaries and interviews for BSB (British Satellite Broadcasting), just three weeks after he closed the *Doctor Who* production office.

In June 1991 Virgin Publishing launched a range of original novels under the umbrella title *Doctor Who: The New Adventures*. The following month, an exhibition entitled Behind the Sofa opened at London's MOMI (the Museum of the Moving Image). The only 'new' episode of *Doctor Who* broadcast in 1991 was the pilot version of *An Unearthly Child*, screened for the first time as part of BBC2's tribute to Lime Grove Studios on 26 August.

January 1992 saw the first BBC documentary about the series since 1977's *Whose Doctor Who*. Director Archie Lauchlan took a rather different approach for *Resistance is Useless*, linking clips from the show with an illuminated anorak that dispensed "curious facts" in a West Midlands accent.

This was a small price to pay for the repeats that followed. Over the course of the next three years BBC2 screened 11 complete stories from all seven Doctors, with average ratings peaking at an impressive 3.6 million for *Planet of the Daleks*.

In May 1993 Radio 5 producer Phil Clarke recorded *The Paradise of Death*, a Barry Letts serial featuring Jon Pertwee, Elisabeth Sladen and Nicholas Courtney. Soon afterwards, BBC Worldwide announced plans to make a 90-minute film – *The Dark Dimension* would have been directed by Graeme Harper from a script by Adrian Rigelsford. All the surviving Doctors were asked to appear, but the project was shelved in July.

Fans were left baffled by Nathan-Turner's *Dimensions in Time*, which was broadcast in 3D as part of BBC1's *Children in Need* on 26 and 27 November. Two days later Kevin Davies' lavish documentary *30 Years in the TARDIS* gave the channel a more satisfying celebration of the anniversary.

In 1994 Radio 2 commissioned *The Ghosts of N-Space*, another Third Doctor serial by Barry Letts, and in 1995 the rumours about the series' television comeback were finally confirmed. Pre-production on an American TV movie was underway...

NEW ADVENTURES

Left: A magazine advertisement for BSB's Doctor Who Weekend in September 1990. The satellite channel had only been on the air since March, and in November would be absorbed by Sky Television.

Below left: A Virgin Megastores advertisement for the VHS box sets of *The Chase/Remembrance of the Daleks* and *The Trial of a Time Lord*. These became some of BBC Video's best-selling *Doctor Who* titles when they were released in autumn 1993. The 'Daleks' box set sold more than 40,000 units, and *The Trial of a Time Lord* sold more than 30,000.

Below: Ace and Davros were added to the Dapol range of action figures in 1990. Unfortunately the initial version of Davros had two arms and was widely circulated before Dapol realised the error.

In the absence of a formal cancellation announcement most fans spent the early 1990s waiting for a new series. A militant minority maintained pressure on the BBC, while others realised that *Doctor Who* could continue in other media.

After much deliberation, John Nathan-Turner had granted permission for a range of original *Doctor Who* novels in October 1989. *Doctor Who: The New Adventures* would be published by Target's new parent company Virgin and overseen by editor Peter Darvill-Evans.

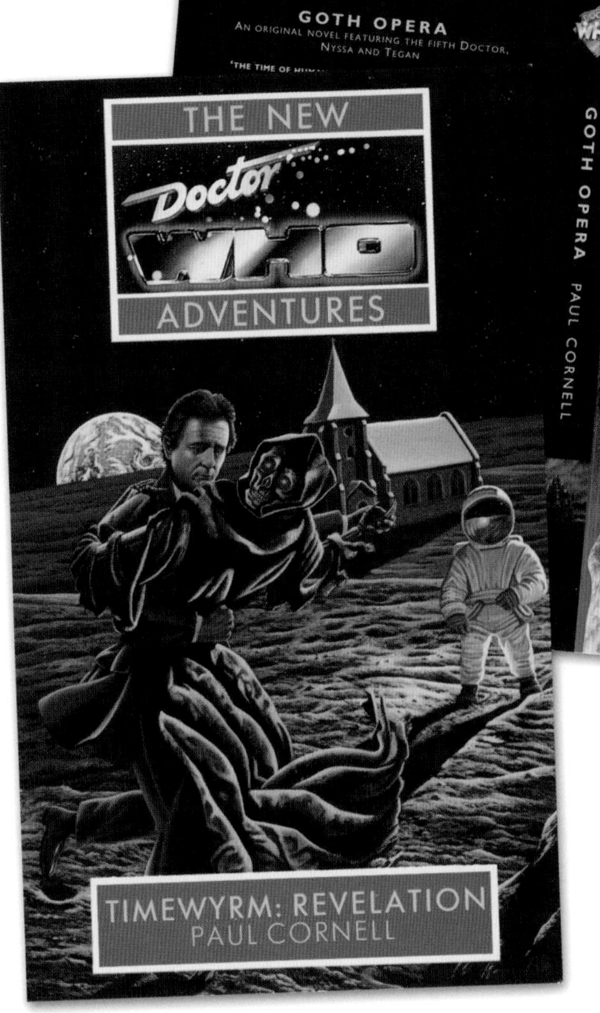

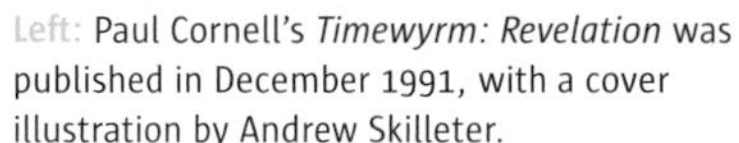

Left: Paul Cornell's *Timewyrm: Revelation* was published in December 1991, with a cover illustration by Andrew Skilleter.

Above: A proof cover for Cornell's *Missing Adventures* novel *Goth Opera*. Virgin asked artist Alister Pearson to remove most of the blood before the book was published in 1994.

Right: The 1993 *Radio Times* promoting *Dimensions in Time* was the first to feature the Sixth and Seventh Doctors on a cover.

Below right: Barry Letts' *The Paradise of Death* was published in 1994 and brought the Target range of *Doctor Who* books to an end. The cover was by Alister Pearson.

Nearly two years later the back cover of John Peel's *Timewyrm: Genesys* announced a new series of "Full-length science fiction novels; stories too broad and too deep for the small screen." *Genesys* was criticised for its efforts to appear more adult than the television series, but the fourth book, Paul Cornell's *Timewyrm: Revelation*, achieved a sophistication that Russell T Davies particularly admired. "I bought a copy of *Revelation* to idle away a train journey, and arrived at Euston three hours later feeling jealous as hell," he told *Doctor Who Magazine* in 2001. "Paul bloody Cornell gave us *Doctor Who*, but he made it real. I mean, real people, laid bare, exposing all their anger, passion and, damn it, nobility. People with histories and hopes and flaws, existing in a world where Chad Boyle, the school bully, is far more terrifying than some super-evil Timewyrm."

Davies would contribute to the range himself, later inviting Cornell and fellow *New Adventures* novelists Mark Gatiss, Gareth Roberts and Matt Jones to write for the revived television series.

Sixty-one *New Adventures* and 33 *Missing Adventures* (depicting past Doctors) were published before Virgin's license was allowed to expire. In June 1997 Terrance Dicks' *The Eight Doctors* became the first BBC novel to feature the latest Doctor. The range was still going strong when, four years later, the book of *Guinness World Records* described the novels as "the largest fictional series built around one principal character," with sales of over eight million copies since the 1970s. The 100th BBC *Doctor Who* book, Paul Magrs' *Mad Dogs and Englishmen*, was published in 2002.

The New Adventures and their BBC Books successors played an important role in the development of *Doctor Who* during its wilderness years. Darvill-Evans' insistence that the stories should forge ahead with the most recent Doctor and companion gave his writers the freedom to develop new concepts, unbound by nostalgia or continuity.

For much of the 1990s, the novels were the only sign that *Doctor Who* had any sort of future at all.

1996

Philip David Segal was born in England and had been a fan of *Doctor Who* since he watched the first episode with his grandfather. Thirteen years later he emigrated to the United States, rising through the ranks of the television industry in Los Angeles until he became the Vice President of Steven Spielberg's Amblin Television. Segal used his executive status and enviable connections to realise a long held ambition. When he told the BBC he wanted to a produce a new series of *Doctor Who*, they listened.

The path from Segal's initial pitch to the broadcast of the *Doctor Who* TV movie was almost as convoluted as the various drafts of the script. But the English-born executive producer, his English writer Matthew Jacobs and English director Geoffrey Sax achieved the near impossible with American money on Canadian soil – they brought the Doctor back from the dead.

In a misjudged attempt to initiate American viewers, the TV Movie unleashed a barrage of hasty exposition that even British fans struggled to assimilate. The Seventh Doctor then regenerated into the Eighth (Paul McGann) who, with the assistance of Grace Holloway (Daphne Ashbrook), fought the Master (Eric Roberts) in San Francisco on 31 December 1999.

The otherwise untitled *Doctor Who* was premiered in the US by Fox on 14 May. Scheduled against a new episode of ABC's sitcom *Roseanne*, the TV Movie drew just 5.6 million viewers. The BBC1 screening took place on 27 May and was dedicated to Jon Pertwee, who had died seven days earlier. The remarkable viewing figure of 9.1 million was of little consequence to the BBC's American co-production partners.

The TV Movie was never formally intended as a pilot, but it was hoped that another movie and maybe a series might result from the deal. Neither would occur. Fans fell hungrily on their favourite aspects of the story while ignoring the controversial innovations and other inconsistencies. They ensured the Eighth Doctor lived on in books and other media, but with a television tenure lasting little more than an hour he never gained a foothold in the public consciousness.

Marcus Berkmann's comments in the *Sunday Express* were typical of reviews that seemed to regard the production as some kind of betrayal. "This was probably not a new beginning at all," he wrote on 2 June. "This was the end."

CULTURE SHOCK

Top: Storyboards detailing the death of ambulance driver Bruce, as his body is possessed by the Master.

Top right: An early concept painting of the TARDIS entrance. The Jules Verne-inspired style was maintained for the finished sets.

Far right: Production designer Richard Hudolin created this concept art for the TARDIS Cloister Room. The Eye of Harmony appears on a raised platform at the centre.

Right: Hudolin's design for the new TARDIS console.

Above: Daphne Ashbrook, Philip David Segal and Paul McGann at the Director's Guild in Los Angeles on 8 May.

The idiosyncratic nature of the TV Movie – quite unlike any *Doctor Who* that went before or came after – makes it the greatest oddity in the canon. Sometimes it isn't even included as canon – Paul McGann was the only surviving Doctor not to be interviewed or even appraised in BBC1's 40th anniversary documentary *The Story of Doctor Who* (2003).

And yet co-executive producer Philip David Segal went to great pains connecting the TV Movie to the series' legacy. He insisted the story should begin with the Seventh Doctor, allowing Sylvester McCoy the belated opportunity to bow out with a regeneration scene. And there were many other historical details for keen-eyed fans to spot: the Sonic Screwdriver, the 900 Year Diary, the John Smith pseudonym, the yo-yo and a copious amount of jelly babies should

Above: In May, the team behind *Doctor Who Magazine* published *The Doctor Who Movie Special* for younger readers.

Right: The new Doctor (Paul McGann) comes alive in the morgue. This scene was filmed at the BC Children's Hospital in Vancouver in late January.

Inset: Sylvester McCoy hands the TARDIS key to McGann in a publicity shot from *Doctor Who*.

Below: The Sonic Screwdriver made its first appearance in the series since the last model was destroyed in *The Visitation* (1982).

have reassured the faithful. But all these details appeared insignificant next to the fundamental changes imposed on the Doctor's background and character.

Matthew Jacobs' script puzzled fans by stating that the Eye of Harmony, a power source previously found on Gallifrey, was now situated inside the TARDIS. More significantly, Jacobs included a revelation about the Doctor's parentage. "I'm half human," he tells Professor Wagg. "On my mother's side." Equally surprising

Right: McGann keeps his lips closed as the Doctor kisses Grace (Daphne Ashbrook). This scene was filmed at Hadden Park, Vancouver in January.

Below: Costume designer Jori Woodman created the Master's robes in the style established by James Acheson for *The Deadly Assassin*.

Inset: The Master (Eric Roberts) tells the Doctor, "I always dress for the occasion".

was the scene where the previously celibate Doctor kisses Grace – this moment of abandon could have been attributed to joyful exuberance, were it not for the fact that he kisses her again, and once more at the end of the story.

Although he was no *Doctor Who* aficionado, McGann knew enough about the series to be concerned about this aspect. "I kept my lips closed," he told *The Daily Telegraph*'s Jan Moir on 30 May 1996. "And the directors were on a long lens, so they didn't notice until it was too late. I know it is 30 years down the line, but the Doctor has always been a very child-like character. If they had asked me to do a bedroom scene I would have said no. What would be the point?"

Interviewed on the Vancouver set in early 1996, Segal explained the rationale behind the Doctor's relationship with Grace. "In terms of tailoring it to an American audience we've really created two movies," he said. "We have a science fiction movie for the fans, which is a wonderful adventure for the Doctor, but we also have a love story wrapped into it, very much reminiscent of *Romancing the Stone* in feel... I think we've done a lovely job of kissing the past, bringing it into the '90s and evolving this concept into a story and show that I think can appeal to a broad audience."

Segal's well-intentioned efforts to revitalise *Doctor Who* had in fact been terminally compromised before a single frame was shot. Interviewed in 2001, he spoke more candidly about the creative compromises he had to make. "I had a US network, Fox, who insisted on it being an Americanised version; I had BBC

Worldwide who wanted something that was terribly commercial; I had Universal who had their own agenda, and then I had the BBC who had their own agenda. So I was being pulled in four different directions."

Segal admitted that Eric Roberts' portrayal of the Master was "interesting", but that hiring the American actor "wasn't a choice I wanted to make."

Roberts' presence was entirely appropriate to the story's San Francisco setting, but casting a new Master was a departure from the approach that maintained continuity with the last television Doctor.

Segal also expressed regret over the general Americanising of the show. "Its Britishness is what makes it work," he said. "The eccentricity of it makes it work. You can take that and plug it into any world but you have to protect that and you have to leave that alone.

"Those American elements and the romance and all of the things that I was being forced to do... I think in some ways actually hurt the character."

The TV Movie's failure to spawn a follow-up has seen it branded a failure, but it's curious to note how much of its content seemed to inspire or filter through to the series of *Doctor Who* that began in 2005. Elements of the title sequence, the orchestral style of theme tune, the Sonic Screwdriver and even the kissing were all present in Series One, along with the single-camera technique and heightened melodrama familiar from the TV Movie.

Doctor Number Eight would not appear in the new series, despite the fact that showrunner Russell T Davies was an admirer. "Paul McGann literally turns in every direction," he said in 2009. "You get the passionate hero and great little moments of comedy. You see literally every facet the man could have brought to it, and he conquers every single one. It's a real shame we didn't see more of that."

Doctor Who
15 January – 21 February
Executive producers: Alex Beaton, Philip David Segal, Jo Wright
Producer: Peter V. Ware
written by Matthew Jacobs
directed by Geoffrey Sax
27 May

Above left and left: From June 1996 to March 1997 the *Radio Times* ran an Eighth Doctor comic strip written by Gary Russell and illustrated by Lee Sullivan. These instalments from the first story, *Dreadnought*, show Sullivan's original inks and a coloured version as printed in the magazine.

Above: In March this issue of the *Radio Times* included a 16-page *Doctor Who* supplement.

Below: The 55th book in Virgin's *New Adventures* series was written by Russell T Davies and published in October.

1997–2004

Vhile BBC Video continued the restoration and elease of old stories, some felt that the TV Movie epresented their last chance to see any new *Doctor Who.*

The TV Movie's popularity in the UK prompted BC Books to bring *Doctor Who* back in house, and hey began publishing original novels from 1997.

Doctor Who and the Curse of Fatal Death was roadcast as part of BBC1's Red Nose Day charity vent on 19 March 1999. Written by Steven Moffat, irected by John Henderson and executive produced y Richard Curtis, the four mini-episodes poked entle fun at the programme's clichés. Rowan tkinson, Richard E Grant, Jim Broadbent, Hugh Grant and Joanna Lumley all got to play future egenerations of the Doctor. Julia Sawalha was his/ er assistant Emma, and Jonathan Pryce tackled the ole of the Master with considerable relish.

Big Finish, a company that took *Doctor Who* rather nore seriously, began releasing original audio dramas n July. Their first production, *The Sirens of Time,* tarred Peter Davison, Colin Baker and Sylvester IcCoy. Subsequent serials would be regarded by nany as the closest thing to an official continuation f *Doctor Who.*

On 1 November 1999 the DVD release of *The Five Doctors* Special Edition provided a glimpse of the new format's potential for presenting archive programmes and supplemental material.

BBC2's Doctor Who Night was presented by To Baker on 19 November and included a number of incisive comedy sketches written and performed by Mark Gatiss and David Walliams. The evening's viewing was promoted with a *Radio Times* cover featuring a portrait of a Dalek by photographer Da Bailey. An article inside confirmed that the BBC w (still) in talks to produce a feature film, and mention a rumour that writer Russell T Davies "had been involved in developing a new *Doctor Who* TV serie

The internet was the next medium to offer *Doct Who* a parallel life, as the BBC webcasts *Death Comes to Time* (2001), *Real Time* (2002), *Shada* (2003) and *Scream of the Shalka* (2003) developed increasingly sophisticated animation to accompany the audio drama.

On 30 December 2003 the BBC1 documentary *The Story of Doctor Who* was a brisk but sincere overview of glories past. The timing of its producti only allowed a brief opportunity to mention the programme's suddenly promising future...

DRAMATIC LICENCE

Right: Most of the *Doctor Who* merchandise released in the years immediately following the TV Movie adopted variations on its basic logo. In June 1997 *The Devil Goblins from Neptune* was the first novel published in BBC Books' 'Past Doctor Adventures' series.

Far right: Although it suffered from technical problems, the 1999 DVD release of *The Five Doctors* was an important step forward in the BBC's presentation of archive *Doctor Who*.

Below: This pewter TARDIS was released in 1997, and was the first item in the Danbury Mint's World of Doctor Who collection.

The best of the fan-led enterprises in the wilderness years created material that wasn't just an approximation of their favourite programme but a valid extension of it.

Big Finish took their name from an episode of Steven Moffat's children's drama *Press Gang* (1989-1993), but are now best known for their ongoing association with *Doctor Who*.

The company was established in 1998 with the ambition to produce *Doctor Who* audio dramas, but managing director Jason Haigh-Ellery and producer Gary Russell initially found BBC Worldwide reluctant to accord them authorised status. Undaunted, they instead turned to Virgin Books, who had maintained their *New Adventures* brand with a range of science fiction novels that didn't feature the Doctor. They produced five adaptations based on these books before the BBC, impressed by the professional quality of their full-cast productions, offered them a licence to produce audio *Doctor Who*.

Peter Davison, Colin Baker and Sylvester McCoy reprised their respective Doctors from the outset, joined by many of the actors who had played their companions. *Storm Warning*, the first story starring Paul McGann, went on sale in 2001 and, four years later, became the first Big Finish serial to be broadcast on the digital radio channel BBC7. Tom Baker finally joined this repertory company of *Doctor Who* veterans in 2011, and in 2013 it was announced that some of his stories would be created in association with Season 12-14 producer Philip Hinchcliffe.

Outstanding examples of the company's prolific output include *Spare Parts* (2002) by *Ghost Light* author Marc Platt, a Cyberman origin

story that stands as one of the most chilling examples of *Doctor Who* in any medium. Another acclaimed production, *Jubilee* (2003), was adapted by its author Robert Shearman for his 2005 television story *Dalek.*

"All drama works better on audio," says Colin Baker, whose Sixth Doctor has certainly benefited from Big Finish's sympathetic scripting. "You're providing the pictures, and I can look exactly how you want me to. I expect that my Doctor works better on audio than he would if I were still playing him on television. I look completely different, after all, but I'm much the same, vocally; you can hardly tell the difference. I may have sunk into decreptitude, but on audio I'm still the lusty young male that I once was."

Top left: Paul McGann, India Fisher and Nicholas Courtney record *Minuet in Hell*, a Big Finish serial released in 2001.

Top right: The cover of *Spare Parts* (2002) was designed by Clayton Hickman, at that time the editor of *Doctor Who Magazine*.

Above left: A BBC postcard promoting the webcast *Scream of the Shalka* (2003). The animated story was said to feature the official Ninth Doctor (voiced by Richard E Grant) but this claim was soon forgotten when the return of the television series was announced.

Above: A draft cover for Philip David Segal and Gary Russell's non-fiction book *Doctor Who: Regeneration* (2000).

Right: The silver version of the Talking Daleks produced by Product Enterprise from 2001.

2005

In September 2003 it was announced that *Doctor Who* would finally return to television. Under the patronage of BBC1 Controller Lorraine Heggessey and Controller of Drama Commissioning Jane Tranter, head writer Russell T Davies and his fellow executive producers, Julie Gardner and Mal Young, began redeveloping the programme for a modern audience. Christopher Eccleston and Billie Piper would star in a season largely comprising self-contained 45-minute episodes. Their recording would be overseen by producer Phil Collinson.

'Series One' began with *Rose* on Saturday 26 March 2005. The action was conducted at a giddying pace but found time to evoke previous debut episodes *Spearhead from Space* and *Robot*.

The series was an instant success, attracting 10.8 million viewers on its opening night. A few days later, fans old and new were astonished to discover that Eccleston had quit, but there were still 12 more episodes for his Doctor's shadowy history to unfold.

The End of the World showcased state of the art CGI in a story set far in the future, while the Doctor took Rose to the Victorian era to meet Charles Dickens (Simon Callow) in *The Unquiet Dead*, written by Mark Gatiss.

The Doctor again met Rose's boyfriend Mickey (Noel Clarke) and her mother Jackie (Camille Coduri) as together they fought the flatulent Slitheen in *Aliens of London* and *World War Three*.

In Robert Shearman's *Dalek* the ultimately moving confrontation between the last surviving Dalek and the last surviving Time Lord revealed more about the apocalyptic Time War that seemingly destroyed both races.

The Long Game was part of Davies' meticulous set-up for the season finale, and Rose tried to rewrite her family's tragic past in *Father's Day*. In *The Empty Child* and *The Doctor Dances* writer Steven Moffat took *Doctor Who* back to its disturbing best, introducing the omnisexual Captain Jack Harkness (John Barrowman) as a regular. *Boom Town* was recorded at landmarks in Cardiff, the show's new production base. The season ended with an invasion force of revitalised Daleks in *Bad Wolf* and *The Parting of the Ways*.

By June David Tennant had enthusiastically inherited the lead role. In *The Christmas Invasion* Doctor Number Ten emerged from the TARDIS with a question that could have been directed at viewers everywhere: "Did you miss me?"

SATURDAY NIGHT FEVER

Rose's new life against the ordinariness of her mum, and to touch base with something the Doctor can never have.

THE EPISODES

The plots aren't worked out in any detail yet. But this is a good example of the style, and the scale, and the pace, and the fun.

EPISODE ONE

Rose meets the Doctor, and the journey begins.

2005. The ordinary world. Houses and shops and telly and cars. Rose Tyler is busy, and hassled, and late, she's got work to do and her mum's lost her keys, and who the hell put that blue box in the middle of the street? She hurries to her boring job – in a big, city centre department store - and Mobbsy, her boyfriend, is on the mobile, asking about tonight, and she's got to go down to the stockroom and –

...hold on. Did that shop-window-dummy just *move..?*

It's not just moving, it's walking towards her, and then suddenly, this strange man appears, and he grabs her hand and they run – somehow, she trusts him, right from the start – and they're running for their lives and all the dummies are moving, they're sinister and faceless and coming to life, all around her - this is impossible, they're men dressed up, they must be –

The strange man saves Rose's life, then disappears, like he was never there. And the dummies have stopped moving, they're just plastic. She pulls the arm off a dummy, just to prove it to herself; ordinary, solid plastic. Surely… that didn't really happen, did it?

But Rose takes the arm home. Like in some vague way, she might investigate this… although the further away she gets, the dafter it seems, and mum's made chips and it's time for EastEnders, and everything's normal…

But the next thing you know, the strange man's turning up at her front door… and the plastic arm is moving! And Rose's boyfriend is acting odd, almost like he's been replaced by a copy… and time is running out…

As the episode hurtles toward a climax, Rose discovers that a box can be bigger inside than outside. And her mind expands with it. Joyously! Aliens, monsters, invasion, danger – all true! And in a brand-new, bigger, madder world, where nothing makes sense any more, and everything is dangerous, she has to make a decision: can she trust the Doctor?

5

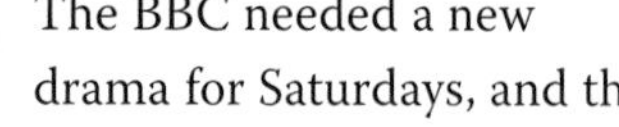

Top: A page from Russell T Davies' 2003 pitch, detailing his proposals for a new series of *Doctor Who*.

Top right: The TARDIS prop and the new control room set at the Unit Q2 studio. When production of *Doctor Who* began all interior sets were built at this facility in Newport, Wales.

Above right: The fifth version of the Sonic Screwdriver made its debut in *Rose*. This is one of several prop screwdrivers created for the series by Millennium FX.

Above: Construction work begins on the standing set of the TARDIS interior. The scale set model sits in the foreground as reference for its shape and size.

The BBC needed a new drama for Saturdays, and the pendulum swung in *Doctor Who*'s favour. "If *Doctor Who* hadn't come back, we'd have had to reinvent it as something else," said Lorraine Heggessey in 2005. What actually happened amounted to both a revival *and* a reinvention.

The first series of the new *Doctor Who* was arguably the most accessible in its history. It had to be – events in the 1980s had shown that the faithful old audience simply wasn't large enough to sustain the programme by itself. The multi-channel environment was now even more competitive, and the cost of producing *Doctor Who* with the single-camera techniques and computer-generated visual effects expected by viewers made it more expensive than ever.

What the programme needed was someone brave enough to give it a new identity without betraying its past. Heggessey and Jane Tranter trusted Russell T Davies to walk that tightrope, and commissioned a new season of *Doctor Who* without even asking to see a formal proposal.

Since contributing *Damaged Goods* to the *New Adventures* range Davies had become one of the most influential television scriptwriters in the country. The highlights of his recent work included the groundbreaking *Queer as Folk* (1999)

Far left: Costume designer Lucinda Wright's sketch of the Ninth Doctor's battered costume. Davies' script for *Rose* described him as looking "Like Terence Stamp if he worked on a market stall."

Left: This early concept sketch for Rose Tyler emphasised a different style to that eventually adopted on screen.

Bottom left: Once Billie Piper had been cast and the script finalised, Rose's look reflected a more urban feel.

Bottom centre: Shop dummy Autons launch their attack in *Rose*. These scenes were recorded at the Queens Arcade in Cardiff.

Below: A partially completed sculpt by Millennium FX for a male Auton head. This would later be used to create a mould from which numerous masks were produced.

for Channel 4 and *The Second Coming* (2003), a controversial religious drama for ITV. Lighter fare such as *Bob & Rose* (2001) and *Mine All Mine* (2004) proved that Davies was equally attuned to a more mainstream audience.

These were the viewers that Davies had in mind when he began devising the new series of *Doctor Who* in autumn 2003. He decided to reassure Heggessey and Tranter by preparing a detailed 'pitch' document which could also be shared with the other writers of Series One. The broad content was largely consistent with what reached the screen in 2005, and reveals a strict revisionism.

Davies' introduction began with the words "A girl meets an alien, and together they travel the universe." The alien was the Doctor: "Your best friend. Someone you want to be with, all the time. He's wise and funny, fast and sarky, cheeky and brave. And considering he's an alien, he's more human than the best human you could imagine. So full of compassion, his heart could burst, and his head's jam-packed with science and art and history...

"The Doctor is lacking one thing. Family. He's a loner. Sometimes, this distance gives him a vital edge – when the world's in danger, he doesn't waste time saving the bloody cat – but sometimes, when he looks at humans, and their mums and dads and lovers and mates, it's like he knows nothing."

Mal Young was the first to suggest Christopher Eccleston for the role. Eccleston had recently given a compelling performance in *The Second Coming*, and when he heard that *Doctor Who* was going to return he asked Davies for an audition.

Above left: Published on 19 May, Justin Richards' *The Clockwise Man* was the first novel based on the new series.

Far left: The BBC Model Unit recreated the upper tower of Big Ben for *Aliens of London*. "It took four weeks to build, half an hour to set up and two seconds to destroy!" recalls model unit supervisor Mike Tucker.

Left: The Slitheen start to panic as their plan unravels in *World War Three*.

Below: A full Slitheen head mask during its latter painting stages at Millennium FX. The mask was worn like a hat, with animatronics operated by remote control.

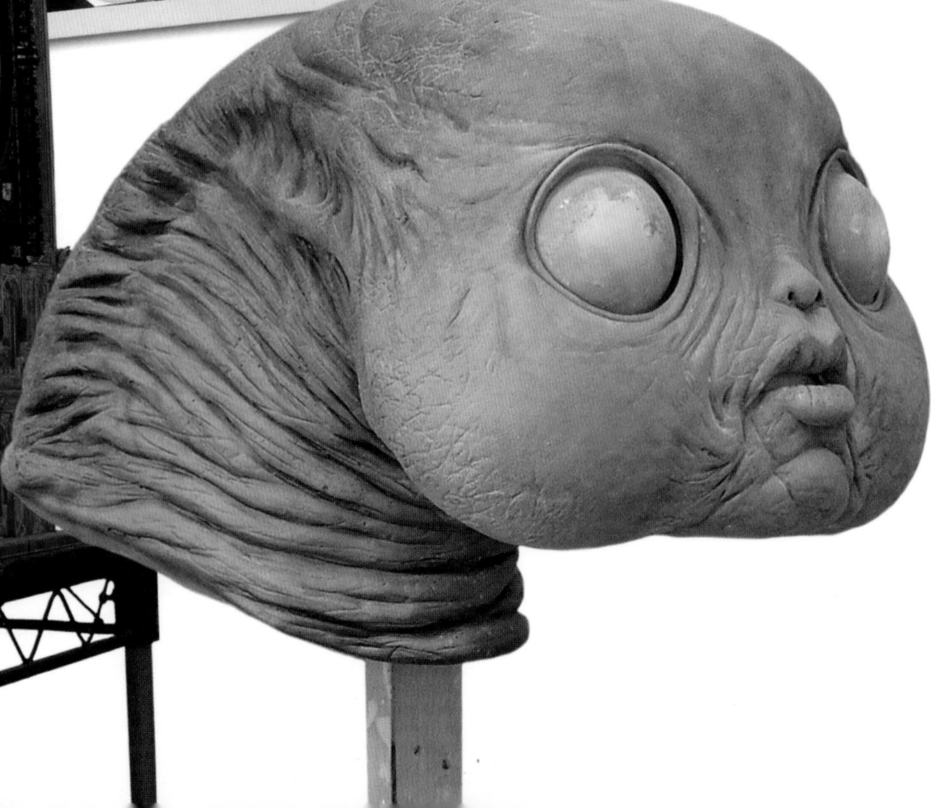

A BBC press release announcing Eccleston as "the ninth television Doctor" was issued in April 2004. Shortly afterwards Davies received a letter from a disgruntled female viewer. "How can you cast a Doctor Who with stubble????" she wrote. "Uggghhh! He looks wretched!!! I will not be watching!"

However, Eccleston perfectly matched Davies' vision of the new-style Doctor. "Not necessarily young, but let's move on from that neutered, posh, public-school, fancy-dress-frock-coat image. He's immediate and tactile. Stand too close to him, you could get burnt."

Lessons had been learned from the alienating references that cluttered the TV Movie. As much as Davies admired Paul McGann's Doctor, the new series would not begin with any sort of handover.

Doctor Number Nine would share his adventures with the 18-year-old Rose Tyler. While this didn't seem like a huge departure from what had gone before, some of the subsequent paragraphs in Davies' pitch would have alarmed more conservative fans.

"The fiction of the Doctor has got 40 years of back-story," he wrote. "Which we'll ignore. Except for the good bits. He's a Time Lord, he's got a Tardis, he's got two hearts, a sonic screwdriver, and he's 900 years old. And, contracts pending, there's an old metal enemy called the Daleks. The Doctor might have a robot dog called K9, because... well, it's a robot dog. I'd have one! But

Top left: Matthew Savage's concept art for *Dalek* showed the exposed creature being tortured.

Top right: The *Radio Times* promoting *Dalek* was voted the best British magazine cover of all time in an internet poll organised by the Periodical Publishers Association in 2008. In 2005 the magazine sold 20,000 posters of the fold-out image, which recreated an iconic photoshoot on Westminster Bridge from the 1964 story *The Dalek Invasion of Earth*.

Above: Rose's sympathy for Van Statten's prisoner initiates a terrible sequence of events in *Dalek*.

Right: A 3D schematic used by The Mill for rendering a CGI Dalek in later sequences of the episode.

Above: The rendered CGI model of Satellite Five, the space station featured in *The Long Game*, *Bad Wolf* and *The Parting of the Ways*.

Top centre: A *Doctor Who* exhibition ran at Brighton Pier from May to November. This was the first opportunity the public had to see props and costumes from the new series.

Top right: A reference photograph of Simon Pegg as the Editor from *The Long Game*. Make-up designer Davy Jones gave the character an almost albino look and fitted Pegg with special contact lenses.

Above right: The key to accessing Floor 500 on Satellite Five in *The Long Game*.

let's see how the scripts work, he can stay in his kennel for now.

"And that's enough. The important thing is, this mythology is discovered, as new, by Rose... The rest of the series' continuity is absolutely irrelevant. I don't care that in 1973 he used gadget X to escape from planet Y. And regeneration isn't important – let's discover it if we need to see it, one day. It's vital to build a new mythology, on screen."

Davies cast off the shackles of continuity by abandoning a central part of the show's mythology. His Doctor would be the last of the Time Lords. "His people are gone. An entire civilisation, mysteriously destroyed... How? Why? The Doctor won't say. But this emphasises his loneliness, and his need for a companion... Whether we develop it or not, 'Last Of The Time Lords' has got a certain ring to it. It makes the Doctor unique. And it guarantees that no old men in silly hats are going to turn up."

In a further effort to capture a wide audience, Davies decided that the programme would never stray far from Earth. "If the Zogs on planet Zog are having trouble with the Zog-monster... who gives a toss?" he wrote. "But if a human colony on the planet Zog is in trouble, a last outpost of humanity fighting to survive... then I'm interested. Every story, somehow, should come back to Earth, to humanity, its ancestors and its descendants."

The episodic structure that had been *Doctor Who*'s format for 26 years was finally discarded.

Right: One of the Art Department's designs for a World War Two-style propaganda poster featured in *The Empty Child*.

Far right: A publicity postcard from the first season of *Doctor Who Confidential*, a BBC3 series that went behind the scenes on each episode.

Below: Mike Smith doubled for Richard Wilson in the scenes where Dr Constantine's face morphs into a gas mask in *The Empty Child/ The Doctor Dances*.

Below right: One of the original gas mask props used in the story.

Bottom: John Barrowman as Captain Jack, sitting astride the German bomb dropped in *The Doctor Dances*. Recording for this sequence took place at the Vale of Glamorgan Railway on Barry Island in late January. Never clearly seen on screen, Schlechter Wolf means 'Bad Wolf'.

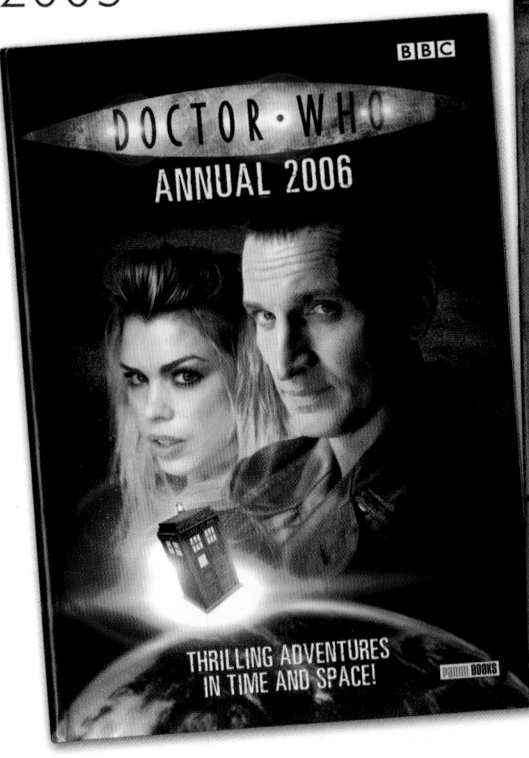

Above left: Panini relaunched the *Doctor Who Annual* with this edition. Subsequent annuals were published by BBC Children's Books.

Above: *Doctor Who: The Shooting Scripts* was published by BBC Books in October. It included all the final scripts for Series One and notes by the episodes' writers.

Left: The Doctor Who Interactive Electronic Board Game was released in September. A battery-powered TARDIS issued pre-recorded phrases.

Below: The TARDIS Talking Money Bank was issued in July and was one of the first items of merchandise featuring the Ninth Doctor.

Davies' producer, Phil Collinson, was a fellow fan who expressed concern that having largely standalone stories would compromise the series' character. In 2008 he told *Doctor Who Magazine*: "I always thought that the cliffhanger was incredibly important to the fabric of the show... But Russell convinced me, and he was right."

Not that the cliffhanger entirely disappeared from the new *Doctor Who*; Series One would contain three two-part stories, and elsewhere the compressed nature of the storytelling moved the cliffhanger to the end of a pre-credit 'hook'.

For all these innovations, much would stay the same. The audience would discover the Doctor through his companions, just as the show's original viewers had done in 1963. The TARDIS' police box exterior was retained, and for one season only the name 'Doctor Who' would be restored to the end credits (it had been 'The Doctor' since Peter Davison's first episode in 1982).

For a while Davies considered resurrecting a 1970s version of the theme tune, only commissioning a new arrangement when it was decided that the old one sounded too 'sparse' when married to the computer graphics created by effects house The Mill. The new recording was by Murray Gold, the composer who would also score the incidental music for every episode.

When pre-production began, Davies instigated tone meetings in order to create a consistency for each episode, and within the series in general. "What should this show look like, feel like, aspire to?" he said in 2004. "Colour, sound, pitch, key? In a year's time, when an eight-year-old kid says 'I like *Doctor Who*,' what picture will be in his head?"

In March 2005 the debut episode *Rose* triumphed against its principal opposition, ITV's light entertainment show *Ant & Dec's Saturday Night Takeaway*. When that series ended, *Doctor Who* proved even more effective against its successor. By the middle of May, *Celebrity Wrestling* could only muster

Above: The *Daily Express' Saturday* supplement for 24 to 30 December previewed *The Christmas Invasion*.

Above centre: One of the Roboform Santa masks created by Millennium FX for *The Christmas Invasion*.

Above right: The disguised Roboforms record a scene on location at The Hayes in central Cardiff in August.

Far right: This Welsh-language postcard was part of a set issued to promote *The Christmas Invasion*. The caption on the front translates as "Sycorax Strong. Sycorax Sound. Sycorax Cool."

2.3 million viewers opposite *Father's Day*, which totalled 8.1.

"I knew that people were expecting [*Doctor Who*] to fail, because that's what people are like," said Davies in 2008. "I *always* knew it would work. I never knew it would work to this extent – I have to say the success of it has gobsmacked all of us – but I knew as a drama it worked."

ITV made one last attempt to topple Series One by scheduling a number of Hollywood blockbusters on Saturday evenings. On 21 May *Star Wars: Episode One – The Phantom Menace* was seen by four million viewers, while more than seven million watched *The Empty Child* on BBC1.

The success of the original *Star Wars* had created expectations of television science fiction that the BBC struggled to meet. Nearly 30 years later, Russell T Davies' extraordinary reinvention put *Doctor Who* back on top.

SERIES ONE

Executive producers: Russell T Davies, Julie Gardner, Mal Young
Producer: Phil Collinson

Rose
20 July – 11 September, 18 October, 10 November 2004
written by Russell T Davies
directed by Keith Boak
26 March

The End of the World
7, 23 September, 4-22 October, 9, 26 November 2004, 18-19 February
written by Russell T Davies
directed by Euros Lyn
2 April

The Unquiet Dead
19 September – 2 October, 19-20, 22 October 2004
written by Mark Gatiss
directed by Euros Lyn
9 April

Aliens of London/World War Three
18 July – 11 September, 4 October, 9-10, 22, 24 November 2004
written by Russell T Davies
directed by Keith Boak
Aliens of London 16 April
World War Three 23 April

Dalek
25 October – 8 November, 23, 26 November 2004
written by Robert Shearman
directed by Joe Ahearne
30 April

The Long Game
30 November – 15 December 2004
written by Russell T Davies
directed by Brian Grant
7 May

Father's Day
11-26 November 2004
written by Paul Cornell
directed by Joe Ahearne
14 May

The Empty Child/The Doctor Dances
17-18 December 2004, 4 January – 11 February, 25 February
written by Steven Moffat
directed by James Hawes
The Empty Child 21 May
The Doctor Dances 28 May

Boom Town
19 January, 1-18 February
written by Russell T Davies
directed by Joe Ahearne
4 June

Bad Wolf/The Parting of the Ways
16 February – 14 March, 21 April
written by Russell T Davies
directed by Joe Ahearne
Bad Wolf 11 June
The Parting of the Ways 18 June

SERIES TWO

Executive producers: Russell T Davies, Julie Gardner
Producer: Phil Collinson

The Christmas Invasion
22 July – 22 August, 8, 22 September, 8 October, 3, 10 November
written by Russell T Davies
directed by James Hawes
25 December

2006

Following the departure of Mal Young, Russell T Davies and Julie Gardner continued to steer the phenomenally successful *Doctor Who*. The second series seemed as fresh as its predecessor, and was now accompanied by a proliferation of books, toys and other merchandise.

New Earth marked the return of computer-generated villain Cassandra (Zoë Wanamaker) and a memorable new race in the feline Sisters of Plenitude. Even more ferocious was the werewolf that threatened the Doctor, Rose and Queen Victoria (Pauline Collins) in *Tooth and Claw*.

Long-standing fans were delighted to see Sarah Jane Smith (Elisabeth Sladen) and K-9 (voiced by John Leeson) in Toby Whithouse's *School Reunion*. With Mickey joining the TARDIS, *The Girl in the Fireplace* continued the bitter-sweet exploration of the Doctor's longevity compared with mere humans.

A classic foe was given a new origin story in *Rise of the Cybermen* and *The Age of Steel*. This two-parter was handled by Graeme Harper – the only director with links to the original run of the series. Events took place on a parallel version of Earth where Rose's father was still alive. Back on the real Earth in 1953, writer Mark Gatiss cleverly used *The Idiot's Lantern* – television itself – as a threat.

The Ood continued *Doctor Who*'s tradition of memorable creatures, first appearing in *The Impossible Planet* and *The Satan Pit*, before *Love & Monsters* made Elton Pope (Marc Warren) the focus of an unconventional tale produced while David Tennant and Billie Piper were recording other stories. Comedian Peter Kay played the Abzorbaloff, a creature designed and named by the winner of a *Blue Peter* competition.

Fear Her presented a more subdued tale of suburban horror before *Army of Ghosts* and *Doomsday* pitted the Daleks against the Cybermen for the first time in the show's history. After the devastating Battle of Canary Wharf the Doctor bid an emotional farewell to Rose as different universes separated them.

The final moments of *Doomsday* led into this year's Christmas episode. *The Runaway Bride* introduced Donna (Catherine Tate), an outspoken companion quite unlike any that had gone before.

The show was branching out in other ways: in October Captain Jack starred in *Torchwood*, *Doctor Who*'s first spin-off drama series. Davies had already developed similar plans for the eternally popular Sarah Jane...

ALSO STARRING…

Doctor Who's capacity for self-renewal – changing both its lead actor and narrative direction – possibly explains why it took so long for a spin-off to appear. *Torchwood*, the first full series to be inspired by *Doctor Who*, began on BBC3 in October 2006.

There would have been a spin-off series much sooner, had William Hartnell's idea for 'The Son of Doctor Who' been taken up by Verity Lambert. Hartnell's accomplished portrayal of dual roles in *The Massacre of St Bartholomew's Eve* illustrates

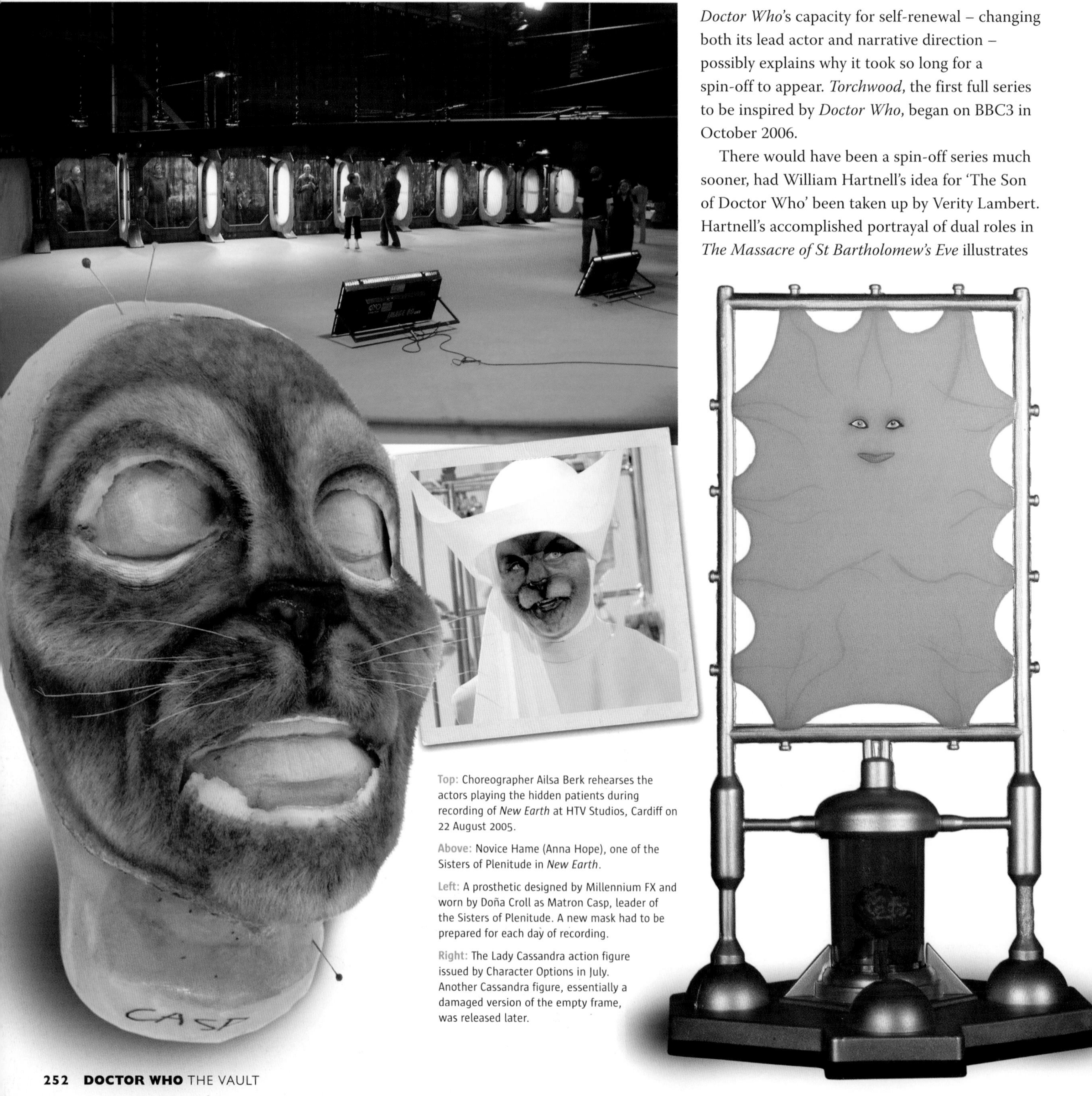

Top: Choreographer Ailsa Berk rehearses the actors playing the hidden patients during recording of *New Earth* at HTV Studios, Cardiff on 22 August 2005.

Above: Novice Hame (Anna Hope), one of the Sisters of Plenitude in *New Earth*.

Left: A prosthetic designed by Millennium FX and worn by Doña Croll as Matron Casp, leader of the Sisters of Plenitude. A new mask had to be prepared for each day of recording.

Right: The Lady Cassandra action figure issued by Character Options in July. Another Cassandra figure, essentially a damaged version of the empty frame, was released later.

how he might have played both the Doctor and his evil offspring.

In 1981 another attempt at a spin-off got a bit further when the pilot episode of *K-9 and Company* went into production. Featuring the character of Sarah Jane Smith as well as K-9, the 50-minute episode was written by Terence Dudley and subtitled *A Girl's Best Friend*. It placed the robot dog in a rural setting that seemed to have been as much inspired by *The Wicker Man* (1973) as *Doctor Who*. It was not an environment that played to K-9's strengths, either as a character or a notoriously unpredictable prop.

Right: The original version of the Tenth Doctor's costume, designed by Louise Page. In the 2007 episode *Gridlock* the Doctor claims that his coat was a present from Janis Joplin.

Bottom right: "This room's the greatest arsenal we could have." The Doctor (David Tennant) arms himself with books in *Tooth and Claw*.

Below: Peter McKinstry joined *Doctor Who*'s Art Department in 2005. This is his interpretation of the multi-prismed telescope featured in *Tooth and Claw*.

Bottom: Showrunner Russell T Davies, co-executive producer Julie Gardner and producer Phil Collinson.

Below: Issue 369 of *Doctor Who Magazine*, published in April, celebrated the return of Sarah (Elisabeth Sladen) and K-9 (voiced by John Leeson) in *School Reunion*.

Below right: "Goodbye, my Sarah Jane." The Doctor and Sarah embrace at the end of *School Reunion*.

Bottom: The version of K-9 fixed by the Doctor in *School Reunion* was built by Nick Kool and Alan Brannan. The prop was operated by Colin Newman during recording.

Bottom right: The date on this clapperboard shows that it was used during the recording of *School Reunion* at Duffryn High School, on the outskirts of Newport.

In 2007 *K-9 and Company* proved painful for Elisabeth Sladen to recall. "What we couldn't ever do with the dog was have a scene with it where it was moving," she said. "It had to be static. It brought everything to a halt. The dog can move on its own, but if you want to have a conversation because something is important then you've got to crouch, and it kind of takes the momentum out of it."

Most of her regrets, however, were reserved for the title sequence that attempted to evoke the pace and sophistication of *Hart to Hart* on a damp Gloucestershire lane in November. "It does my head in," said Sladen, cringing at the memory. "I looked like such an omelette."

Billed as a Christmas special, *K-9 and Company* was watched by 8.4 million when it was first broadcast by BBC1 on 28 December 1981. Its quaint dialogue and traditional styling proved popular with BBC management as well. On 5 January 1982 Graeme McDonald, now Head of BBC Drama Group, sent a memo to David Reid, the Head of Series and Serials, saying that *K-9 and Company* was significantly better than *Castrovalva* Part One, the latest episode of its 'parent' programme.

"The discrepancy of production standards between the K9 show and last night's *Dr Who* make me think you should have a word with John Nathan-Turner about the latter," he wrote. "*Dr Who* is too valuable a show to us to let its standard slip so dramatically and I think we need to be sure that John is aware of the shortcomings of that opening episode."

Despite this enthusiasm, logistics and changes in BBC management meant that a series was never commissioned and *Doctor Who* would have to wait 25 years for its next spin-off.

Soon after starting production on Series One of *Doctor Who*, Russell T Davies realised the

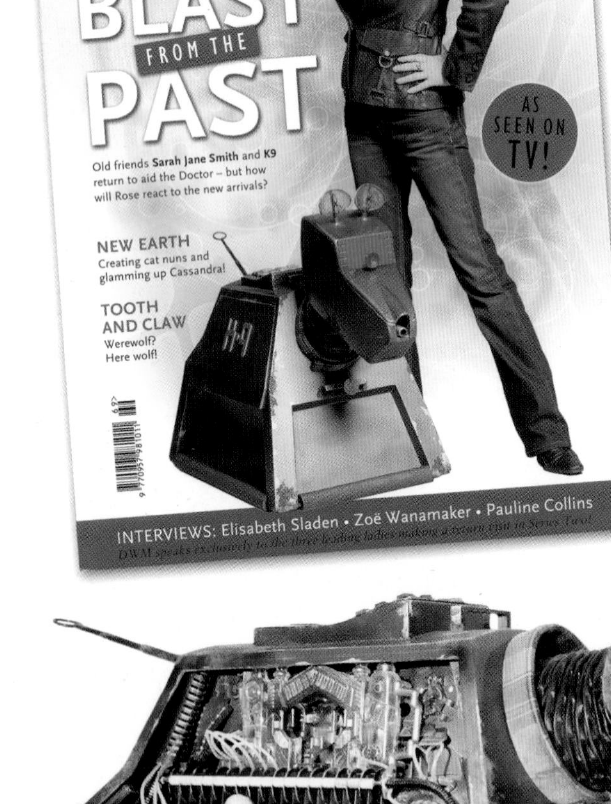

potential of spin-offs, not just in storytelling terms but as a way of exploiting the production talent at BBC Wales.

"It's no good having one hit show," he said in 2008. "That's the foundation of nothing, because when that show ends everything's gone and everyone's out of work. And when you see what the crews can do here [at BBC Wales], and the design teams and the level of expertise, you think 'It's stupid to just make one show.'"

Davies was keen not to produce 'clone' series, which he considered the various *Star Trek* franchises to be. *Torchwood* was to be *Doctor Who*'s older brother, aimed squarely at an adult audience. The mysterious (and apparently immortal) main character,

Above: Rose and Mickey (Noel Clarke) are sedated by Clockwork Robots on board the spaceship *SS Madame de Pompadour* in *The Girl in the Fireplace*.

Right: One of the elaborate masks created by Millennium FX for the Clockwork Robots in *The Girl in the Fireplace*.

Above right: Ellen Thomas wears the mask during location recording at Ragley Hall, Warwickshire in October 2005.

Below: One of the freezing guns from *The Girl in the Fireplace*. The props were rigged to blast dry ice via a pipe that was carefully threaded out of shot for each scene.

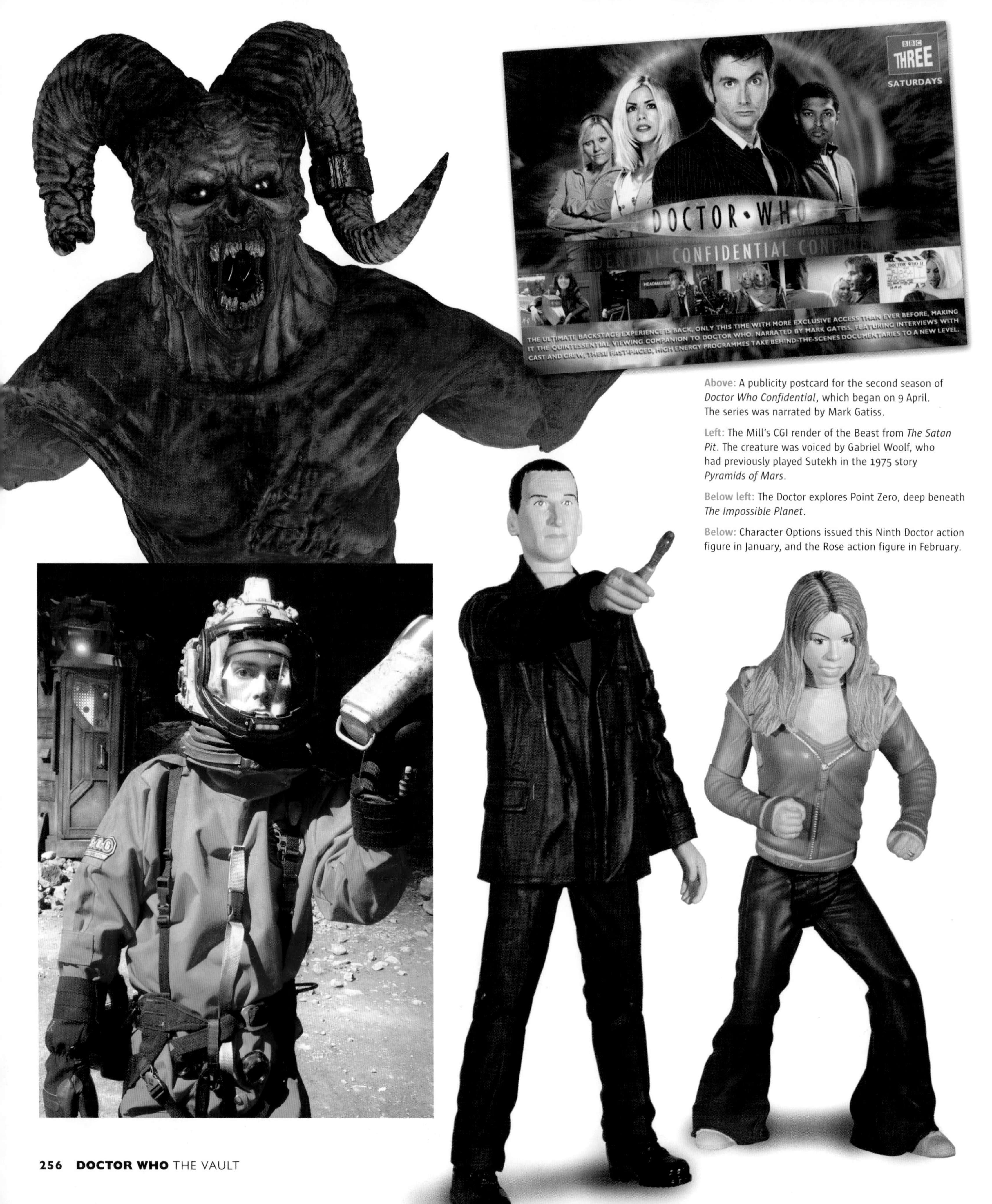

Above: A publicity postcard for the second season of *Doctor Who Confidential*, which began on 9 April. The series was narrated by Mark Gatiss.

Left: The Mill's CGI render of the Beast from *The Satan Pit*. The creature was voiced by Gabriel Woolf, who had previously played Sutekh in the 1975 story *Pyramids of Mars*.

Below left: The Doctor explores Point Zero, deep beneath *The Impossible Planet*.

Below: Character Options issued this Ninth Doctor action figure in January, and the Rose action figure in February.

Captain Jack, had been firmly established in the main series. The charisma John Barrowman brought to the role made Jack an obvious choice to lead the Torchwood team as it investigated alien and paranormal phenomena from a secret base in Cardiff.

The Torchwood Institute itself had been established in the 2006 *Doctor Who* series, its origins depicted in *Tooth and Claw*, and the clandestine organisation it would become was a main component of season finale *Army of Ghosts* and *Doomsday*. The name 'Torchwood' was also derived from the parent show. When boxes of master tapes of the revived series were shipped from Cardiff to London they were labelled 'Torchwood', an anagram of 'Doctor Who', in order to disguise their contents.

Torchwood was not conceived in isolation. To balance this adult spin-off, Davies was keen to create something for the younger audience too. *School Reunion* became a template for *The Sarah Jane Adventures*, a series that exploited the continuity established in *K-9 and Company* (and ratified in *The Five Doctors*) – namely, that the Doctor gave Sarah her own version of the robot dog.

"This is very carefully designed," said Davies in 2008, "as *Doctor Who* in the middle as a family show, *Sarah Jane* for children and *Torchwood* for adults. They're very very different shows, all nonetheless feeding into each other in a very complicated way. Even the children's audience likes that complication... It's a chance to complicate the mythological world of *Doctor Who*, without making *Doctor Who* itself too complicated."

Davies was confident in the appeal of K-9. As well as featuring the dog in his own drama *Queer as Folk*, he had toyed with the idea of including K-9 in the main series right from the start. Bringing

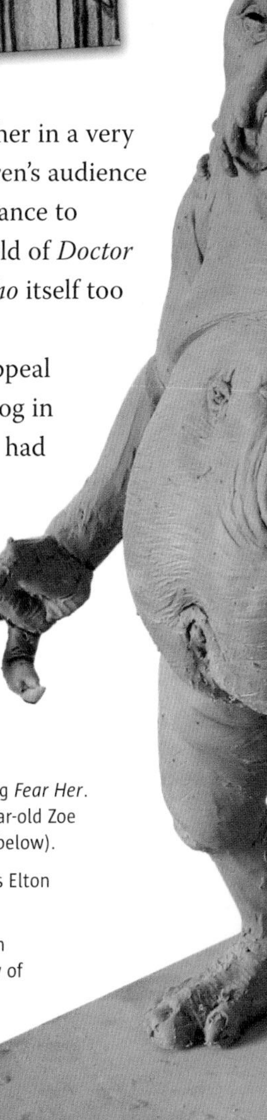

Top left: The *Battles in Time* trading card game and fortnightly magazine was published by GE Fabbri and launched nationally on 20 September.

Top centre: BBC publicity postcards promoting *Fear Her*. The artwork on these examples was by six-year-old Zoe Barkes (top) and 11-year-old Jeremy Barclay (below).

Top right: The Abzorbaloff (Peter Kay) pursues Elton in *Love & Monsters*.

Right: The Abzorbaloff was created by William Grantham, the winner in the 8 to 10 category of *Blue Peter*'s latest competition to design a *Doctor Who* monster. Millennium FX based this clay maquette on his original picture.

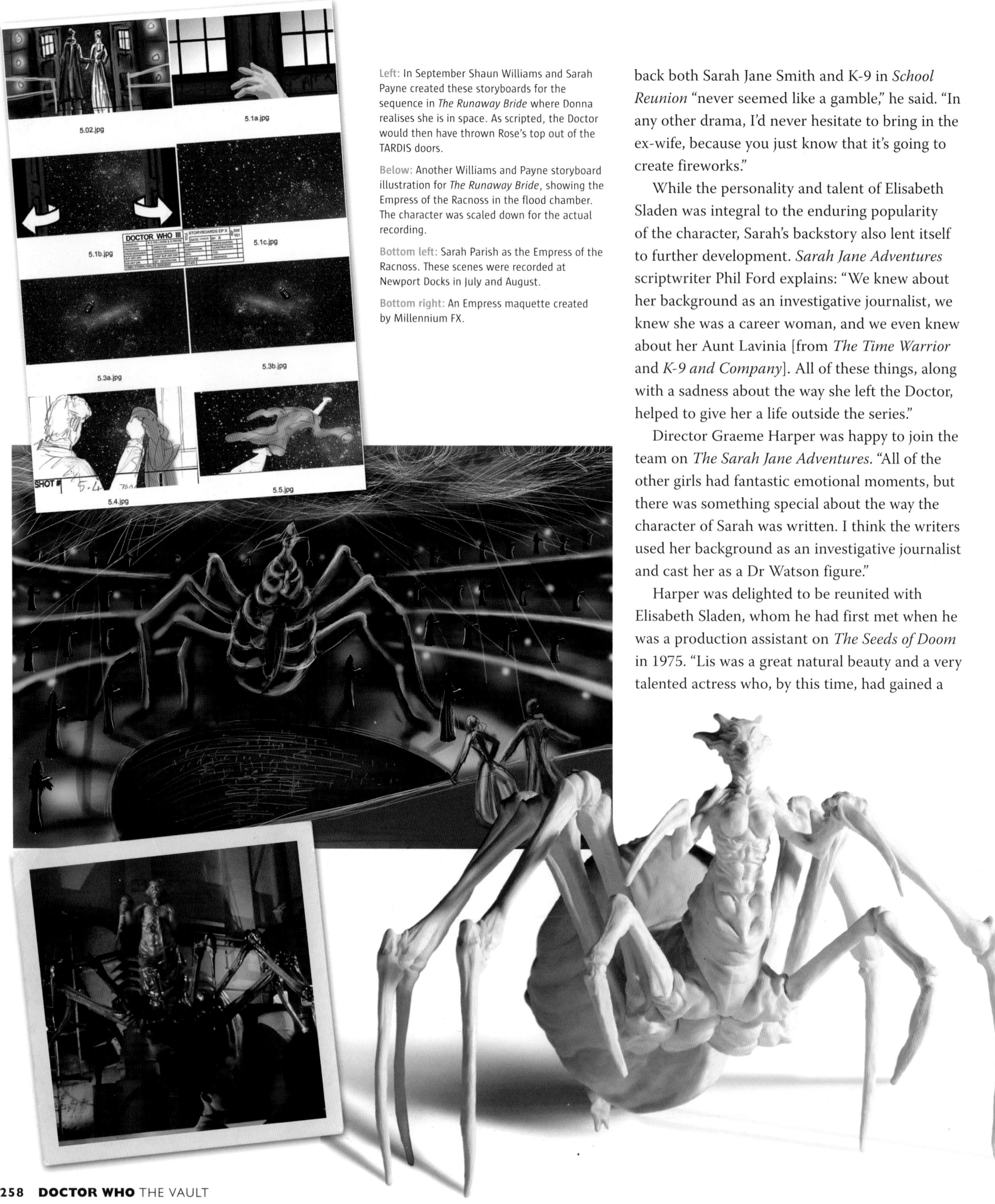

Left: In September Shaun Williams and Sarah Payne created these storyboards for the sequence in *The Runaway Bride* where Donna realises she is in space. As scripted, the Doctor would then have thrown Rose's top out of the TARDIS doors.

Below: Another Williams and Payne storyboard illustration for *The Runaway Bride*, showing the Empress of the Racnoss in the flood chamber. The character was scaled down for the actual recording.

Bottom left: Sarah Parish as the Empress of the Racnoss. These scenes were recorded at Newport Docks in July and August.

Bottom right: An Empress maquette created by Millennium FX.

back both Sarah Jane Smith and K-9 in *School Reunion* "never seemed like a gamble," he said. "In any other drama, I'd never hesitate to bring in the ex-wife, because you just know that it's going to create fireworks."

While the personality and talent of Elisabeth Sladen was integral to the enduring popularity of the character, Sarah's backstory also lent itself to further development. *Sarah Jane Adventures* scriptwriter Phil Ford explains: "We knew about her background as an investigative journalist, we knew she was a career woman, and we even knew about her Aunt Lavinia [from *The Time Warrior* and *K-9 and Company*]. All of these things, along with a sadness about the way she left the Doctor, helped to give her a life outside the series."

Director Graeme Harper was happy to join the team on *The Sarah Jane Adventures*. "All of the other girls had fantastic emotional moments, but there was something special about the way the character of Sarah was written. I think the writers used her background as an investigative journalist and cast her as a Dr Watson figure."

Harper was delighted to be reunited with Elisabeth Sladen, whom he had first met when he was a production assistant on *The Seeds of Doom* in 1975. "Lis was a great natural beauty and a very talented actress who, by this time, had gained a

vast amount of experience. She had also gained a huge amount of real-life experience as a wife and mother and all that was brought to *The Sarah Jane Adventures*."

A 60-minute episode, *Invasion of the Bane*, was broadcast on 1 January 2007, reintroducing Sarah and establishing her adopted son Luke (Tommy Knight) and his friends. A full series had already been commissioned and duly followed in the autumn. Terrance Dicks, who as script editor was instrumental in the creation of the character in the first place, enjoyed what he saw. "*The Sarah Jane Adventures* was more like the *Doctor Who*s I used to do than many of the new episodes of *Doctor Who* – the episodes were half an hour long with a reasonably comprehensible and straightforward plot and a cliffhanger!"

Contractual issues prevented K-9 playing a major part in *The Sarah Jane Adventures*. A concurrently developed Australian series, simply titled *K-9*, made its British debut on Channel 5 in December 2010, but aside from giving the (now flying) dog a starring role it had only tenuous links to *Doctor Who*.

Torchwood ran for four series. The fifth series of *The Sarah Jane Adventures* was in production when Elisabeth Sladen sadly died in April 2011.

There have been no full-series spin-offs from *Doctor Who* since the US-based *Torchwood: Miracle Day* ended in September 2011, followed by the final *Sarah Jane Adventures* the following month. But other official spin-offs have continued online. In 2006, each story was preceded by a short 'TARDISode' prequel. While not every story has got that treatment since, there is a plethora of 'mini-sodes' like *Pond Life*, *Night and Day*, *The Great Detective*, Strax's idiosyncratic reports back to his native Sontar and other extensions to *Doctor Who*.

The Doctor's story remains too big for just one series.

Above left: The Doctor and Rose say goodbye at Bad Wolf Bay in *Doomsday*.

Above centre: Issue 7 of *Doctor Who Adventures*. This fortnightly title for younger fans was published by BBC Magazines and launched in March.

Above: Edited by Clayton Hickman and published by Panini in September, the first *Doctor Who Storybook* included contributions from Russell T Davies, Mark Gatiss, Tom MacRae, Steven Moffat and Robert Shearman.

SERIES TWO
(continued)
Executive producers:
Russell T Davies, Julie Gardner
Producer: Phil Collinson

New Earth
1, 22 August, 5-26 September, 7, 8 October, 3 November 2005
written by Russell T Davies
directed by James Hawes
15 April

Tooth and Claw
26 September – 27 October 2005
written by Russell T Davies
directed by Euros Lyn
22 April

School Reunion
23 August – 8 September, 8 October 2005
written by Toby Whithouse
directed by James Hawes
29 April

The Girl in the Fireplace
6 October, 12-27 October 2005
written by Steven Moffat
directed by Euros Lyn
6 May

Rise of the Cybermen/The Age of Steel
1-28 November, 5-6, 15-16 December 2005, 7, 11-13, 18 January, 18, 22 February, 9 March
written by Tom MacRae
directed by Graeme Harper
Rise of the Cybermen 13 May
The Age of Steel 20 May

The Idiot's Lantern
23 January, 7-23 February
written by Mark Gatiss
directed by Euros Lyn
27 May

The Impossible Planet/The Satan Pit
28 February – 11 April
written by Matt Jones
directed by James Strong
The Impossible Planet 3 June
The Satan Pit 10 June

Love & Monsters
19-31 March
written by Russell T Davies
directed by Dan Zeff
17 June

Fear Her
24 January – 6 February, 10, 15, 22-23 February
written by Matthew Graham
directed by Euros Lyn
24 June

Army of Ghosts/Doomsday
2 November, 16 November – 15 December 2005, 3-19 January, 20, 27 January, 9, 31 March, 11 April
written by Russell T Davies
directed by Graeme Harper
Army of Ghosts 1 July
Doomsday 8 July

SERIES THREE
Executive producers:
Russell T Davies, Julie Gardner
Producer: Phil Collinson

The Runaway Bride
4 July – 3 August, 19 October
written by Russell T Davies
directed by Euros Lyn
25 December

2007

On New Year's Day *The Sarah Jane Adventures* joined the ongoing *Torchwood*, BBC3's documentary series *Doctor Who Confidential* and children's magazine programme *Totally Doctor Who* in a diverse stable of programmes sharing the same inspiration. Later in 2007 there were yet more spin-offs in the animated serial *The Infinite Quest* and the mini-episode *Time Crash*, featuring Peter Davison alongside David Tennant.

The third season of the series proper began with the Doctor travelling alone. In *Smith and Jones* he met Martha (Freema Agyeman), a medical student whose fondness for him would soon become an infatuation.

In *The Shakespeare Code* the Doctor helped the playwright exorcise the alien Carrionites. *Gridlock*'s futuristic traffic jam highlighted Russell T Davies' talent for anchoring outlandish narratives in readily identifiable scenarios. Aficionados were amused by a cameo from the crab-like Macra, last seen in 1967.

A theme familiar from that year's *The Evil of the Daleks* was given a fresh interpretation by the Cult of Skaro in the impressively realised *Daleks in Manhattan* and *Evolution of the Daleks.*

Mark Gatiss played a character whose own genetic meddling had terrible consequences in *The Lazarus Experiment*. *Torchwood* writer Chris Chibnall made his *Doctor Who* debut with *42*, its title a reference to the real-time countdown dictating the Doctor and Martha's efforts to save a stricken spaceship.

While Chibnall had picked up on elements of the popular US drama *24*, Paul Cornell's subsequent story adapted his 1995 *Doctor Who* novel *Human Nature*. This memorable two-parter came to a conclusion with *The Family of Blood.*

Steven Moffat's *Blink* was a landmark for *Doctor Who*, bending narrative form as never before. Its time-jumping Weeping Angels would become arguably the most successful monsters created for the new series.

The harrowing trilogy *Utopia* and *The Sound of Drums/Last of the Time Lords* brought the season to a crescendo, reintroducing not only Captain Jack but also the Master. Played initially by Derek Jacobi, the evil Time Lord soon regenerated into a younger form played with startling insanity by John Simm. In the last episode Martha decided to stay on Earth.

The 72-minute *Voyage of the Damned* was lighter fare for Christmas. The casting of guest star Kylie Minogue helped attract 13.3 million viewers – a remarkable feat, even for the indefatigable *Doctor Who.*

A BRIEF HISTORY OF TIME

Right: Neill Gorton with one of the Judoon masks his company Millennium FX created for *Doctor Who* and *The Sarah Jane Adventures*. On the right is a Weevil mask from *Torchwood*.

Below: *The Sarah Jane Adventures* began with *Invasion of the Bane*, an episode written by Russell T Davies and Gareth Roberts. Terrance Dicks' novelisation of the story was published by BBC Children's Books on 1 November.

Bottom: Character Options' K-9 Supermag Construction Kit was a 75-piece magnetic playset released in March.

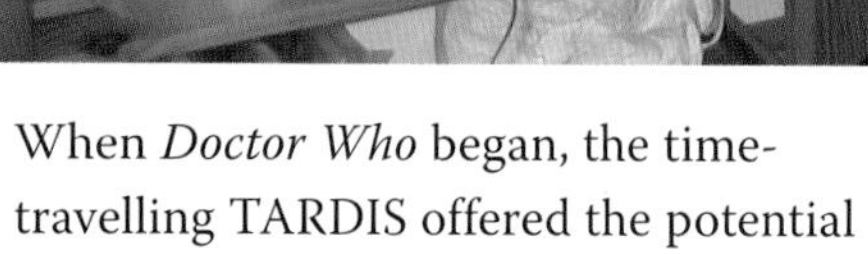

When *Doctor Who* began, the time-travelling TARDIS offered the potential for limitless adventure.

In *The Cave of Skulls*, the series' second episode, a cynical Ian Chesterton says, "Time doesn't go round and round in circles. You can't get on and off whenever you like in the past or the future."

"Really?" says the Doctor. "Where does time go, then?"

"It doesn't go anywhere," states Ian confidently. "It just happens and then it's finished."

Moments later Susan opens the TARDIS doors and Ian is proved wrong.

The first restrictions on time travel were introduced by story editor David Whitaker as early as 1964. In *The Aztecs* the Doctor is appalled by Barbara's intention to persuade the 15th century civilisation to abandon its practise of human sacrifice. "What you are trying to do is utterly impossible," he tells her in the first episode, *The Temple of Evil*. "I know. Believe me, *I know*."

The programme's caution about such historical milestones was partly an effort to appease Sydney Newman, who wanted *Doctor Who* to maintain a semi-educational stance wherever possible. *The Aztecs* presents a relatively modest drama against a cultural backdrop that remains unaltered at the end. Not every historical story worked so well within this dramatic straitjacket.

A pitfall of initiating, or otherwise significantly altering, chapters of text-book history was illustrated in the 1965 serial *The Myth Makers*. In the third episode, *Death of a Spy*, we learn that the Doctor is familiar with the Trojan

Right: The Macra were given a CGI make-over by The Mill for their fleeting return in *Gridlock*.

Below: Two storyboard illustrations showing the Doctor's jump from Brannigan's car in *Gridlock*, and a composite image from the finished sequence.

Below right: Martha wakes up to life in the slow lane when she is kidnapped by Milo and Cheen in *Gridlock*.

Bottom: *Gridlock* was the final story to feature the enigmatic Face of Boe. This deluxe figure featured a movable mouth and was released by Character Options in March.

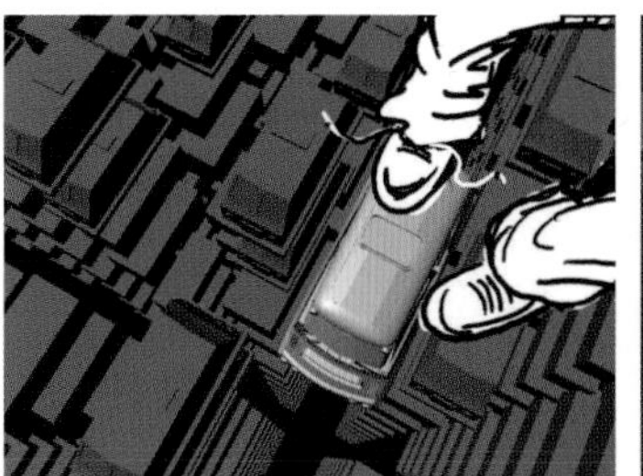

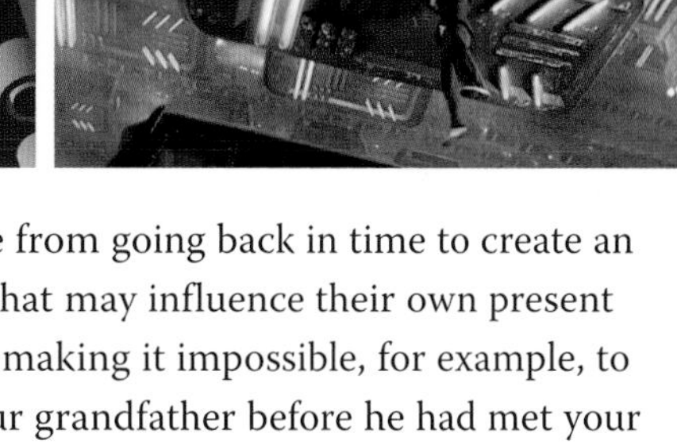

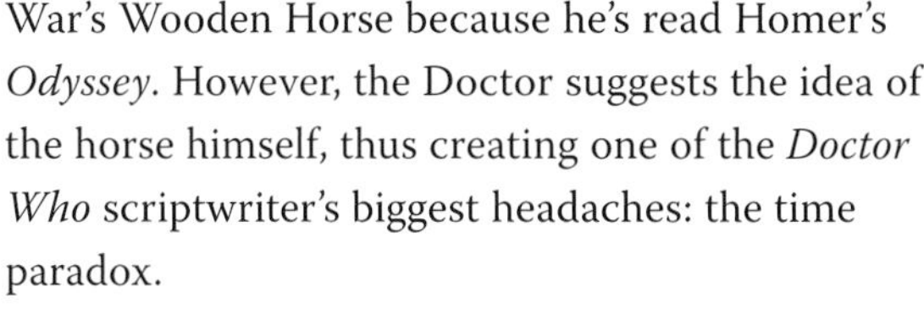

War's Wooden Horse because he's read Homer's *Odyssey*. However, the Doctor suggests the idea of the horse himself, thus creating one of the *Doctor Who* scriptwriter's biggest headaches: the time paradox.

How could Homer have written about the Wooden Horse if the Doctor hadn't made it happen? And how could the Doctor have suggested the horse if he hadn't remembered Homer writing about it?

By 1966 the programme's policy of non-interference in history was forgotten. The Doctor spends much of the fourth episode of *The Gunfighters* trying to *prevent* the gunfight at the OK Corral, surely one of the best-known events in 19th century American history.

What Whitaker called 'sideways' travel through time has been explored in stories that depict alternative events, such as *The Space Museum* (where the travellers jump a 'time track' and see their future selves as exhibits), *Pyramids of Mars* (where the Doctor shows Sarah the devastation Sutekh would wreak without their intervention) and 2008's *Turn Left* (where a Time Beetle influences Donna's choice on a car journey, resulting in the Doctor's death). These stories prove that the future, as well as the past, can be manipulated.

The tricky question of time paradoxes was eventually tackled in the 1972 story *Day of the Daleks*. In Episode Two the Doctor tells Jo that the 'Blinovitch Limitation Effect' would prevent anyone from going back in time to create an event that may influence their own present – thus making it impossible, for example, to kill your grandfather before he had met your grandmother. Unfortunately the Doctor never

Right: The Mill's CGI render of the mutated Professor from *The Lazarus Experiment*.

Below: Character Options' Martha Jones action figure was released in June.

Bottom centre: One of the prop guns used in *Evolution of the Daleks*. This combination of human and Dalek technology was partly inspired by the Thompson submachine gun that is synonymous with America's Prohibition era.

Bottom right: The Dalek-humans begin the invasion of Manhattan in *Evolution of the Daleks*.

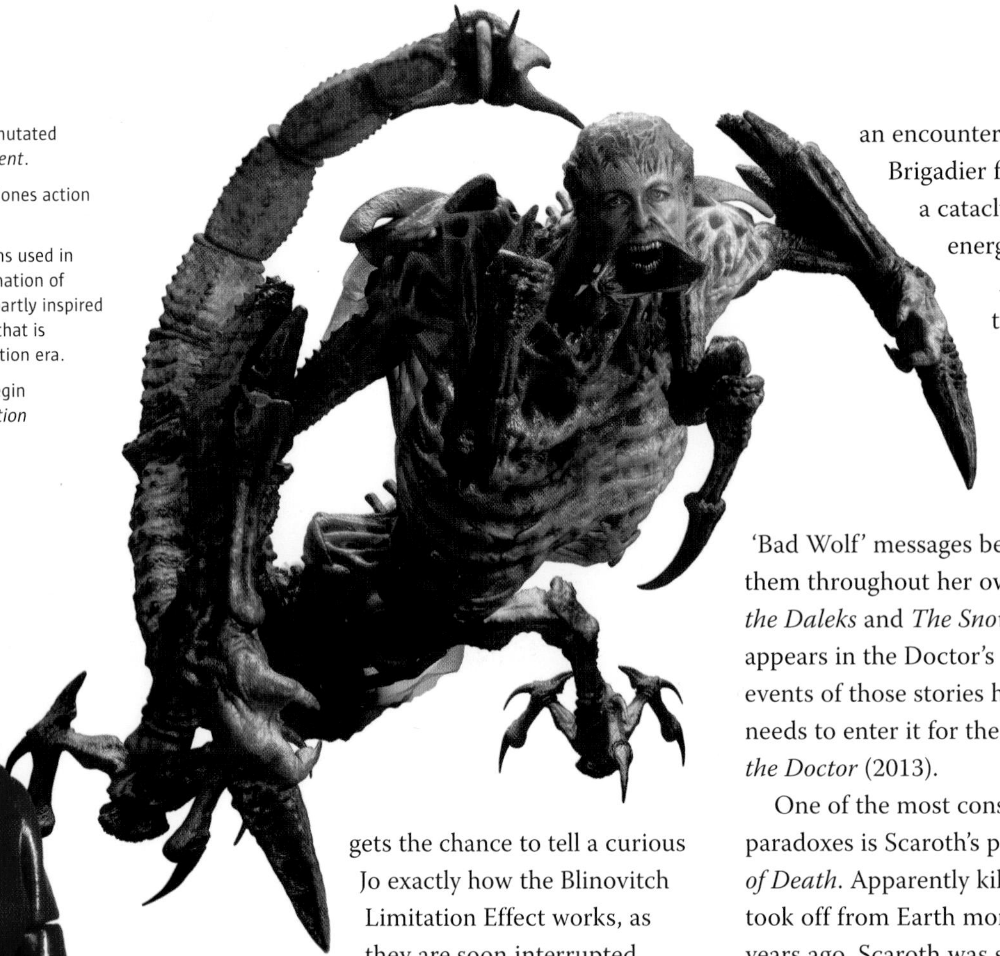

gets the chance to tell a curious Jo exactly how the Blinovitch Limitation Effect works, as they are soon interrupted.

Blinovitch's unexplained phenomenon would become ingrained in *Doctor Who* continuity, referred to in *The Three Doctors* as the First Law of Time, and more commonly as the inability to cross your own 'timeline'.

Only the Time Lords had the power to manage such transgressions when they engineered events in *The Three Doctors, The Five Doctors* and *The Two Doctors*. The consequences of unauthorised timeline activity were depicted in *Mawdryn Undead*, when an encounter between versions of the Brigadier from 1977 and 1983 caused a cataclysmic release of temporal energy.

Doctor Who has better tolerated another type of paradox, one that allows you to seed events on your own timeline. In Series One, Rose learns the implications of her 'Bad Wolf' messages before she decides to scatter them throughout her own history. In *Asylum of the Daleks* and *The Snowmen* (both 2012) Clara appears in the Doctor's time stream before the events of those stories help her to realise why she needs to enter it for the first time in *The Name of the Doctor* (2013).

One of the most conspicuous self-negating paradoxes is Scaroth's painstaking scheme in *City of Death*. Apparently killed when his spaceship took off from Earth more than 400 million years ago, Scaroth was splintered throughout time, advancing human civilisation to the point when the furthest 'splinter' could develop a time machine to travel back to a point just before take-off and prevent the accident. Thanks to the Doctor's intervention the ship explodes, as before, providing the spark that leads to the creation of life on Earth. Without the Doctor's intervention Scaroth would have got away, and there would have been no life on Earth to help him build his time machine in the first place.

Since the destruction of Gallifrey in the Time War, the programme has had to find new ways to manage the contradictions of time travel. The Blinovitch Limitation Effect is largely still observed – in *The End of Time* Part One (2009) the Doctor tells Wilf: "I can't go back inside my own timeline. I have to stay relative to the causal nexus." And in *The Name of the Doctor* the Doctor's visit to his own tomb triggers a debilitating paradox.

While Russell T Davies was showrunner, the idea of 'fixed points in time' was introduced to curb the Doctor's omnipotence in the absence of

Right: "Is that fair… that we stop?" John Smith (David Tennant) and Joan Redfern (Jessica Hynes) share their first kiss in *Human Nature*.

Below: One of the Scarecrow costumes from *Human Nature/The Family of Blood*. The masks were sculpted in clay by Rob Mayor, the co-director of Millennium FX.

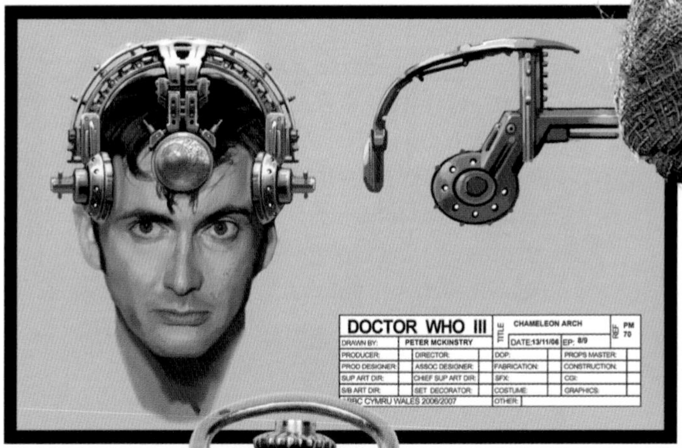

Below left: Peter McKinstry's concept art for the Chameleon Arch, the device that rewrites the Doctor's biology in *Human Nature*.

Bottom left: The Doctor's fob watch, which restores his memories and biological data in *The Family of Blood*.

Right: In *Blink*, Sally Sparrow (Carey Mulligan) searches an abandoned house for Kathy, unaware that her friend has been sent back in time by the Weeping Angels.

Below: The mask made by Millennium FX for the Weeping Angel played by Aga Blonska.

Below centre: In November Character Options released this action figure of Professor Yana from *Utopia*.

Below right: The *Radio Times* of 30 June-6 July came with two different covers, one featuring David Tennant as the Doctor and the other with John Simm as the regenerated Master, first seen at the end of *Utopia*.

the Time Lords. Even by going back in time, Rose was unable to prevent the death of Pete Tyler in *Father's Day*, and the Doctor ensured that Mount Vesuvius erupted in *The Fires of Pompeii* (2008). Despite the Doctor's best efforts to save her, Adelaide Brooke took her own life at the end of *The Waters of Mars* (2009); her inevitable death was almost certainly another fixed point in time.

Steven Moffat has done more than any other *Doctor Who* writer to embrace the complexities of time travel and its ensuing paradoxes. In both his 1996 short story *Continuity Errors* and 1999 spoof *The Curse of Fatal Death* the Doctor found

time-jumping resolutions to his problems. Series Two episode *The Girl in the Fireplace* saw him visit various points in the life of Madame de Pompadour. His trips to 18th century France were made through 'time windows' on a 51st century spaceship.

For *Blink*, first screened in 2007, Moffat attempted his boldest experiment yet. The story showed how the Doctor as a time-travelling protagonist could govern a non-linear narrative in which he barely appeared. Sally Sparrow (Carey Mulligan) compiles a dossier about the Weeping Angels – stone statues that send their victims back in time. When the Doctor is sent back by the Angels he uses Sally's information to warn her via DVD 'Easter eggs' and other hidden messages, allowing her to create the dossier that ultimately saves them both.

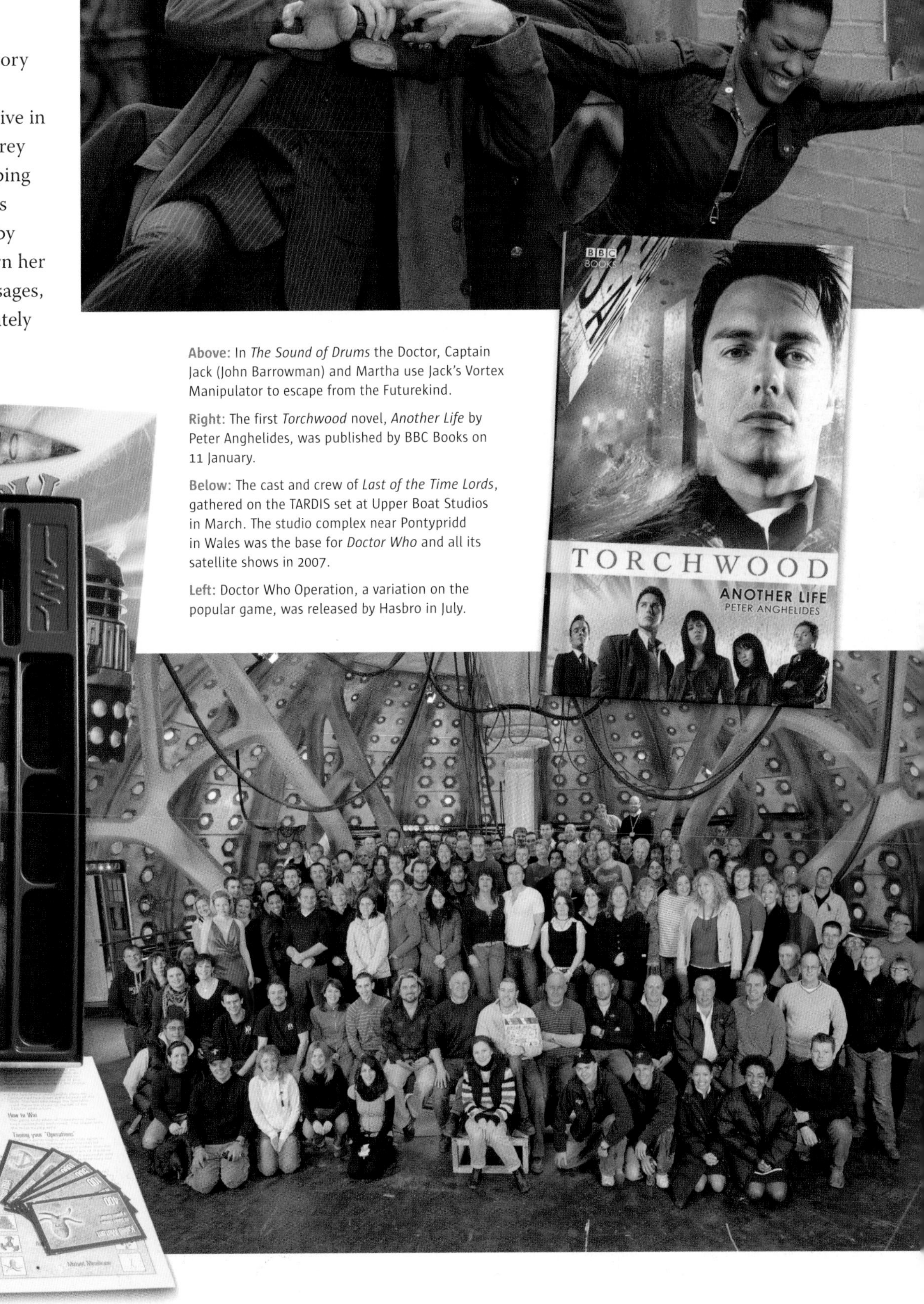

Above: In *The Sound of Drums* the Doctor, Captain Jack (John Barrowman) and Martha use Jack's Vortex Manipulator to escape from the Futurekind.

Right: The first *Torchwood* novel, *Another Life* by Peter Anghelides, was published by BBC Books on 11 January.

Below: The cast and crew of *Last of the Time Lords*, gathered on the TARDIS set at Upper Boat Studios in March. The studio complex near Pontypridd in Wales was the base for *Doctor Who* and all its satellite shows in 2007.

Left: Doctor Who Operation, a variation on the popular game, was released by Hasbro in July.

Left: The costume designed by Louise Page for Kylie Minogue in *Voyage of the Damned*.

Above: This illustration by production designer Edward Thomas shows his original intention for the computer-generated chasm beneath the damaged floor of Deck 31 in *Voyage of the Damned*.

Right: A publicity photo of the Doctor and plucky waitress Astrid Peth (Kylie Minogue), taken during the sequence where they face the robotic Host aboard the *Titanic*.

Below: A Peter McKinstry concept illustration of the *Titanic*, a luxury cruise-liner from the planet Sto.

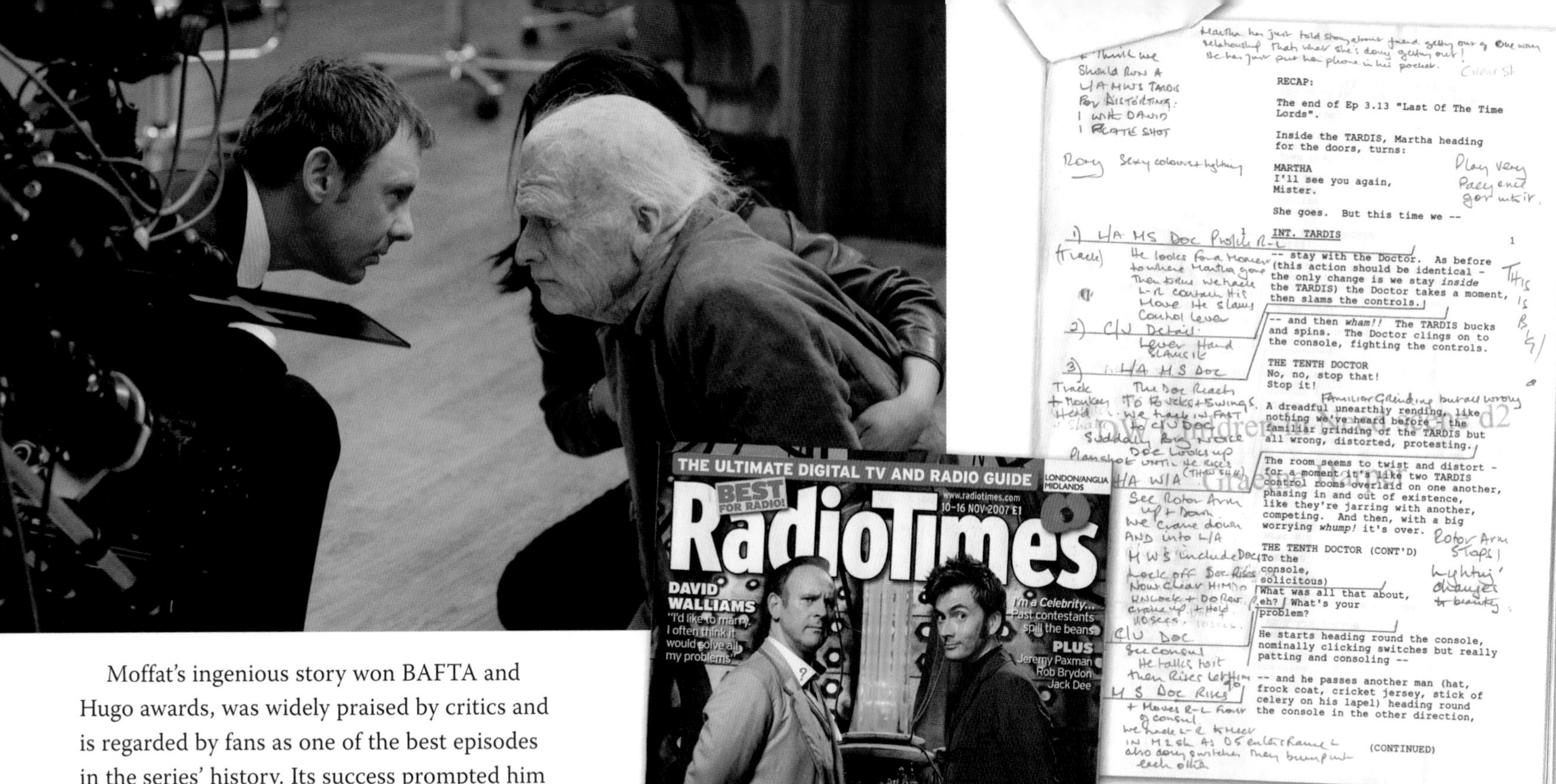

RECAP:

The end of Ep 3.13 "Last Of The Time Lords".

Inside the TARDIS, Martha heading for the doors, turns:

MARTHA
I'll see you again, Mister.

She goes. But this time we --

INT. TARDIS 1

-- stay with the Doctor. As before (this action should be identical - the only change is we stay *inside* the TARDIS) the Doctor takes a moment, then slams the controls.

-- and then *wham!!* The TARDIS bucks and spins. The Doctor clings on to the console, fighting the controls.

THE TENTH DOCTOR
No, no, stop that! Stop it!

A dreadful unearthly rending, like nothing we've heard before - the familiar grinding of the TARDIS but all wrong, distorted, protesting.

The room seems to twist and distort - for a moment it's like two TARDIS control rooms overlaid on one another, phasing in and out of existence, like they're jarring with another, competing. And then, with a big worrying *whump!* it's over.

THE TENTH DOCTOR (CONT'D)
(To the console, solicitous)
What was all that about, eh? What's your problem?

He starts heading round the console, nominally clicking switches but really patting and consoling --

-- and he passes another man (hat, frock coat, cricket jersey, stick of celery on his lapel) heading round the console in the other direction,

(CONTINUED)

Moffat's ingenious story won BAFTA and Hugo awards, was widely praised by critics and is regarded by fans as one of the best episodes in the series' history. Its success prompted him to attempt increasingly complex narratives in subsequent seasons.

"It's no big deal," he says. "It's just that I quite like time travel stories. *Doctor Who* is a show about a character with a time machine, so it's kind of irresistible."

Moffat took over from Russell T Davies as showrunner in 2010. His tenure is partly defined by the sort of paradox stories that script editors used to warn their writers about. Viewers needed to pay very careful attention to his Series Five finale *The Pandorica Opens/The Big Bang* (2010) and the Series Six arc that began with the Doctor's apparent death in *The Impossible Astronaut* (2011). As per one of the Doctor's most famous quotes from *Blink*, fans came to expect "a big ball of wibbly-wobbly, timey-wimey stuff" from Moffat's scripts.

"If you had a Time Lord brain, as opposed to your boring little human brain, you would understand the delicate surgery that can sometimes be done to alter an outcome in the way you'd like," he says. "When everyone else is shaking their fists and saying television is dumbing down I quite like the fact that there's a Saturday evening series that some people think is too complicated – and yet madly popular."

Which just goes to show that some time-travelling rules are made to be broken.

Isn't that where we came in?

Above left: John Simm, Freema Agyeman and David Tennant during recording of *The Sound of Drums* at Upper Boat Studios on 27 February.

Above: Director Graeme Harper's annotated camera script for *Time Crash*. This *Children in Need* mini-episode was written by Steven Moffat and recorded on 7 October.

Left: Peter Davison told the *Radio Times* that *Time Crash* "gives me such credibility with my children."

SERIES THREE
(continued)
Executive producers: Russell T Davies, Julie Gardner (unless otherwise stated)
Producer: Phil Collinson (unless otherwise stated)

Smith and Jones
8-25 August, 12-13 September, 2, 13 October, 7 November 2006, 17 January
written by Russell T Davies
directed by Charles Palmer
31 March

The Shakespeare Code
23 August – 15 September, 2, 13 October 2006
written by Gareth Roberts
directed by Charles Palmer
7 April

Gridlock
18-29 September, 2, 18 October, 7 November 2006
written by Russell T Davies
directed by Richard Clark
14 April

Daleks in Manhattan/Evolution of the Daleks
13 October – 23 November, 5, 8 December 2006
written by Helen Raynor
directed by James Strong
Daleks in Manhattan 21 April
Evolution of the Daleks 28 April

The Lazarus Experiment
3-19 October, 7 November 2006
written by Stephen Greenhorn
directed by Richard Clark
5 May

42
15-30 January, 8-9, 20 February, 1, 6, 13 March
written by Chris Chibnall
directed by Graeme Harper
19 May

Human Nature/The Family of Blood
27 November – 15 December 2006, 3-11, 17 January, 5, 23 February
written by Paul Cornell
directed by Charles Palmer
Producer: Susie Liggat
Executive producers: Phil Collinson, Russell T Davies, Julie Gardner
Human Nature 26 May
The Family of Blood 2 June

Blink
7, 20 November – 2, 9 December 2006, 9 January
written by Steven Moffat
directed by Hettie Macdonald
9 June

Utopia
15, 30 January – 23 February, 1 March
written by Russell T Davies
directed by Graeme Harper
16 June

The Sound of Drums/Last of the Time Lords
7 February – 19 March
written by Russell T Davies
directed by Colin Teague
The Sound of Drums 23 June
Last of the Time Lords 30 June

SERIES FOUR
Executive producers: Russell T Davies, Julie Gardner
Producer: Phil Collinson

Voyage of the Damned
9 July – 8, 21 August, 20 October
written by Russell T Davies
directed by James Strong
25 December

2008

The year began with the surprise return of a one-off companion and concluded with a story packed full of friends and allies.

The unique chemistry between David Tennant and Catherine Tate was reignited when the Doctor and Donna belatedly continued their association as *Partners in Crime*. The sobering truth behind *The Fires of Pompeii* gave way to the icy *Planet of the Ood*, in which the creatures' tragic back story was told.

The Sontaran Stratagem and its concluding episode *The Poison Sky* brought the war-obsessed clones back to Earth and a confrontation with UNIT (now renamed the Unified Intelligence Taskforce).

In *The Doctor's Daughter* the Time Lord's genetically engineered offspring Jenny was played by a real-life Doctor's daughter, Georgia Moffett, whose father is Peter Davison. In 2011 Georgia would become David Tennant's wife.

The Unicorn and the Wasp centred on a young Agatha Christie (Fenella Woolgar), before *Silence in the Library* and *Forest of the Dead* introduced the unfathomable Professor River Song (Alex Kingston). In another time-bending narrative from writer Steven Moffat, the character was shown at the end of her life, after many as yet unseen adventures with the Doctor.

The action in the next story confined a solo Doctor to a besieged tour bus on the planet *Midnight*, while *Turn Left* placed Donna in an alternate dystopian reality where the Doctor had died defeating the Racnoss in *The Runaway Bride*.

The final story of the series – *The Stolen Earth* and *Journey's End* – reunited former companions Rose, her mother Jackie, Mickey, Sarah Jane Smith, Martha and Captain Jack. The epic struggle with Davros (Julian Bleach) and his Daleks was almost too much for Donna, who returned to her family with no memory of the Doctor.

An Appreciation Index of 91 per cent – one of the greatest approval ratings in television history – meant that Phil Collinson left the series on a high. Susie Liggat was the first of numerous subsequent producers to fulfil the role from 2008 onwards.

On 29 October, David Tennant chose the National Television Awards to announce that he would be leaving *Doctor Who*. On Christmas Day *The Next Doctor* was revealed to be the deluded Jackson Lake (David Morrissey), who nevertheless joined forces with the real Doctor to defeat the Cybermen. An audience of 13.1 million forgave the teasing title.

CARNIVAL OF MONSTERS

Below: *Partners in Crime* featured the Adipose, creatures made from human body fat. The Mill created the Adipose using the MASSIVE (Multiple Agent Simulation System In Virtual Environment) program. Russell T Davies asked for a single fang to be added to the design.

Below centre: Character Options' 8" Pyrovile from *The Fires of Pompeii* was issued in August.

Below right: Ood Sigma from *Planet of the Ood* was released as a Character Options action figure in July, initially only available from Argos stores.

Doctor Who has always thrived on the imagined horrors that lurk in the shadows of a child's bedroom. It takes the mundane, like a doll, or the repulsive, like maggots and spiders, and makes them deadly.

Russell T Davies knew that the public wanted to see monsters in *Doctor Who*, and the revitalised series didn't disappoint. As well as the returning Ood, the 2008 episodes featured the Adipose, the Pyroviles and the Hath. A giant wasp and the Time Beetle played on insect phobias, while the Sontarans, Daleks and Cybermen appeared in the same season for the first time since 1975.

Doctor Who and its aliens were already indivisible when newspapers in the mid-1960s eagerly nominated a whole host of new creatures to challenge the Daleks' supremacy. Perhaps mindful of her battle with Sydney Newman over the first Dalek story, Verity Lambert recoiled from journalists' use of the word 'monsters'. On 18 January 1965, L Marsland Gander of the *Daily Telegraph and Morning Post* suggested that she

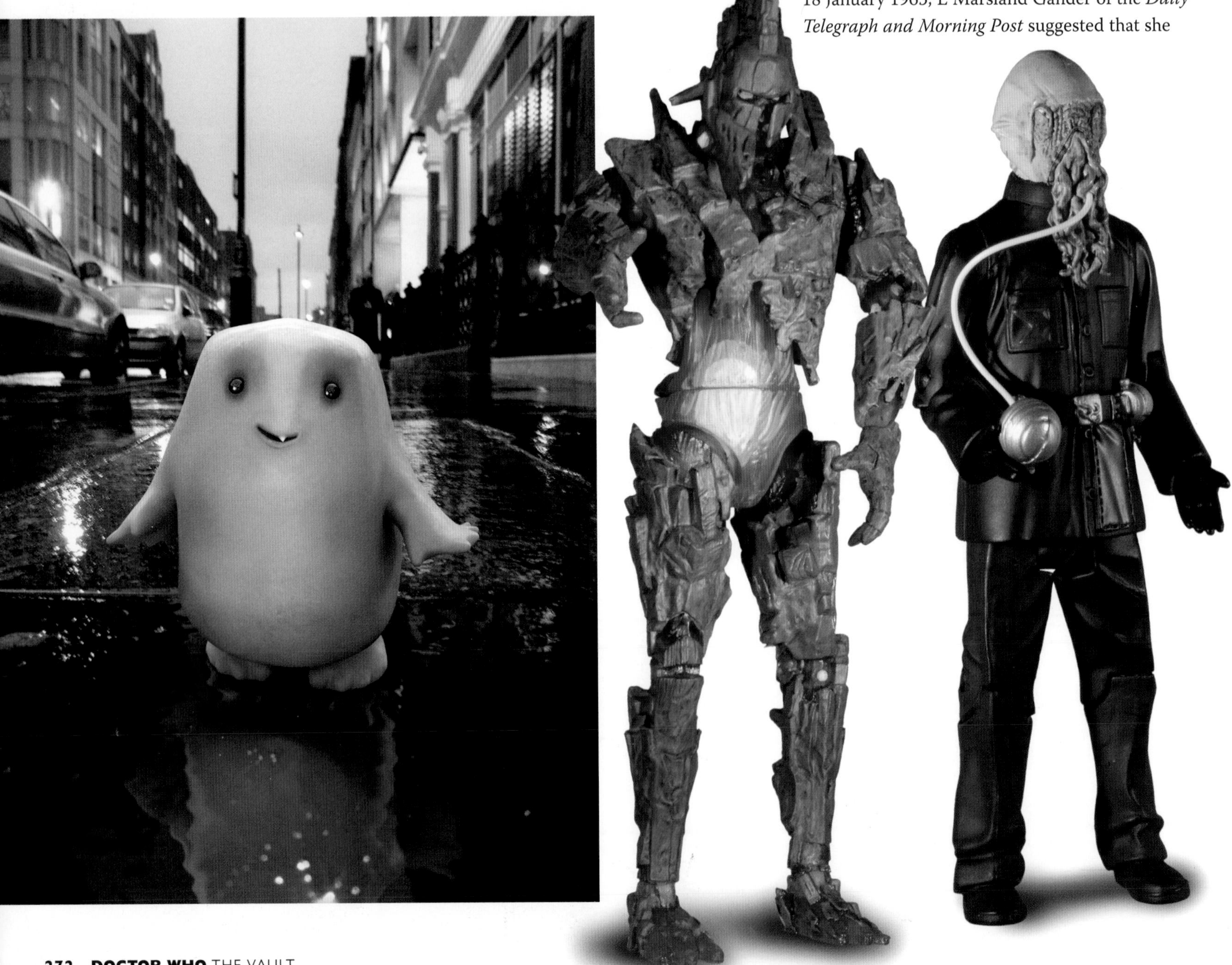

was trying to be kind to the actors inside the various costumes. "*Dr Who*, in fact, has created a new class of employment," he wrote. "To be the hind legs of an elephant may be the nadir of theatrical ambition, but to be a creature of another world is a loftier proposition."

At around the same time, Zarbi operator Jack Pitt took a pragmatic view of playing a Zarbi in *The Web Planet*. "It's hard work," he told the *Daily Sketch*, "but it pays the rent."

When the Doctor himself tried to justify his actions to the Time Lords at the end of *The War Games*, he presented images of Daleks, Cybermen, Ice Warriors and Yeti. His defence neglected the humanoid Dominators in favour of their robot servants the Quarks.

Yet ironically, the most successful and memorable monsters were often not monsters at all but more accurately *villains*. Part of that attraction, aside from an innate fascination with the frightening, lies in the dramatic possibilities offered by the articulate monster.

For Philip Hinchcliffe and Robert Holmes, the move towards the lone villain was a creative as well as a budgetary decision. In 1983 Hinchcliffe recalled: "It became apparent to me that the success of the programme depended to a large extent on getting good character actors, so we aimed at writing parts that would attract better actors

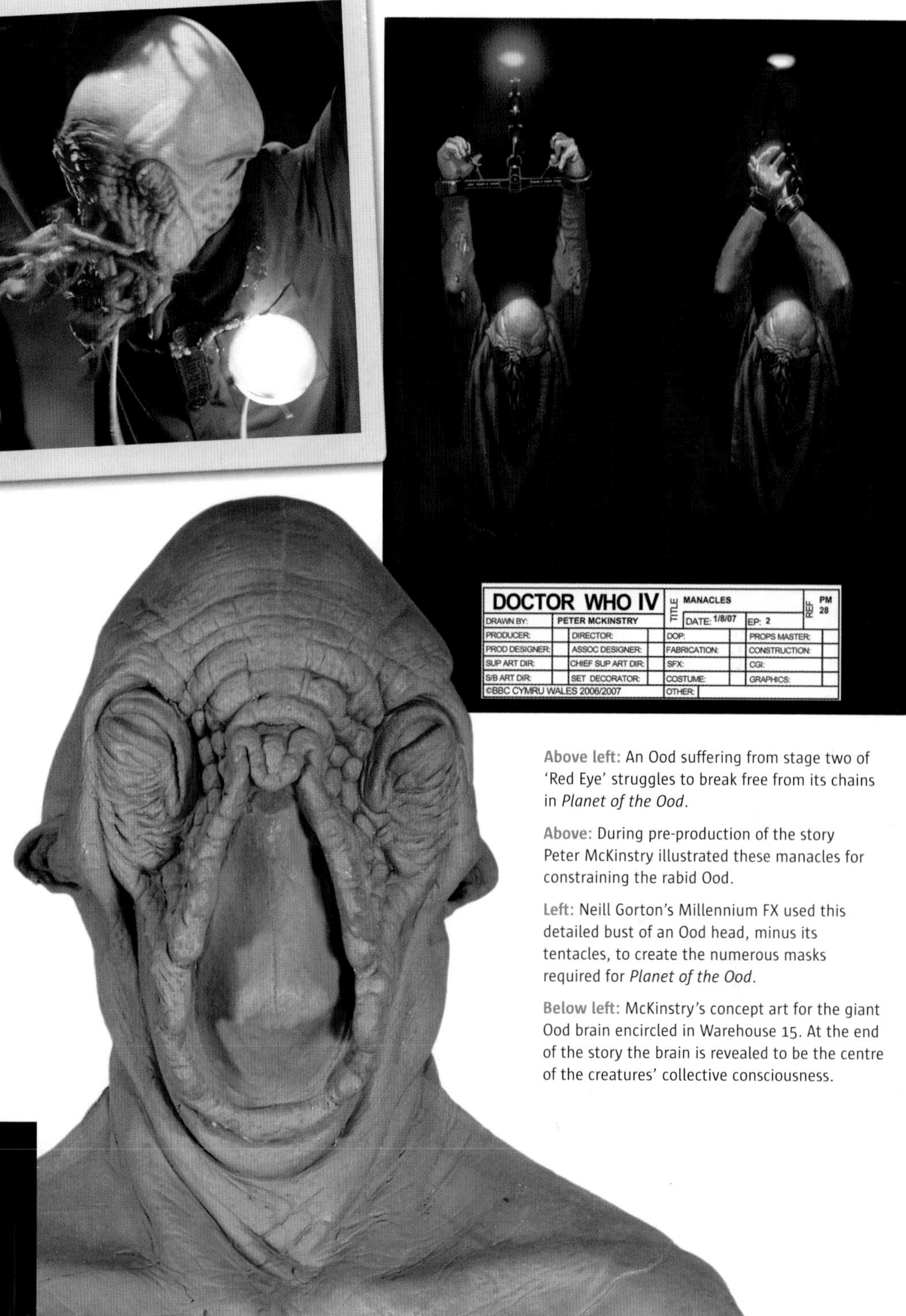

Above left: An Ood suffering from stage two of 'Red Eye' struggles to break free from its chains in *Planet of the Ood*.

Above: During pre-production of the story Peter McKinstry illustrated these manacles for constraining the rabid Ood.

Left: Neill Gorton's Millennium FX used this detailed bust of an Ood head, minus its tentacles, to create the numerous masks required for *Planet of the Ood*.

Below left: McKinstry's concept art for the giant Ood brain encircled in Warehouse 15. At the end of the story the brain is revealed to be the centre of the creatures' collective consciousness.

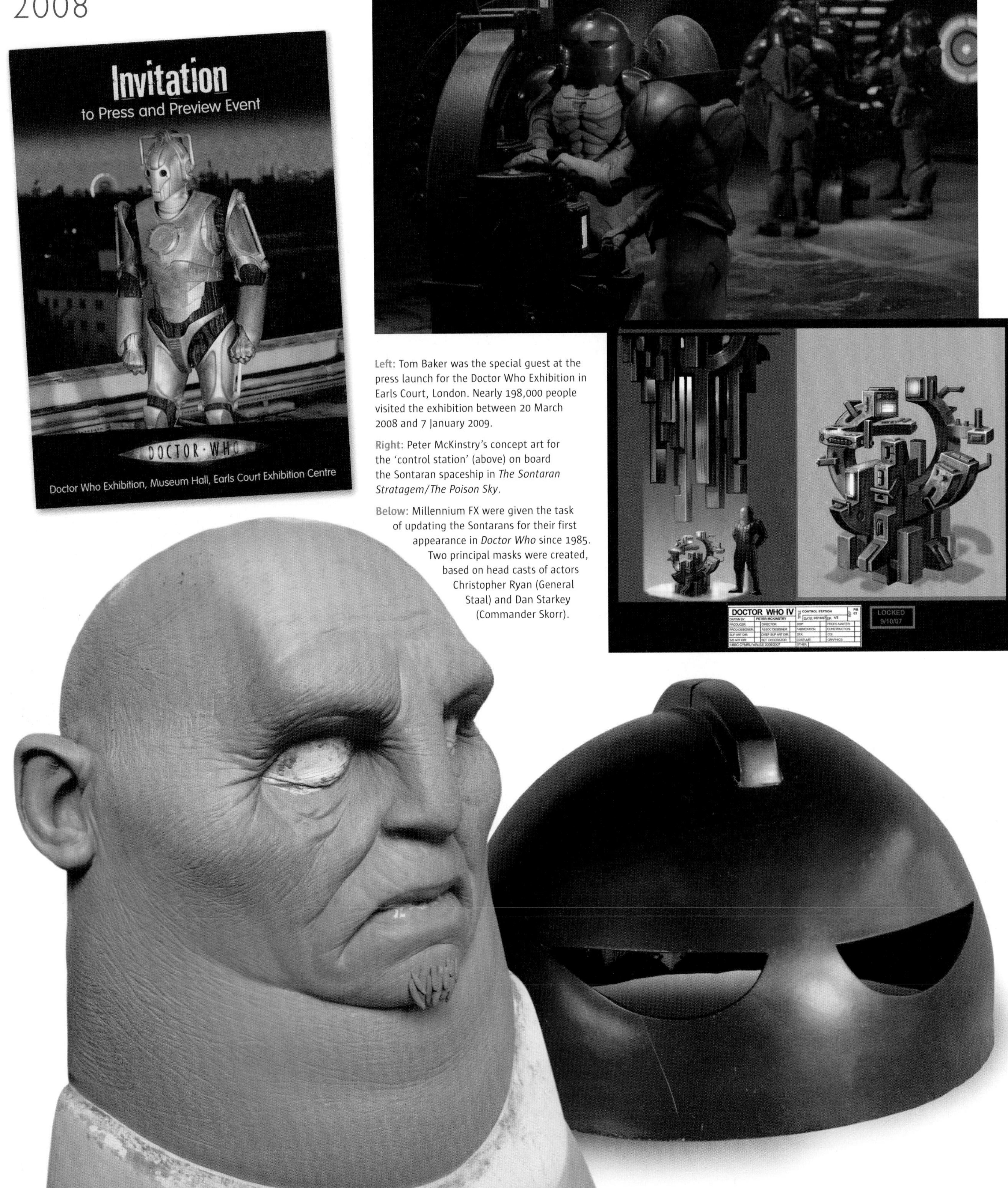

Left: Tom Baker was the special guest at the press launch for the Doctor Who Exhibition in Earls Court, London. Nearly 198,000 people visited the exhibition between 20 March 2008 and 7 January 2009.

Right: Peter McKinstry's concept art for the 'control station' (above) on board the Sontaran spaceship in *The Sontaran Stratagem/The Poison Sky*.

Below: Millennium FX were given the task of updating the Sontarans for their first appearance in *Doctor Who* since 1985. Two principal masks were created, based on head casts of actors Christopher Ryan (General Staal) and Dan Starkey (Commander Skorr).

Far left: Once removed from the printed polythene bag it came in, Issue 397 of *Doctor Who Magazine* (cover dated 23 July) reflected the return of Russell T Davies' Bad Wolf story arc in *Turn Left*.

Left: The first edition of *Doctor Who: The Writer's Tale* was published by BBC Books on 25 September. The 512-page book comprised highlights of the illuminating e-mail correspondence between Russell T Davies and *Doctor Who Magazine* writer Benjamin Cook.

Below: Character Options' Hath action figure was released in December.

Bottom: Martha and the Hath get acquainted in *The Doctor's Daughter*.

who could see a thumping good role for them to play. When that happens you can raise the level of illusion in the programme almost to infinity."

Examples of this approach to casting chief monsters and villains are evident not only in stories that Hinchliffe produced (*Genesis of the Daleks, Pyramids of Mars, The Brain of Morbius, The Talons of Weng-Chiang* and many others) but throughout the programme's history. Julian Glover's Scaroth matches Tom Baker's Doctor for wit and presence in *City of Death*, just as Anthony Stewart Head's Krillitane Mr Finch makes a powerful foil for David Tennant in *School Reunion*.

Andrew Cartmel, one of Robert Holmes' successors as script editor, regrets what he considered to be *Doctor Who*'s over-reliance on less sophisticated monsters in the late 1980s. "John Nathan-Turner's philosophy on monsters was self-evident. He said it was what the audience expected – he might actually have said it was what the kiddies expected. I didn't agree with this. There weren't any monsters in *Ghost Light* so that's why he asked for the husks to menace Ace. And of course they're the worst thing in it. They're terrible."

The Haemovores in *The Curse of Fenric* are surely amongst the series' most successful creature designs, but Cartmel feels this is another instance where the monsters compromised a story. "You needed a big bad in *Fenric*, you couldn't just have had the girl vampires," he says. "I don't remember the exact discussions we had on that, but I do remember the way my heart sank when I saw the Haemovores. I thought, 'Oh well, here we go again...' If other people like them then I'm delighted, but they don't work for me."

Cartmel reserves his harshest criticism for the Cheetah People in the original series' final televised story, *Survival*. "We were pushing for them to be entirely humanoid," he says. "We just wanted the actors to wear contact lenses and fangs, partly because we knew that if we went that way there'd be less chance of them somehow becoming laughable. At John's insistence they had to be made full-blown creatures. The result is the debacle we see before us today."

The obligation felt by successive producers and their writers to come up with new monsters has, on occasions, overstretched even the most resourceful designer. In 1979 the Mandrels might have prompted *The Sun* to ask "Is This Dr Who Monster Far Too Scary?" but that wasn't the question in most viewers' minds when they lumbered on screen in *Nightmare of Eden*. In his survey of *Doctor Who*'s monsters for the BBC's *Did You See..?* on 14 March 1982, Gavin Scott rather charitably put the Mandrels' failure down to their walk.

"I suspect the time will come when the psycho-historians of the future will regard the BBC's *Doctor Who* programmes as a goldmine for discovering the true nature of late 20th century neuroses," said Scott. "What other cultural enterprise has worked so hard and so long... to come up with every conceivable variation on the theme of alien beings who are both repulsively horrible and irresistibly attractive?"

Scott's comparison between the Sontarans and Humpty Dumpty was far from original, but his description of their appearance as a "skilled distortion of the comfortable images of childhood" went some way towards explaining the enduring appeal of the diminutive brutes.

The idea that a Sontaran's head would be much the same shape as its domed space helmet was initially a joke on the part of costume designer James Acheson. However, the troll-like appearance he devised was kept not only for the creatures' first appearance in *The Time Warrior* but also their subsequent stories.

"In all the four years, that's the only time I've dug out an old episode to remind me of what a

Top: The Firestone, the most prized jewel of Lady Clemency Eddison's collection in *The Unicorn and the Wasp*.

Top centre: Lady Eddison (Felicity Kendal) is unaware that the Firestone is a Vespiform telepathic recorder.

Top right: Panini's *Doctor Who Storybook* was published in August and included the Ice Warrior story *Cold* by Mark Gatiss. The cover illustration was by Alister Pearson.

Above: The Mill's CGI render of the shape-shifting Vespiform, created for *The Unicorn and the Wasp*.

Above: The Doctor prepares to run from "the shadows that melt the flesh" in *Silence in the Library*.

Left: "Spoilers!" Professor River Song (Alex Kingston) refuses to let the Doctor read her diary in *Silence in the Library*.

Below: River Song's diary is a detailed prop filled with sketches and notes supplied by the *Doctor Who* Art Department.

Right: One of the spacesuit costumes designed by Louise Page for *Silence in the Library/Forest of the Dead*.

Far right: The animatronic Time Beetle prop, created for *Turn Left* by Millennium FX.

Right: On the planet Shan Shen Donna falls under the spell of a fortune teller and the Time Beetle leaps onto her back.

Below: Character Options' plush Adipose was released in October and became one of the most popular *Doctor Who* toys of Christmas 2008.

Below right: In July the *Radio Times* hailed the return of Davros (Julian Bleach) in *The Stolen Earth*. An eight-page *Doctor Who* feature included interviews with Neill Gorton, who created the new Davros mask, and Bleach, who described the character as "a cross between Hitler and Stephen Hawking."

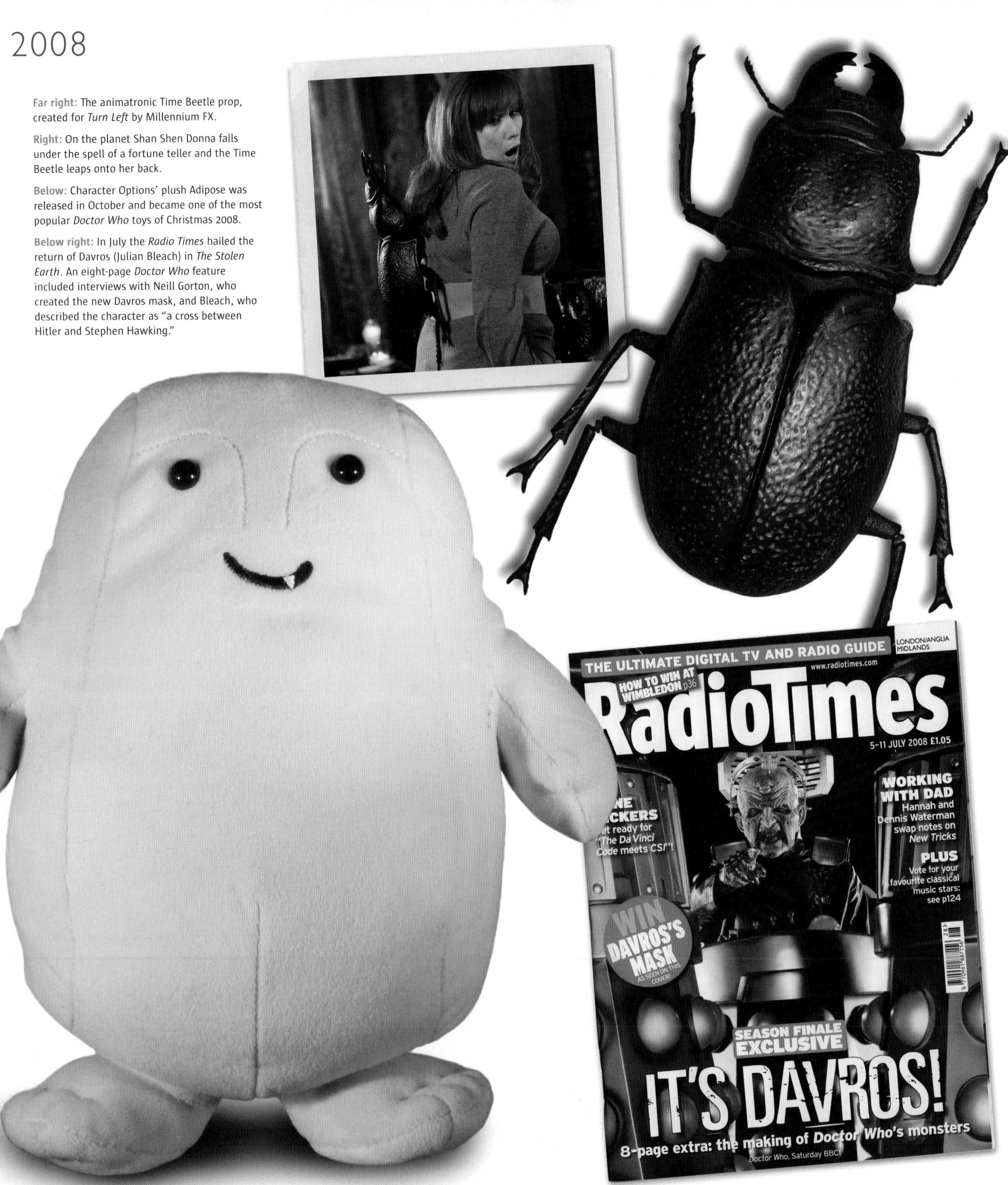

Far left: Issue 400 of *Doctor Who Magazine* was published in September and included a free poster of David Tennant reading Issue 1 of *Doctor Who Weekly* on the TARDIS set.

Left: Russell T Davies had first described the Cybershades in *The Writer's Tale*: "Cyberman heads in flowing black robes, like wraiths, sort of creepy half-Cybermen." The stooped bio-mechanoids appeared in the Christmas special, *The Next Doctor*.

Below: An original Cybershade costume, designed by Louise Page for *The Next Doctor*.

monster is like," said Russell T Davies in 2008. "*The Time Warrior* is great. It's exactly the sort of story we'd do now."

The Sontarans of Series Four were led by General Staal (Christopher Ryan). His costume retained the distinctive collar and helmet from Acheson's design, and Helen Raynor's script for *The Sontaran Stratagem*/*The Poison Sky* honoured Robert Holmes' original conception of the creatures as militaristic monomaniacs. In *The Poison Sky* their unforgiving code of conduct and fanatical pursuit of glory leads the Doctor to describe them as "The finest soldiers in the galaxy, dedicated to a life of warfare."

For a race of clones grown in batches of millions, the Sontarans' appearance was curiously inconsistent throughout the 1970s and '80s. With advances in all areas of alien manufacturing, both physical and computer-generated, monster-making in the 21st century is less of a lottery. Since 2005, Neill Gorton's company Millennium FX has produced new creatures and updated versions of classic monsters – including the Cybermen, Sontarans, and Zygons – while The Mill has animated creatures such as the werewolf in *Tooth and Claw* and the army of Adipose for *Partners in Crime*.

However sophisticated the make-up and visual effects, the most successful monsters have been those whose condition or ambitions are somehow relevant to our society. The most potent example remains the terrible price the Daleks paid for surviving a nuclear war. The Cybermen echoed Kit Pedler's fear of extreme spare-part surgery, before *Rise of the Cybermen* reconfigured them as a warning about homogeneity in the digital age. The Sontarans – warriors who all look and think and act the same – exploit a similar anxiety about the loss of identity.

All of these, and so many other *Doctor Who* monsters, offer a distorted reflection of our own worst fears.

SERIES FOUR (continued)
Executive producers: Russell T Davies, Julie Gardner (unless otherwise stated)
Producer: Phil Collinson (unless otherwise stated)

Partners in Crime
4-23 October, 19, 20-29 November, 18 December 2007
written by Russell T Davies
directed by James Strong
5 April

The Fires of Pompeii
13 September – 2 October, 20 October 2007
written by James Moran
directed by Colin Teague
12 April

Planet of the Ood
21 August – 7 September, 16 November 2007
written by Keith Temple
directed by Graeme Harper
Producer: Susie Liggat
19 April

The Sontaran Stratagem/The Poison Sky
23 October – 22 November, 5, 18 December 2007, 24 January, 29 February
written by Helen Raynor
directed by Douglas Mackinnon
Producer: Susie Liggat
Executive producers: Phil Collinson, Russell T Davies, Julie Gardner
The Sontaran Stratagem 26 April
The Poison Sky 3 May

The Doctor's Daughter
11-21 December 2007, 7-24 January
written by Stephen Greenhorn
directed by Alice Troughton
10 May

The Unicorn and the Wasp
8-21 August, 6-7 September, 16 November 2007
written by Gareth Roberts
directed by Graeme Harper
Producer: Susie Liggat
Executive producers: Phil Collinson, Russell T Davies, Julie Gardner
10 May

Silence in the Library/Forest of the Dead
15 January – 14 February, 19-20 March
written by Steven Moffat
directed by Euros Lyn
Silence in the Library 31 May
Forest of the Dead 7 June

Midnight
27 November – 11 December 2007
written by Russell T Davies
directed by Alice Troughton
14 June

Turn Left
22 November – 8 December 2007, 18, 24, 31 January, 20 March
written by Russell T Davies
directed by Graeme Harper
Producer: Susie Liggat
Executive producers: Phil Collinson, Russell T Davies, Julie Gardner
21 June

The Stolen Earth/Journey's End
31 January, 18 February – 31 March, 1 May
written by Russell T Davies
directed by Graeme Harper
The Stolen Earth 28 June
Journey's End 5 July

THE SPECIALS
Executive producers: Russell T Davies, Julie Gardner

The Next Doctor
7 April – 3 May
written by Russell T Davies
directed by Andy Goddard
Producer: Susie Liggat
25 December

2009

David Tennant's final year as the Doctor comprised four specials, beginning in Easter 2009 with *Planet of the Dead*, written by Russell T Davies and Gareth Roberts. The first episode of *Doctor Who* to follow *Torchwood* into high definition featured extensive location recording in Dubai. Michelle Ryan guest starred as Lady Christina de Souza, a thief trapped on a double-decker bus with the Doctor as it travelled to an arid planet devastated by giant flying ray creatures.

For *The Waters of Mars*, Davies was credited alongside *Sarah Jane* and *Torchwood* writer Phil Ford. This episode swapped the desert of San Helios for the confines of Bowie Base One in 2059. Having defeated relentless water creature the Flood, the uncharacteristically arrogant Doctor fails in his attempt to change the destiny of Captain Adelaide Brooke (Lindsay Duncan). Time itself put history back on track and the Doctor firmly in his place.

The Waters of Mars was broadcast in November, the same month that viewers of digital television could watch *Dreamland*. This animated story consisted of six short instalments written by Ford and set in 1950s America.

Uniquely for the current series, the two episodes of *The End of Time* kept the same title qualified by 'Part One' and 'Part Two'. The Master returned from the dead, and so did Gallifrey. The Doctor's planet had been destroyed in the Time War, but now Rassilon (Timothy Dalton) and the Time Lords launched a ruthless scheme to cheat death at the expense of all creation.

Part Two was broadcast on 1 January 2010. At its conclusion, the Doctor sends the Time Lords to oblivion before sacrificing himself to save Donna's grandfather Wilf (Bernard Cribbins). Dying from radiation poisoning, the Tenth Doctor visits his past friends and companions for the last time before returning to the TARDIS to regenerate.

David Tennant was not the only departing member of the team. Executive producer Julie Gardner left BBC Wales (where she was Head of Drama) to take up a post at BBC America. Russell T Davies, the chief architect of *Doctor Who*'s revival and incredible five-year success, also moved on. For the first time in the show's history the core production team as well as the regular cast were all about to change.

The End of Time was the end of an era.

PICTURE PERFECT

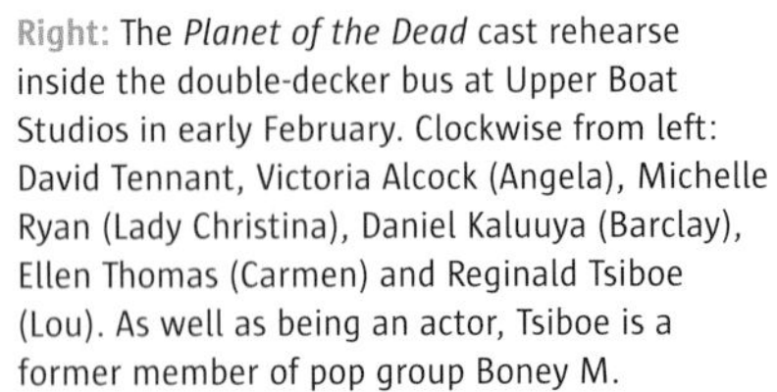

Right: The *Planet of the Dead* cast rehearse inside the double-decker bus at Upper Boat Studios in early February. Clockwise from left: David Tennant, Victoria Alcock (Angela), Michelle Ryan (Lady Christina), Daniel Kaluuya (Barclay), Ellen Thomas (Carmen) and Reginald Tsiboe (Lou). As well as being an actor, Tsiboe is a former member of pop group Boney M.

Below right: The Doctor puts a hammer to the Cup of Athelstan, the key to getting back through the wormhole in *Planet of the Dead*.

Below: The Cup of Athelstan, stolen from London's International Gallery by Lady Christina. Made of pure gold, it proved to be ideal for conducting electricity between an anti-gravity clamp and the steering wheel of a London bus.

Planet of the Dead, the first of the 2009 specials, was a milestone for several reasons. *Doctor Who*'s 200th story was also the first to be recorded in high definition.

In February, co-executive producer Julie Gardner told *Doctor Who Magazine*: "This is something we've been moving towards ever since *Doctor Who* resumed production in 2005, and we've been planning this for many months now, in conjunction with the Art Department, The Mill and all the design teams."

High definition was the latest stage in a technological journey that began in 1963 when *Doctor Who* was produced with facilities that were inadequate even by the standards of the day. In convention appearances and interviews during the early years of fandom, some actors recalled that the programme had been a live production in the 1960s. This was an easy mistake to make: while never actually broadcast live, *Doctor Who* was largely performed 'as live'. Video editing was so expensive and clumsy that only the most serious line fluffs or collapsing scenery warranted an unscheduled break in recording.

"You tried to get the stuff onto a telerecording in a fully edited form," says Frank Cox, who directed episodes of *Inside the Spaceship* and *The Sensorites* in 1964. "It was rather like shooting a stage show, in that the cameras were ranged around the outside of the set, so you couldn't

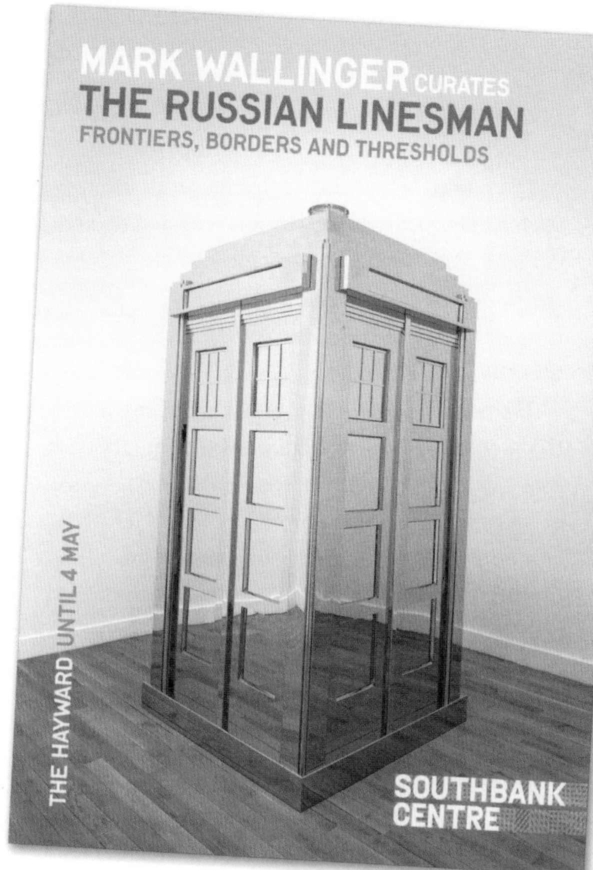

Far left: Mark Wallinger's mirrored 'TARDIS' art installation was first exhibited at the Venice Biennale in 2001. In February 2009 it became part of his latest exhibition, at London's Hayward Gallery. "The works in the exhibition use illusion, artifice and dislocating devices to look at our accidental time and place in the world afresh," he explained.

Left: The second instalment of *Hornet's Nest*, a five-part audio serial that was released by BBC Audiobooks between September and December. This was Tom Baker's first dramatic portrayal of the Doctor since *Logopolis* in 1981.

Below: The bus used in *Planet of the Dead* was given the number 200 in recognition of the story's status.

really get in for the sort of impact shots that the thing probably needed. Looking back on that first episode, I was quite horrified at the naïveté of the shooting. It was very simple. I think we rehearsed from about ten in the morning, and at seven in the evening we went for it!"

The relatively primitive quality of 1960s broadcast standards probably did *Doctor Who* a few favours. When the series began, viewers would have watched it on television screens no larger than 23 inches – most made do with a considerably smaller picture. An analogue black and white image composed of just 405 horizontal lines hid a multitude of deficiencies in costume, make-up and set design.

Doctor Who's first 625-line recording was Episode 3 of *Enemy of the World*, broadcast in January 1968. From *Spearhead from Space* in January 1970 viewers could watch the series in colour, although the exorbitant price of television sets meant that many took their time to upgrade to these innovations. Viewers in London could hear the Sylvester McCoy episodes in stereo from *Remembrance of the Daleks* onwards, and in the show's enforced break digital television enabled the next advancements.

The TV Movie was shot on 35mm film, and from this point *Doctor Who* was exclusively made using the single-camera technique, a movie industry method that allowed a greater flexibility and fluidity than was possible with the relatively static cameras common to television studios.

"The way *Doctor Who* used to be made with the old multi-camera system was a financial choice," says Graeme Harper. "To make it single-camera, with the technology that's available now, is much more expensive. Nowadays multi-camera studios are relatively rare. Soaps still do things that way, as a means of getting though the material quickly, but it's not the best way to make a programme. Soaps compete with each other, and they all use the same format, but *Doctor Who* has to compete with other drama programmes such as *Downton Abbey*, which is also shot single-camera. When you shoot single-camera you have the opportunity to create something much more interesting – the chance to get each shot looking as good as you can."

Harper first worked in high definition, a system that generally enables between 720 and 1080 scan lines, in 2006. "It didn't excite me at first, because of the problems it presented," he says. "The cameras showed up all the blemishes, every piece of background detail."

The technology rapidly improved, offering more control over depth of field, and in 2009 James Strong became the first director to apply the technology to an entire episode of *Doctor Who*. *Planet of the Dead* was broadcast simultaneously on BBC1 and BBC HD on 11 April. The BBC1 audience of 9.5 million dwarfed the 205,000 on BBC HD, but even this modest figure made it the channel's most watched programme to date. The audience for high definition was growing quickly, and when *The End of Time* Part Two was broadcast on BBC HD on 1 January 2010 the audience for *Doctor Who* had more than doubled to 480,000.

"Like it or not, episodes of *Doctor Who* made in the old style just aren't going to cut it," says showrunner Steven Moffat. "The gap between *Star Trek* and *Doctor Who* really widened with the new films, which have extraordinary production values and incredible CGI. Kids will watch those *Star Trek* films on their big widescreen HD televisions, and the next thing they watch could well be one of our episodes of *Doctor Who*. That's what we have to try to compete with."

Top left: The final *Doctor Who Storybook* from Panini was published on 31 August, and included contributions from television scriptwriters Mark Gatiss, Matt Jones, James Moran and Keith Temple.

Above left: The Doctor and Adelaide (Lindsay Duncan) use the robot Gadget to escape from Andy and Tarak in *The Waters of Mars*.

Left: The Gadget prop was designed and operated by Colin Newman and Lynn Walters.

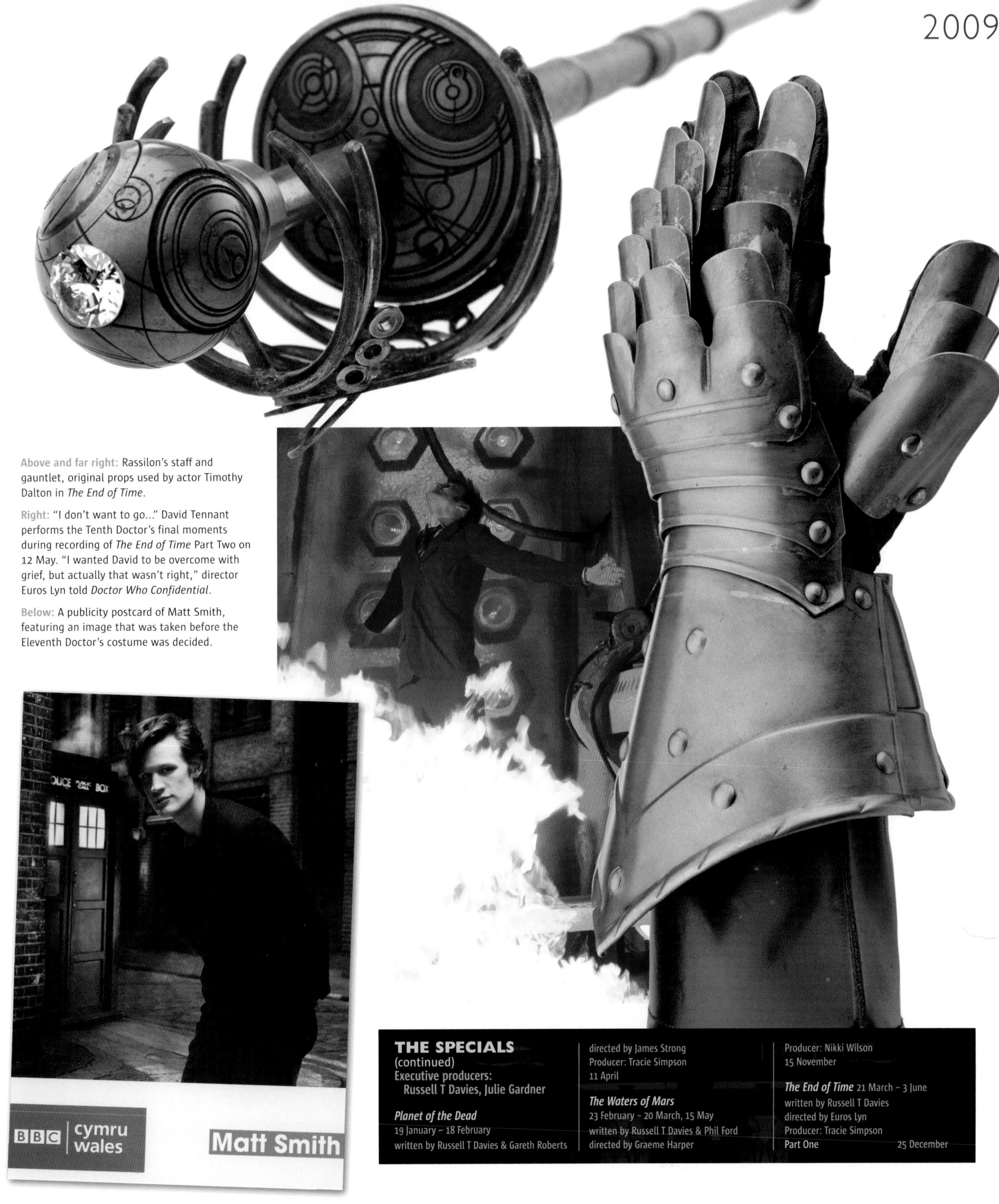

Above and far right: Rassilon's staff and gauntlet, original props used by actor Timothy Dalton in *The End of Time*.

Right: "I don't want to go..." David Tennant performs the Tenth Doctor's final moments during recording of *The End of Time* Part Two on 12 May. "I wanted David to be overcome with grief, but actually that wasn't right," director Euros Lyn told *Doctor Who Confidential*.

Below: A publicity postcard of Matt Smith, featuring an image that was taken before the Eleventh Doctor's costume was decided.

THE SPECIALS
(continued)
Executive producers:
Russell T Davies, Julie Gardner

Planet of the Dead
19 January – 18 February
written by Russell T Davies & Gareth Roberts
directed by James Strong
Producer: Tracie Simpson
11 April

The Waters of Mars
23 February – 20 March, 15 May
written by Russell T Davies & Phil Ford
directed by Graeme Harper
Producer: Nikki Wilson
15 November

The End of Time 21 March – 3 June
written by Russell T Davies
directed by Euros Lyn
Producer: Tracie Simpson
Part One 25 December

2010

An entire edition of *Doctor Who Confidential* was devoted to revealing the Eleventh Doctor. It was broadcast by BBC1 on 3 January 2009 and featured *Doctor Who*'s new head writer and showrunner Steven Moffat. For Series Five he would be joined by two other executive producers, Piers Wenger and Beth Willis.

At 26 Matt Smith was the youngest actor ever cast as the Doctor, but he studied the series' legacy while preparing for the role. Patrick Troughton's performance in *The Tomb of the Cybermen* would inform the geekish eccentricity that Smith applied to his own portrayal.

At the beginning of 2010 the new Doctor, and his newly regenerated TARDIS, arrived in *The Eleventh Hour*. The story also introduced Amy Pond (Karen Gillan) and her fiancé Rory Williams (Arthur Darvill), as well as a narrative arc about a crack in time and space.

The Beast Below was set on a huge starship carrying the migrating population of the UK. *Victory of the Daleks* depicted the Doctor's devious foes as apparent allies of Winston Churchill and culminated with the unveiling of the new paradigm Daleks – a bold and colourful reworking of the classic design that divided opinion.

The Weeping Angels and the inscrutable River Song returned in *The Time of Angels* and *Flesh and Stone*. *The Vampires of Venice* was a more traditional story, with Croatia standing in for 16th century Italy.

The surreal *Amy's Choice* included an unsettling performance by Toby Jones as the Dream Lord. Two-parter *The Hungry Earth* and *Cold Blood* resurrected the Silurians (glossing over their inaccurate nomenclature) in a homage to the best of early '70s *Doctor Who*. Screenwriter Richard Curtis then contributed *Vincent and the Doctor*, which sensitively depicted the inner torment of Van Gogh (Tony Curran).

Matt Smith showed off his football skills in *The Lodger*, a whimsical tale set in contemporary Colchester. *The Pandorica Opens* and *The Big Bang* brought the season to a densely plotted finale. Amy finally married Rory and, although they stayed with the Doctor, neither played a prominent role in the next special, *A Christmas Carol*.

A few months earlier Smith made a virtual appearance in the arena show *Doctor Who Live*. The rapturous reception he received proved that the Eleventh Doctor had already earned a secure place in the affections of viewers.

SOUNDS UNFAMILIAR

Right: The young Amelia Pond (Caitlin Blackwood) plays host to the newly regenerated Doctor (Matt Smith) in *The Eleventh Hour*. The Doctor's meal was not all it seemed – the custard was real, but the fish fingers were disguised coconut cakes.

Below: A poster promoting *Doctor Who Live: The Monsters are Coming!*

Below right: The Mark VII Sonic Screwdriver was created by the TARDIS and given to the Doctor in *The Eleventh Hour*.

Bottom: Composer Murray Gold pictured at AIR Studios in Hampstead, London.

It might have been called *Doctor Who Live* and subtitled *The Monsters are Coming!* but the 2010 arena tour was essentially a showcase for the incidental music of *Doctor Who* composed by Murray Gold. Ever since the programme's return in 2005, music had been an especially prominent part of its presentation.

The first of several *Doctor Who at the Proms* concerts was hosted by Freema Agyeman in 2008. Catherine Tate was also present, and David Tennant appeared in a mini-episode called *Music of the Spheres,* specially shot on the TARDIS set.

Apart from the TARDIS and – arguably – the Doctor himself, the original theme music remains one of the series' few recognisable elements from 1963. Composed by Australian Ron Grainer, it was realised by the BBC's Radiophonic Workshop, the organisation that provided 'special sound' effects right up until 1989. Grainer's composition

Left: Shot on 19 November 2009, the 40-second trailer for Series Five was directed by Michael Geoghegan and gave audiences their first glimpse of the new Doctor and Amy Pond (Karen Gillan). The following March a 3D version accompanied cinema screenings of Tim Burton's *Alice in Wonderland*.

Below: The Smiler androids from *The Beast Below* were created by Rob Mayor and the team at Millennium FX. The fibreglass masks were painted and glazed to create the appearance of a crazed surface.

Bottom left: Aboard the *Starship UK* children line up to be graded by a Smiler.

was commissioned in August 1963 and brought to life by the Workshop's Delia Derbyshire. As a 1964 BBC press release explained, "the music was constructed note by note, with infinite patience, and without the use of any live instrumentalists whatsoever."

Derbyshire recorded the notes she needed on magnetic tape from sine and square wave oscillators, a white noise generator and a frequency oscillator nicknamed the 'Wobbulator'. The tape was then cut into small pieces equating to musical notes of the desired lengths before being spliced back together to create the various tracks that were assembled to form the theme. The result was groundbreaking, disconcerting and impossible to imitate, although numerous cover versions were attempted. It's perhaps unsurprising, then, that the very first piece of *Doctor Who* merchandise was a 7" single of Derbyshire's arrangement, issued by Decca in February 1964. In a sense, the record pre-dated the programme itself, as it was produced from an early version of the theme which was later adjusted to better match the opening titles created by Bernard Lodge.

Over the years, Ron Grainer's music has been through several regenerations – some more successful than others. Although a synthesiser version was created for the tenth anniversary, it was rejected by producer Barry Letts and only appeared on two early edits of episodes sent in error to ABC Television in Australia. It wasn't until 1980 that a notably different version,

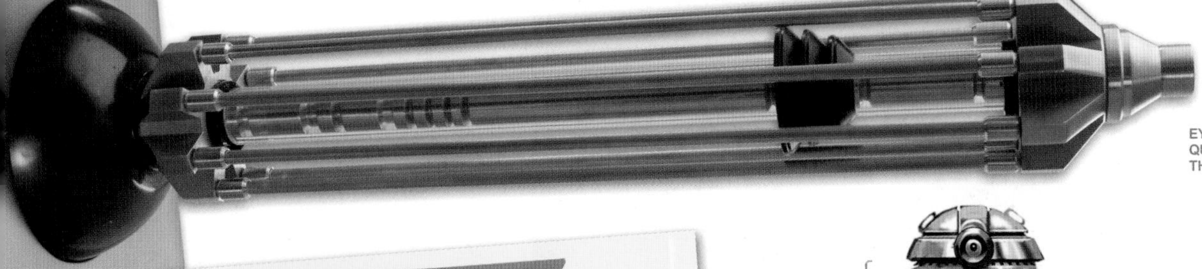

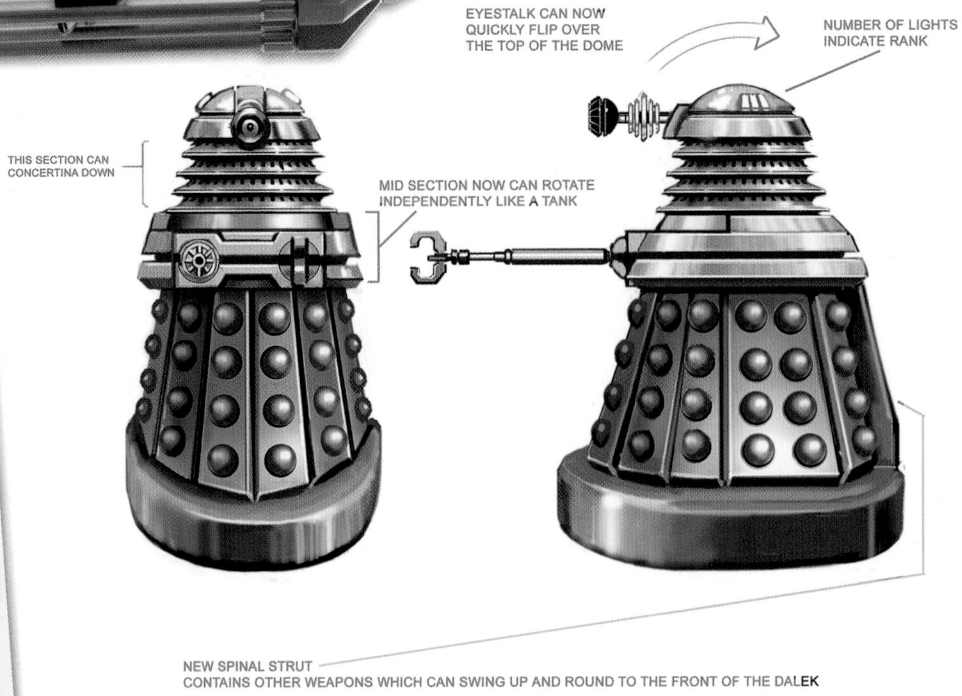

EYESTALK CAN NOW QUICKLY FLIP OVER THE TOP OF THE DOME

NUMBER OF LIGHTS INDICATE RANK

THIS SECTION CAN CONCERTINA DOWN

MID SECTION NOW CAN ROTATE INDEPENDENTLY LIKE A TANK

NEW SPINAL STRUT
CONTAINS OTHER WEAPONS WHICH CAN SWING UP AND ROUND TO THE FRONT OF THE DALEK

—TO VICTORY!

Printed by the MOD Ministry of Defense Room 73,Public Relations Office, Fillongley Warks .Test Print awaiting approval,Not for public display until further notice.
Ref: 20773/04021985

the first not derived from the original 1963 recordings, made its debut with Part One of *The Leisure Hive*. This was arranged by Peter Howell of the Radiophonic Workshop, and was in turn replaced by versions from Dominic Glynn (for Season 23) and Keff McCulloch (Seasons 24-26).

From *Rose* onwards, the various versions of the theme have all been arranged by incidental music composer Murray Gold.

When *Doctor Who* began, there was no facility to match incidental music to the recorded pictures. Like sound effects, directors decided what music they needed from the script, and briefed the composer accordingly. The music was composed to an estimated length and recorded ahead of the programme. It was then played in at the appropriate point while the production was being recorded in the studio.

For a key sequence in *Devil's Planet*, the third episode of *The Daleks' Master Plan*, director Douglas Camfield's brief to composer Tristram Cary read: "As the Dalek Supreme says 'You threaten our unity!' I would like to cut to the figure of Zephon. When that statement is made he is doomed – and he knows it. Everyone else gets out of the way prior to the execution. Could we have a piece of music which I shall bang in on the cut to Zephon and keep going until the Daleks fire, then I'll cut it out. (Duration: 15 secs.)"

Cary had created the bizarre musique concrète for the first Dalek story, and was one of a number of film composers that worked on the series; others

Top: One of the new paradigm gun props from *Victory of the Daleks*.

Top right: Collaborating with production designer Edward Thomas, Peter McKinstry created these front and side elevations of the new Dalek paradigm. This was the most radical Dalek redesign in the show's history.

Above: McKinstry's propaganda poster from *Victory of the Daleks*.

Below: A selection of the Character Options Time Squad figures that were released from 2009 to 2010.

Left: Matt Smith holds his position during recording of *Flesh and Stone* at Upper Boat Studios in August 2009.

Below left: "Where are we up to?" A barefoot River Song meets Amy on the planet Alfava Metraxis. This sequence for *The Time of Angels* was recorded at Southerndown Beach, Dunraven Park on 20 July 2009.

Below centre: Steven Moffat's script for *The Time of Angels* described River Song as looking like a "sexy, dangerous, 1940s' femme fatale". These are the shoes chosen for her by costume designer Ray Holman.

Below: The Projected Weeping Angel action figure was released by Character Options in September.

included Richard Rodney Bennett, who wrote a delicate flute-based score to accompany *The Aztecs*. By far the most prolific composer for the original series was Australian Dudley Simpson, whose first *Doctor Who* score was for *Planet of Giants* in 1964. By the end of the decade Simpson was composing most of the music for the show, and throughout the Third and Fourth Doctor's time it was unusual for anyone else to be commissioned.

Some directors did favour other composers – Carey Blyton, for example, provided distinctive scores for *Doctor Who and the Silurians*, *Death to the Daleks* (featuring a quirky Dalek 'theme' played by the London Saxophone Quartet) and *Revenge of*

Right: Restac (Neve McIntosh) orders her Silurian warriors to execute the Doctor, Nasreen (Meera Syal), Mo (Alun Raglan) and Amy as war with the human race looms in *Cold Blood*.

Below: The creation of the Silurians in *The Hungry Earth/Cold Blood* was supervised by Rob Mayor at Millennium FX. This headpiece and armoured faceplate saved the extras from having to undergo complex procedures in make-up. The use of a translucent silicone called Plat-Gel allowed more natural expressions from the actors playing the lead Silurians.

Below right: Character Options' Amy Pond figure was released in June, while Series Five was still being shown.

the Cybermen. Tristram Cary returned to score *The Mutants* in 1972, the same year the Radiophonic Workshop's Malcolm Clarke composed an avant-garde score for *The Sea Devils*. This complemented the synthesised music Dudley Simpson had been experimenting with the previous year – which, ironically, has dated the sound of Season Eight more than most of *Doctor Who*'s music.

Douglas Camfield declined to use Dudley Simpson after they worked together on *The Crusade* in 1965. He turned to a variety of other sources, including composer Geoffrey Burgon, who provided atmospheric scores for both *Terror of the Zygons* and *The Seeds of Doom* several years before his work on *The Life of Brian* and *Brideshead Revisited* brought him international recognition.

An alternative to employing a composer was to use commercially available tracks from music libraries. This was the option Camfield took for *The Web of Fear*, for example. Throughout the 1960s a large number of *Doctor Who* stories were accompanied by library tracks that may also have been used in many other productions. Most worked perfectly, and some are more memorable than the specially composed music of the era. It's hard to believe, for example, that the strident theme that accompanies the Cyberman assault in *The Moonbase*, and their awakening in *The Tomb of the Cybermen*, was not written to match the pictures. It is, in fact, a Martin Slavin composition called 'Space Adventure', which had already appeared in *The Tenth Planet*.

By the early 1970s library music had largely been phased out and incidental scores were written to the complete, edited episodes. In 1972 Simpson told the *Radio Times*, "I compose the music to picture and normally have two days to write two episodes. What I do is mix 'live' music with synthesised electronic sounds. I use about five musicians usually, who sometimes double up on instruments."

Simpson's own compositions for *Doctor Who* run to many hours of bespoke music. His epic score for *The Evil of the Daleks* used a version of the theme music bass-line to accompany appearances of the Daleks, while *The Ambassadors of Death* introduced a fondly remembered, if little-used, theme for UNIT. One of Simpson's most melodious scores was one of his last – for the Paris-based *City of Death* in 1979. The following season, new producer John Nathan-Turner took a different approach.

From *The Leisure Hive* onwards, Nathan-Turner decided that musicians from the Radiophonic Workshop would compose all of *Doctor Who*'s incidental scores. Bringing the music in-house would save money, but Nathan-Turner was also keen to give the programme a predominantly electronic sound. In September 1979 he sent two records by French composer Jean Michel Jarre to Graeme McDonald, the BBC's Head of Series and Serials, explaining: "I think this kind of 'synthesiser music' would be very useful in next season's *Doctor Who*."

McDonald was less convinced when he replied the following month: "I wonder though if it isn't too ethereal/floating/romantic/drug-oriented for our purposes. I can see it working well for things floating across a giant screen but not so well for

Left: *Doctor Who: The Brilliant Book 2011* was edited by Clayton Hickman and published by BBC Books on 21 September.

Below: Issue 423 of *Doctor Who Magazine* was edited by Tom Spilsbury and published in June. The cover of this issue, showing the crack in time that haunted the Doctor in Series Five, was only fully revealed when the magazine was removed from its printed polythene bag.

Bottom: The severed Cyberman head that reactivates beneath Stonehenge in *The Pandorica Opens*.

Bottom left: Amy cautiously examines a seemingly dead Cyberman in *The Pandorica Opens*.

our tight, pacy, small screen show. And there's very little that's dramatic about it..."

Nathan-Turner went ahead regardless, establishing a style of music that would endure – though not always produced in-house – until 1989.

The 1996 TV Movie featured a more conventional score, as well as an orchestral realisation of Grainer's theme arranged by John Debney. It was a sign of things to come.

Since 2005 Murray Gold has provided all the incidental music for the programme. "It's a dream job because it's so good and you get to write huge amounts of music," he said in 2005. Initially, all this music was realised by Gold himself, but from *The Christmas Invasion* onwards key tracks became fully orchestral – performed by the BBC National Orchestra of Wales and conducted by Ben Foster.

The show's music now emphasised the moods of each scene as never before, and seemed much

Top: *The Sun* gave away this DVD as part of its Christmas TV listings supplement on 18 December.

Top left: Matt Smith lies outside the Pandorica as a shot is lined up by the camera operator and focus puller during recording of *The Pandorica Opens* at Upper Boat Studios in February.

Far left: In *The Big Bang* the Doctor arrives dressed for the wedding of Amy and Rory (Arthur Darvill).

Left: The Rory and Amy figures specially made for their wedding cake in *The Big Bang*.

Left: Matt Smith, Katherine Jenkins (as Abigail) and Laurence Belcher (as young Kazran) record the mid-air sleigh-ride for *A Christmas Carol* against a green screen at Upper Boat Studios.

Below: This portrait of Kazran's father, Elliot Sardick (Michael Gambon), was painted by production designer Michael Pickwood's eldest daughter, Katie.

Bottom left: The Christmas edition of *The Big Issue* promoted the latest festive episode with a Matt Smith interview.

louder. "If you channel-hop on a Saturday night," said Russell T Davies in 2007, "you're up against the big light entertainment shows, like *Ant & Dec*, with a shiny black floor and a huge audience. With background music behind everything. They're phenomenally loud, those shows, and I believe that's what draws an audience. So we decided to make *Doctor Who* really noisy."

"Regardless of what it actually sounds like," says Gold, "the way the music is most different from *Doctor Who* music of old is in what it does, in the emotional story it tells. But just because I write the music, I don't impose the aesthetic. A decision about how *Doctor Who* is going to sound is too big a decision to be left to the composer, it really must rest in the hands of the executive producers."

Just like every other aspect of the design and realisation of the programme, the music of *Doctor Who* remains an integral part of the show's identity.

THE SPECIALS (continued)

Executive producers: Russell T Davies, Julie Gardner

The End of Time (continued)
Part Two — 1 January

SERIES FIVE

Executive producers: Steven Moffat, Piers Wenger, Beth Willis

The Eleventh Hour
24 September – 19 October, 18 & 20 November 2009, 7 & 12 January
written by Steven Moffat
directed by Adam Smith
Producer: Tracie Simpson
3 April

The Beast Below
24 August, 7-23 September, 2, 4 & 12 November 2009
written by Steven Moffat
directed by Andrew Gunn
Producer: Peter Bennett
10 April

Victory of the Daleks
21 August – 23 September 2009
written by Mark Gatiss
directed by Andrew Gunn
Producer: Peter Bennett
17 April

The Time of Angels/Flesh and Stone
10 July – 19 August, 24 September – 14 October, 12, 20 & 25 November 2009
written by Steven Moffat
directed by Adam Smith
Producer: Tracie Simpson
The Time of Angels — 24 April
Flesh and Stone — 1 May

The Vampires of Venice
25 November – 18 December 2009, 7-13 January, 13 February
written by Toby Whithouse
directed by Jonny Campbell
Producers: Tracie Simpson, Patrick Schweitzer
8 May

Amy's Choice 18 February – 20 March
written by Simon Nye
directed by Catherine Morshead
Producer: Tracie Simpson
15 May

The Hungry Earth/Cold Blood
20 October – 18 November 2009, 13 & 29 January
written by Chris Chibnall
directed by Ashley Way
Producer: Peter Bennett
The Hungry Earth — 22 May
Cold Blood — 29 May

Vincent and the Doctor
24 November – 15 December 2009, 4-13 January, 19 March
written by Richard Curtis
directed by Jonny Campbell
Producers: Tracie Simpson, Patrick Schweitzer
5 June

The Lodger
3-20 March
written by Gareth Roberts
directed by Catherine Morshead
Producer: Tracie Simpson
12 June

The Pandorica Opens/The Big Bang
26, 31 August, 22 September, 3, 15 December 2009, 14 January – 17 February, 3-4, 18 March
written by Steven Moffat
directed by Toby Haynes
Producer: Peter Bennett
The Pandorica Opens — 19 June
The Big Bang — 26 June

A Christmas Carol
12 July – 9 August, 8 October
written by Steven Moffat
directed by Toby Haynes
Producer: Sanne Wohlenberg
25 December

2011

The narrative arc of Series Six was more entrenched in the individual stories than ever, as the Doctor and his friends moved closer to an event that Amy, Rory and River Song witnessed in the first episode *The Impossible Astronaut* – the death of the Doctor beside Lake Silencio in Utah. The second part of the story, *Day of the Moon*, saw them help the Doctor challenge the horrifying Silence in 1969.

Old-fashioned pirates seemed to be the menace in *The Curse of the Black Spot*, but the crew were themselves victims of an alien 'Siren' (Lily Cole). In the 1980s John Nathan-Turner had used *The Doctor's Wife* as a decoy title to mislead office 'spies'. For writer Neil Gaiman the Doctor's other half became a Hugo-winning reality, in which the TARDIS was personified as the captivating Idris (Suranne Jones).

The Rebel Flesh and *The Almost People* presented a more conventional monster story, albeit one in which the Doctor met an ersatz version of himself. It ended with the revelation that Amy was in fact a copy – the real Amy having been kidnapped at some unspecified point in the past.

A Good Man Goes to War saw the Doctor call on many allies, including Madame Vastra (Neve McIntosh), her maidservant Jenny Flint (Catrin Stewart) and the 'reformed' Sontaran Strax (Dan Starkey). The real Amy was duly rescued, and River Song's true identity was finally revealed in a cliffhanger that was not resumed until the second half of the series was broadcast in the autumn.

Let's Kill Hitler continued River's story before *Night Terrors* used a dolls' house as the setting for a disturbing tale with intriguing subtextual possibilities. *The Girl Who Waited* explored the emotional consequences of time travel before both Amy and Rory left the Doctor at the end of *The God Complex*.

In *Closing Time* the Doctor teamed up with *The Lodger*'s Craig Owens (James Corden) to defeat the Cybermen and their Cybermats, the latter redesigned for their first screen appearance since 1975.

Amy, Rory and River returned, together with semi-regular characters Winston Churchill (Ian McNeice) and Madame Kovarian (Frances Barber) for *The Wedding of River Song*. After the temporal convolutions of this season finale, the Christmas episode *The Doctor, the Widow and the Wardrobe* was a more straightforward and sentimental tale.

FRIENDS AND RELATIVES

Above: Alex Kingston and Matt Smith relax during a break in recording *The Impossible Astronaut* at Rock Lake Beach, Utah in November 2010. This was the first time that episodes of *Doctor Who* had taken their principal cast to the United States.

Above right: Karen Gillan, Matt Smith and Arthur Darvill in the Oval Office set constructed at Upper Boat Studios for *The Impossible Astronaut/Day of the Moon*.

Below: The envelopes containing the cryptic invitations in *The Impossible Astronaut*.

Far right: Character Options' action figure of *The Impossible Astronaut* was released in September.

The notion that the Doctor can travel round the universe revisiting old friends and calling in favours is far removed from the series' original premise. The proliferation of familiar faces in *A Good Man Goes to War* and again in *The Wedding of River Song* demonstrates a huge change in the treatment of supporting characters.

The return of a character from a previous adventure – or even of a companion – was rare in the original run of the programme. The First Doctor's inability to steer the TARDIS precluded both this and his ability to return Ian and Barbara home. The only returning enemy was the Daleks, and the only character that the First Doctor encountered more than once was the Monk (Peter Butterworth) from *The Time Meddler*, who resurfaced in several episodes of *The Daleks' Master Plan*.

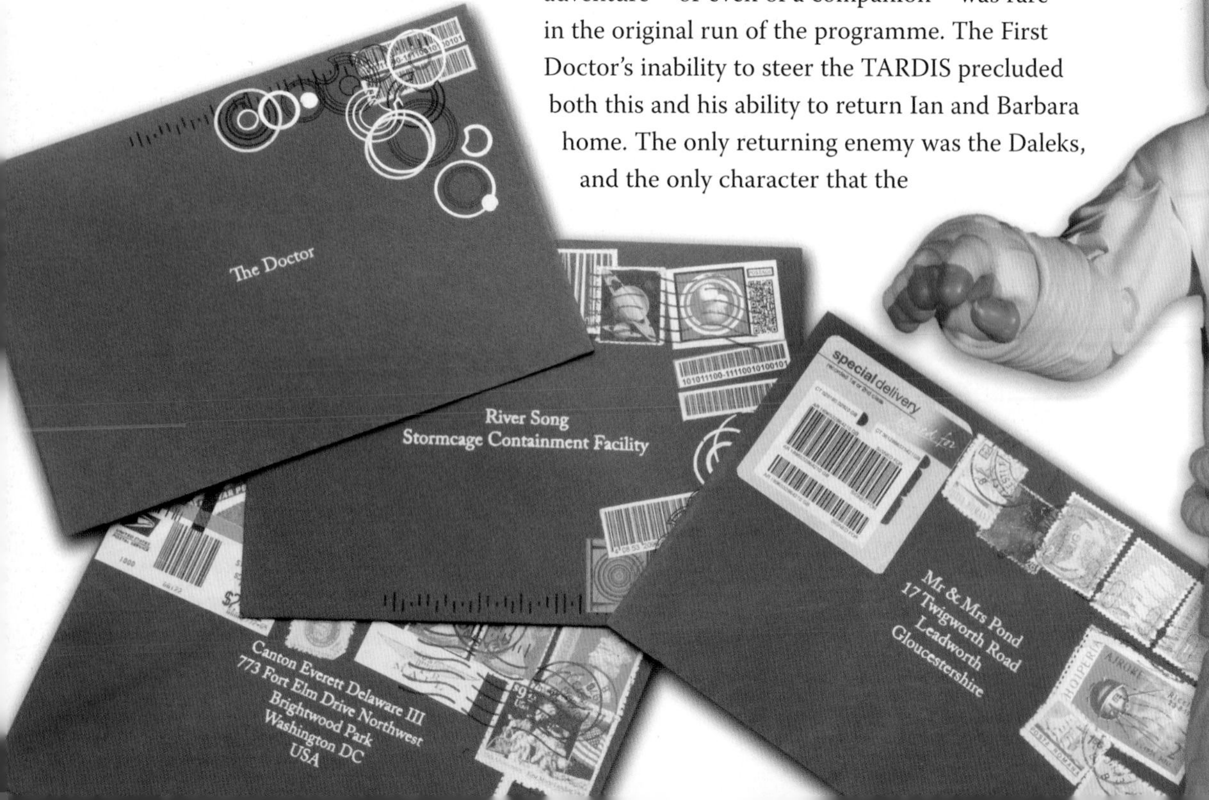

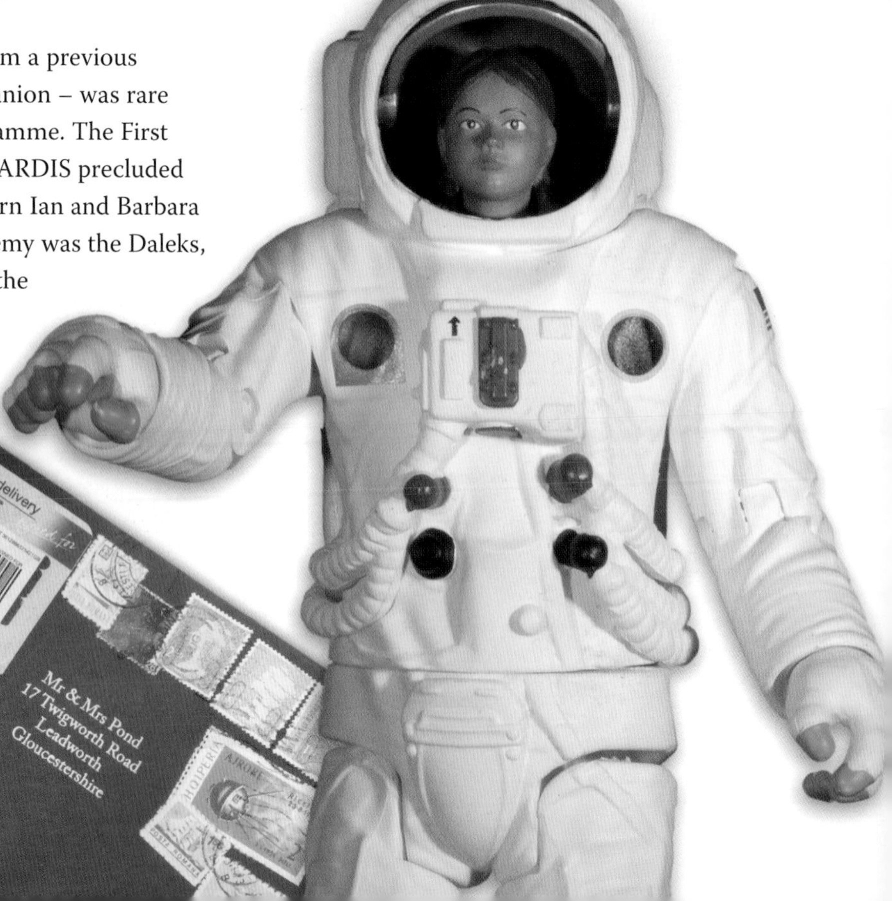

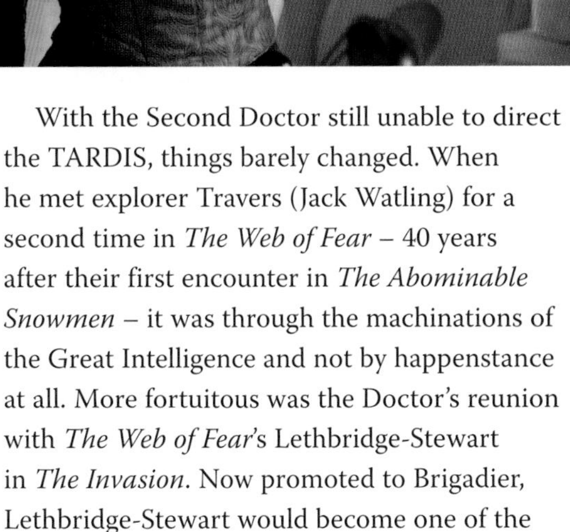

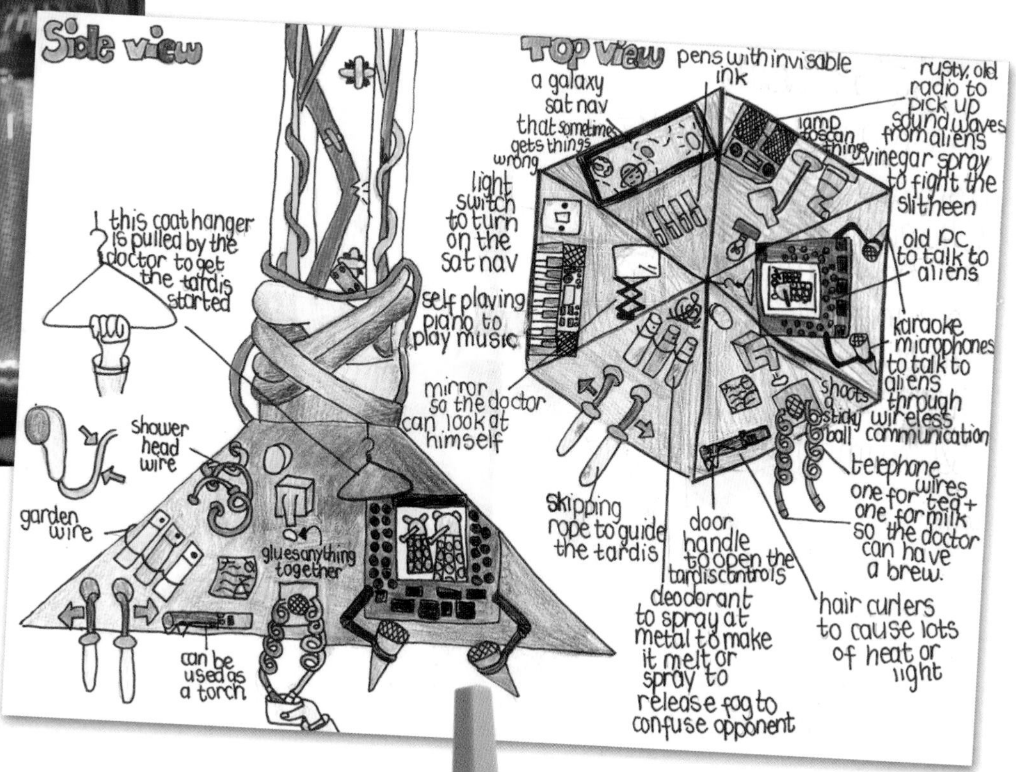

With the Second Doctor still unable to direct the TARDIS, things barely changed. When he met explorer Travers (Jack Watling) for a second time in *The Web of Fear* – 40 years after their first encounter in *The Abominable Snowmen* – it was through the machinations of the Great Intelligence and not by happenstance at all. More fortuitous was the Doctor's reunion with *The Web of Fear*'s Lethbridge-Stewart in *The Invasion*. Now promoted to Brigadier, Lethbridge-Stewart would become one of the series' most important semi-regulars.

Writing to the BBC Copyright Department to enquire about obtaining the rights to the Lethbridge-Stewart character from creators Mervyn Haisman and Henry Lincoln, producer Derrick Sherwin explained, "As you will see from the revised format, we intend to continue with UNIT and consequently Lethbridge-Stewart."

But there was a caveat. "Of course, should the authors prove difficult over granting copyright, we could change our minds..." The reason Sherwin

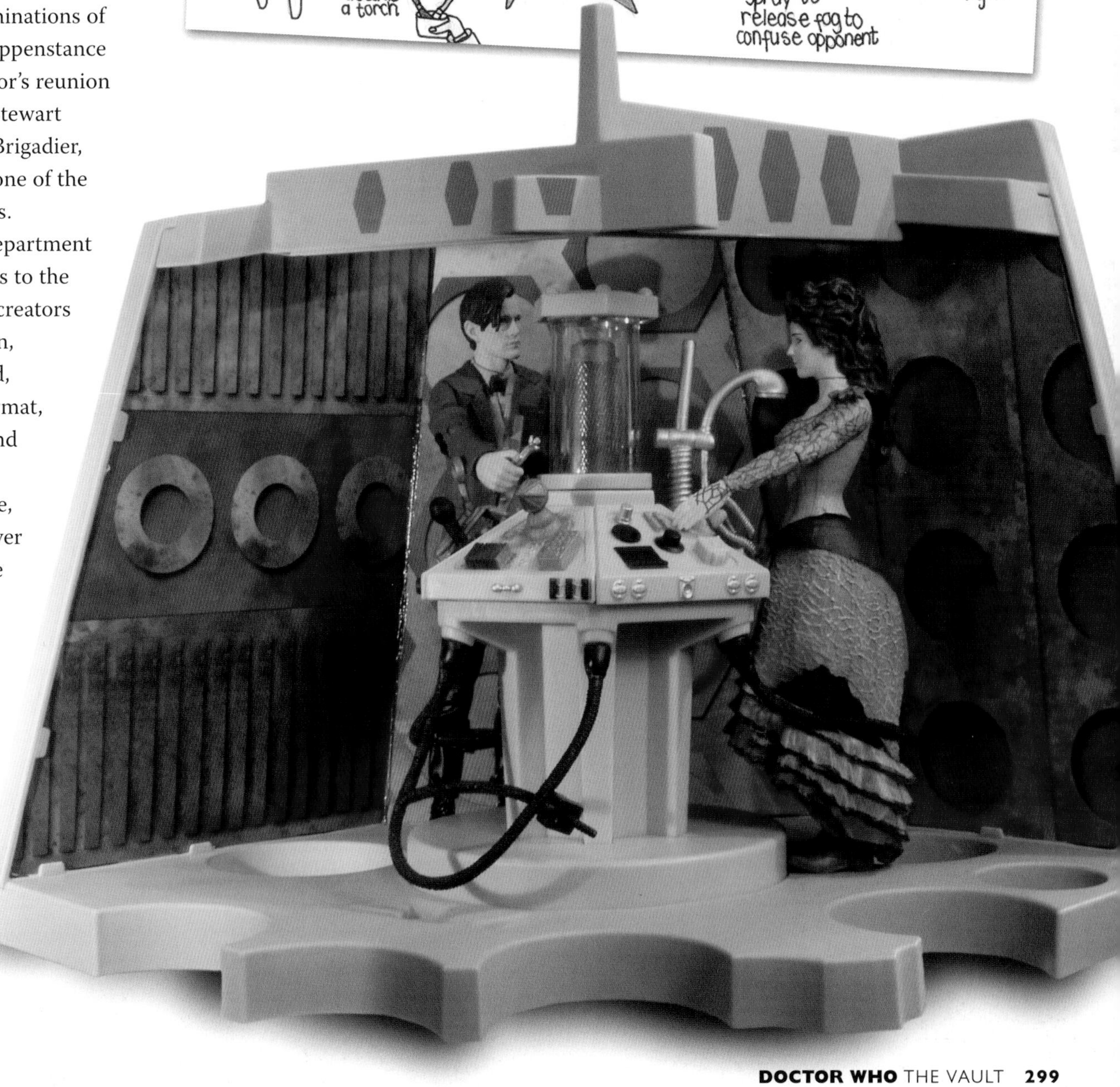

Top: Idris (Suranne Jones) activates the makeshift console in *The Doctor's Wife*.

Top right: On 7 October 2009 *Blue Peter* invited viewers to design a new TARDIS console in their latest *Doctor Who* competition. The winning entry came from the 11 to 12 category and was designed by Susannah Leah. Her picture formed the basis of the console the Doctor and Idris construct in *The Doctor's Wife*.

Right: The Junkyard TARDIS playset, seen here with Eleventh Doctor and Idris action figures, was released by Character Options in June.

Right: Shaun Williams' concept art for the harnesses used to control the Gangers in *The Rebel Flesh/The Almost People*.

Far right: Issue 436 of *Doctor Who Magazine* was published in June and paid tribute to Nicholas Courtney, who died on 22 February. Tom Baker was interviewed inside: "The only consolation we have is that we knew him, and we loved him."

Below right: The Ganger version of Jennifer Lucas (Sarah Smart) in *The Rebel Flesh*.

Below: Character Options released this action figure of the Ganger Doctor in September.

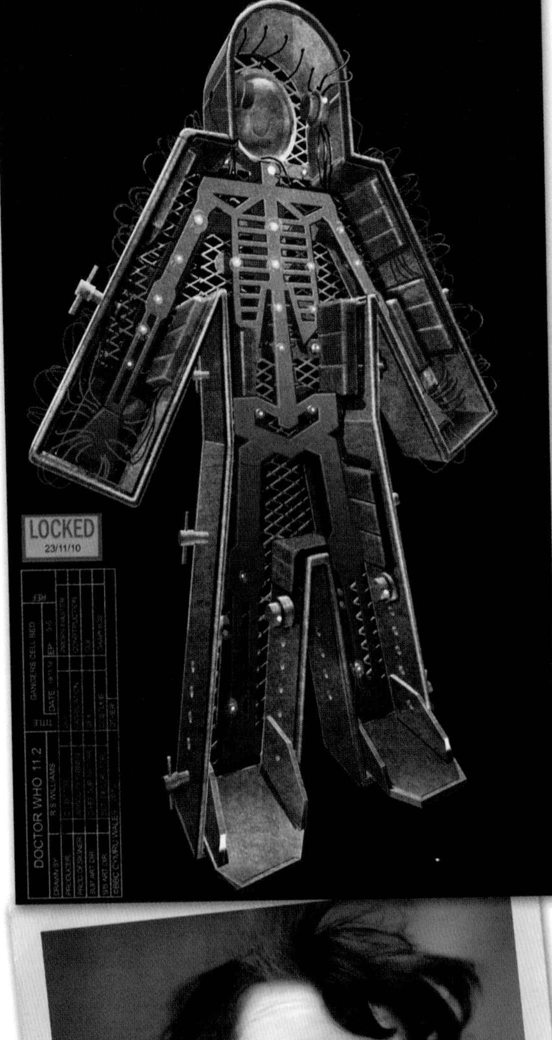

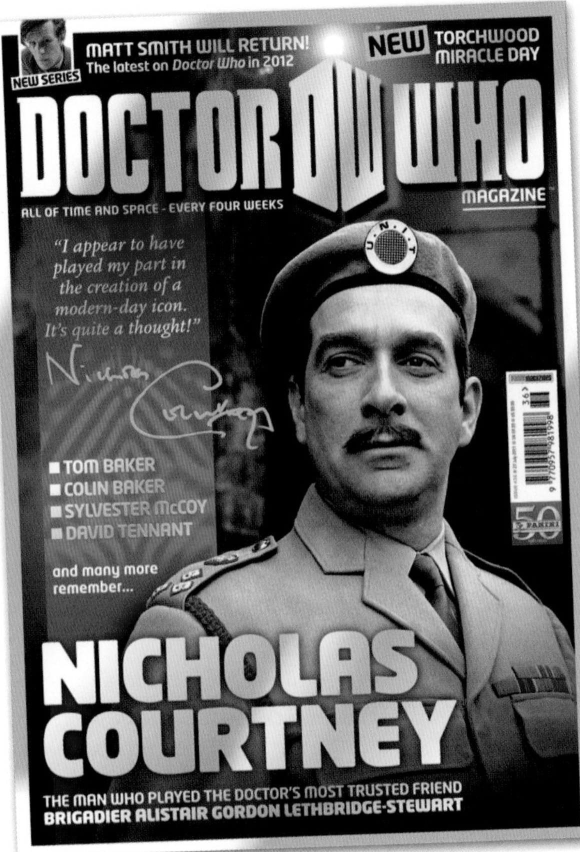

cited for especially wanting to retain the Brigadier was that the production team "very much like the actor who played him." In the event, Haisman and Lincoln agreed to a fee of £5 for every episode in which Lethbridge-Stewart appeared and Nicholas Courtney was offered a two-year contract.

"UNIT, and the Brigadier, were very important to *Doctor Who* in the early 1970s," says Terrance Dicks, who was the show's script editor at the time. "They helped to answer a lot of questions the audience may have asked: Where did the exiled Doctor live? How did he eat? Where did he get his equipment from? Linking UNIT to the Doctor gave him a sort of family, with the Brigadier at its head."

Despite his crucial role in the programme, Courtney never forgot there was a hierarchy, as Dicks recalls: "Jon [Pertwee] was the star of the programme, and he was confident in that position. Nick was not the kind of person who would have considered competing with him, or pushing himself forward. Nick would, however, occasionally suggest lines of dialogue for the Brigadier. Actors are prone to do this sort of thing, and generally speaking I don't approve. You don't want the inmates running the asylum; my job is to provide the lines, your job is to learn them and not bump into the scenery. But I did occasionally indulge Nick, purely because his suggestions were so funny. He was immensely pleased when he got a line in, such as the one in *The Three Doctors* when he refuses to believe he's on an alien planet and declares, 'I'm fairly sure that's Cromer.'"

The Brigadier's troops included Sergeant Benton, promoted from a more junior rank in *The Invasion* on the merits of actor John Levene.

Above: The Doctor lends his childhood cot to Amy and Rory for their baby in *A Good Man Goes to War*.

Right: The original prop, complete with intricate Gallifreyan writing and a suitably space-themed mobile.

Far right: The *Radio Times* featured Steven Moffat's preview of Series Six and a first look at the Silence in the issue dated 16-22 April.

Below: A revised and updated edition of Gary Russell's *Doctor Who Encyclopedia*, covering *Rose* through to *The Wedding of River Song*, was published by BBC Books in October.

Above: The Teselecta featured in *Let's Kill Hitler* and *The Wedding of River Song*. This concept art by Shaun Williams shows various views of the bridge manned by its miniaturised crew.

Above right: One of the Teselecta antibodies built by Millennium FX for *Let's Kill Hitler*. The props were mounted on rods and their tentacles were operated by remote control.

Below: Production designs for the dolls' house in *Night Terrors*.

Below left: A publicity photo of Karen Gillan, Jamie Oram, Matt Smith, Daniel Mays and Arthur Darvill, taken during recording of *Night Terrors* at Dyrham Park, near Bath in September 2010.

Left: The Amy peg doll costume made by Robert Allsopp & Associates for *Night Terrors*.

Several other UNIT soldiers came and went, but more stability was added with Captain Mike Yates (Richard Franklin) and the occasional appearance of Corporal Bell (Fernanda Marlowe). Even so, UNIT was a very fluid organisation, explained in the narrative by its personnel being seconded from other parts of the armed forces.

However much the early 1970s production team admired the actors involved, UNIT was primarily a financial and logistical convenience. "Although I think there are some very good UNIT stories, I never really liked the UNIT set-up," admits Dicks. As soon as they reasonably could, he and producer Barry Letts set about dismantling it. The Doctor regained the use of his TARDIS, and gradually left UNIT behind.

The Doctor was free to roam through all of time and space, now with notional control of his ship. Recurring characters became a rarity again, and it was unusual for him even to return to a previous destination. One exception in the 1970s was the planet Peladon, although only the histrionic Alpha Centauri survived from *The Curse of Peladon* into its sequel, *The Monster of Peladon*.

This lack of recurring characters made the universe seem like a bigger place, and chance meetings with old friends – such as Tegan in *Arc of Infinity*, the Brigadier in *Mawdryn Undead* and *Battlefield*, or Glitz in *Dragonfire* – appeared all the more incredible for it. Exceptions occurred where a story provided a good reason for old friends to reunite, as in *The Five Doctors* and *The Two Doctors*. As was the case before his exile to Earth, it tended to be monsters and villains who crossed the Doctor's path more than once.

This all changed when Russell T Davies took over as showrunner. With Rose Tyler, Davies deliberately created a companion who brought other characters with her. "I knew exactly what I was after, which was female viewers," he explained in 2005. "Rose had to be fully rounded, so I invented the mother, I invented the boyfriend... something that brings it back down to Earth, brings it back to the year 2005, that touches your heart, that gets you crying."

A continuity of characters, and the chance to witness the developing relationships between these characters, suited the more emotional narrative that Davies wanted to explore. In that

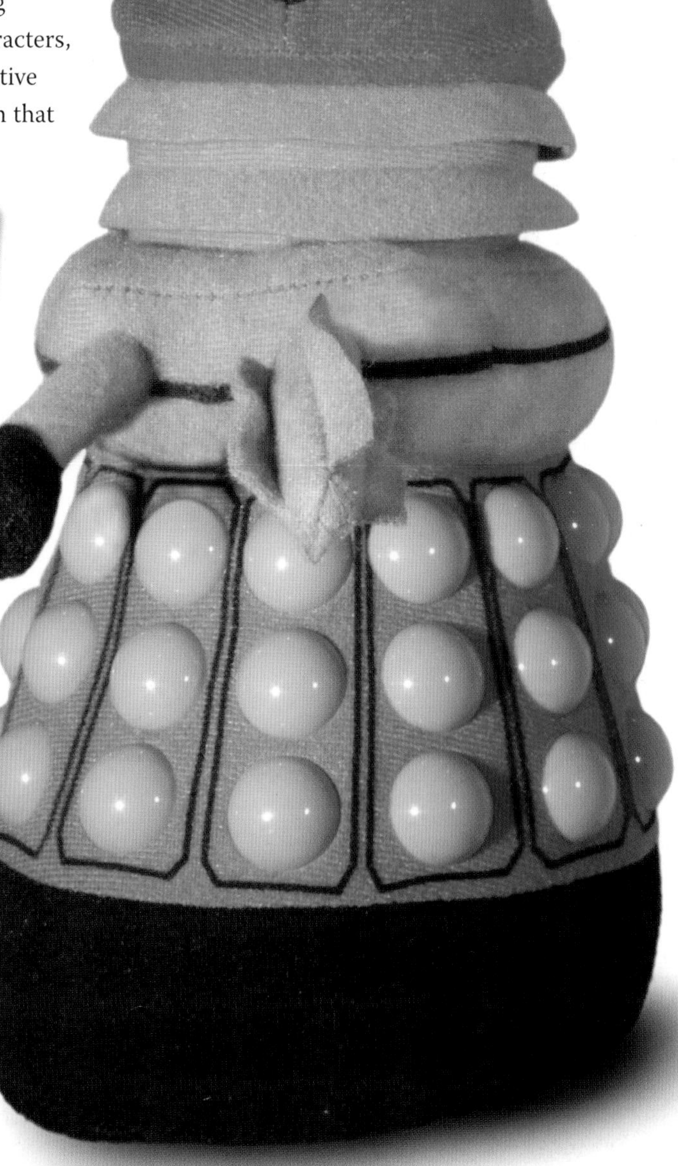

Below: A plush talking Dalek, issued by Underground Toys in July. The red version was also based on the new paradigm design.

Below left: The Doctor, Craig (James Corden) and baby Alfie go Cybermat-hunting at the Sanderson & Grainger department store in *Closing Time*.

Bottom left: A Cybermat from *Closing Time*. Millennium FX built three such props with varying capabilities. These were the first Cybermats seen in *Doctor Who* since *Revenge of the Cybermen* in 1975.

2011

Right: The Doctor's bow tie proves an integral part of *The Wedding of River Song*. Moments later, River learns the Doctor's secret.

Below: The prop distress beacon mounted on the Area 52 pyramid in *The Wedding of River Song*.

Bottom right: In *The Wedding of River Song* the Doctor encounters the severely damaged Supreme Dalek. He ransacks its data core for the Daleks' information on the Silence, then detaches its eye stalk to use as a calling card.

respect Rose formed a template, and each companion who has followed brought with her a family that got caught up in the Doctor's life.

Other characters, like Harriet Jones (Penelope Wilton), recurred because they were based in the same 'real' world where Rose lived. The same world as an older Sarah Jane Smith, whose path crossed the Doctor's again in *School Reunion*. By the same token, Martha and Donna could return, though Rose's reappearance from another universe took a little more explaining. By the time Amy joined the Doctor, the very fact that she didn't seem to have a family aside from her fiancé Rory became an important plot point.

Even the Doctor now has a family of sorts, just as he did back in 1963 (although whether he is actually married to River Song is a moot point). Just like the UNIT era in the early 1970s, the show features a group of characters whom he regularly sees. Only now it's usually through choice rather than a court order.

Such meetings provide multiple points of reference and identification for the audience. There is a sense of satisfaction in spotting the characters who have appeared in *Doctor Who* before, even if it was only in a minor capacity. Winston Churchill, Dorium Maldovar (Simon Fisher-Becker), and Charles Dickens (Simon Callow) all reappear in *The Wedding of River Song* – alongside River Song herself, Amy and Rory,

Left: *Doctor Who Magazine* published another tribute issue in October, this time to Elisabeth Sladen, who died on 19 April. "We all miss her," said Matt Smith inside. "She has a special place in people's hearts."

Above: Cyril Arwell (Maurice Cole) and the Wooden Queen (Paul Kasey) in *The Doctor, the Widow and the Wardrobe*.

Right: The Wooden Queen costume created under Neill Gorton's supervision at Millennium FX.

returning villainess Madame Kovarian, the Silence and the Teselecta robot.

As in *A Good Man Goes to War*, the Eleventh Doctor isn't shy about seeking help from characters he already knows. When he retires from active duty in *The Snowmen*, it is his 'family' – Madame Vastra, Jenny and Strax – who worry about him. When they appear again in *The Crimson Horror*, the story becomes something of an ensemble piece with each adding their own special talents to the mix as they help the Doctor.

The Doctor has always been something of a paternal figure to his friends and companions. But, seeing him together with the same friends, and with their relatives and friends, also serves to demonstrate how different he really is. When he takes a room with Craig Owens in *The Lodger* he remains amusingly ignorant of human social protocol. He revisits Craig in *Closing Time*, and is similarly awkward when he helps with the childcare and takes a job in a shop. His relationship with Idris – his own TARDIS – in *The Doctor's Wife* is perhaps the truest and purest friendship we ever see him form.

Despite an increased tendency to surround himself with familiar people and seek out old friends, the Doctor remains a unique, and very lonely, character.

SERIES SIX

Executive producers: Steven Moffat, Piers Wenger, Beth Willis (unless otherwise stated)
Producer: Marcus Wilson (unless otherwise stated)

The Impossible Astronaut/Day of the Moon
13 October – 19 November 2010, 27 January
written by Steven Moffat
directed by Toby Haynes
The Impossible Astronaut 23 April
Day of the Moon 30 April

The Curse of the Black Spot
27 January – 15 February, 10-11 March
written by Steve Thompson
directed by Jeremy Webb
7 May

The Doctor's Wife
22 September – 12 October 2010
written by Neil Gaiman
directed by Richard Clark
Producer: Sanne Wohlenberg
14 May

The Rebel Flesh/The Almost People
23 November – 16 December 2010, 3-8, 27 January, 14 February, 18 April
written by Matthew Graham
directed by Julian Simpson
The Rebel Flesh 21 May
The Almost People 28 May

A Good Man Goes to War
11-27 January, 8, 14 February, 20, 28 March, 6 April
written by Steven Moffat
directed by Peter Hoar
4 June

Let's Kill Hitler
22 March – 30 April, 10-11 July
written by Steven Moffat
directed by Richard Senior
27 August

Night Terrors
6-23 September, 1, 7 October 2010, 16 April
written by Mark Gatiss
directed by Richard Clark
Producer: Sanne Wohlenberg
3 September

The Girl Who Waited
23 February, 1-23 March, 26 April
written by Tom MacRae
directed by Nick Hurran
10 September

The God Complex
16 February – 18 March
written by Toby Whithouse
directed by Nick Hurran
17 September

Closing Time
3-17 March, 6 April
written by Gareth Roberts
directed by Steve Hughes
Producer: Denise Paul
24 September

The Wedding of River Song
19 November 2010, 4-30 April
written by Steven Moffat
directed by Jeremy Webb
1 October

The Doctor, the Widow and the Wardrobe
12 September – 9 October
written by Steven Moffat
directed by Farren Blackburn
Executive producers: Steven Moffat, Piers Wenger, Caroline Skinner
25 December

2012

Series Seven of *Doctor Who* was again split into two sets of episodes – the first five, together with a Christmas special, broadcast in the second half of 2012. Steven Moffat and his fellow executive producer Caroline Skinner (who had joined the show with *The Doctor, the Widow and the Wardrobe*) decided to structure the season more loosely. The now traditional running themes and character development were still present, but each episode would be a standalone adventure, promoted in the style of a movie blockbuster.

Moffat's *Asylum of the Daleks* cleverly subverted the audience's expectations of the Doctor's oldest foe, sidelining the new paradigm in favour of the 2005 design and a whole host of battle-scarred Daleks from the show's history. The biggest surprise, however, was the introduction of future companion Clara (Jenna Coleman), here depicted as the ill-fated Oswin. The story ended with all knowledge of the Doctor wiped from the Daleks' memory banks.

Despite the impressive CGI and props that created *Dinosaurs on a Spaceship* it was the contest between the Doctor and the pitiless Solomon (David Bradley) that gave this episode its memorable denouement. *A Town Called Mercy* tackled themes of justice and revenge as the Terminator-style Gunslinger (Andrew Brooke) terrorised a frontier town.

The Power of Three returned the TARDIS crew to the present day to work alongside Rory's father Brian (Mark Williams) and UNIT's Kate Lethbridge-Stewart (Jemma Redgrave), daughter of the late Brigadier.

River Song was back in *The Angels Take Manhattan*, the tale that brought Amy and Rory's eventful journey with the Doctor to an enforced end as the Weeping Angels trapped them in 1930s New York.

In Christmas episode *The Snowmen* the melancholy Doctor retreated to his redesigned TARDIS. He was coaxed out of his ennui by Vastra, Jenny and Strax, together with nanny (and part-time barmaid) Clara Oswald. The story's villain was the Great Intelligence, here embodied by Richard E Grant and occasionally voiced by Ian McKellen. Moffat's script not only resurrected a classic adversary but established the logic behind its late 1960s stories *The Abominable Snowmen* and *The Web of Fear*.

Tragically Clara died in the struggle to thwart the Intelligence, but as *Doctor Who*'s 50th anniversary year started it became clear that her story had in fact only just begun...

CINEMA SCOPE

Above: Matt Smith confers with director Nick Hurran on the *Asylum of the Daleks* set at Roath Lock Studios in April. They are joined by (left to right) Arthur Darvill, Karen Gillan and Anamaria Marinca.

Left: The figure-hugging dress worn by Oswin Oswald (Jenna Coleman) in *Asylum of the Daleks* was designed by Howard Burden.

Below: Jenna Coleman was announced as the Doctor's new companion in March, and originally credited as Jenna-Louise Coleman. Her early appearance as Oswin was one of Series Seven's best-kept secrets.

Doctor Who was created for television, based on certain literary antecedents but with no obvious debt to the stage or cinema. In common with its studiobound contemporaries of the 1960s, some early episodes were distinctly theatrical – the claustrophobic angst of *Inside the Spaceship* would have seamlessly translated to the stage – but the series has more frequently harboured cinematic aspirations.

Many episodes of *Doctor Who* have allied themselves to movie genres and narratives in both superficial and fundamental ways. Ian Chesterton's first encounter with the TARDIS in *An Unearthly Child* prompted the exclamation "It's alive!" and 13 years later *The Brain of Morbius* made more conspicuous references to Universal's *Frankenstein*, while its production design owed just as much to the Hammer film adaptations of the same story. The 1996 TV Movie featured an extract from Universal's 1931 classic to underline the similarity between its most famous scene and the events inside the hospital morgue.

The Gunfighters owed much of its style to the cinematic genre of the Western. Director Rex Tucker even staged the

The early Fourth Doctor stories are renowned for plundering horror and science fiction films, but this was a largely superficial exercise. Less analysed is the influence of *The Wizard of Oz* (1939) on *The Three Doctors* and *Dragonfire*.

Of all *Doctor Who*'s writers, it seems Terry Nation was the most consistently enamoured of cinematic conventions. A comprehensive education in vintage Hollywood bore fruit with *The Dalek Invasion of Earth* (any number of wartime invasion films), *The Chase* (*The Perils of Pauline*) and *The Android Invasion* (*Invasion of the Body Snatchers*).

Left: The long-awaited novelisation of Douglas Adams' *Shada* was written by Gareth Roberts and published by BBC Books on 15 March.

Below left: In March the BBC staged an official *Doctor Who* convention at the Millennium Centre in Cardiff. Matt Smith and Steven Moffat were among the guests.

Below: This large Triceratops animatronic was built by Millennium FX and nicknamed 'Tricey' by the crew. A prop dinosaur egg can be seen alongside it.

climactic confrontation at Ealing Studios so he could shoot and edit on film.

Aside from genres, particular films have proved especially inspirational: *The Thing From Another World* (1951), one of the earliest depictions of an unstoppable monster on the loose, had similarly sub-zero tributes in the form of *The Ice Warriors* and *The Seeds of Doom*. And it's hard to imagine that the rampaging *Godzilla* (1954) wasn't somewhere in the back of Malcolm Hulke's mind when he devised *Invasion of the Dinosaurs*. It would be another 38 years before *Doctor Who* could convincingly depict *Dinosaurs on a Spaceship* with visual effects that surpassed the groundbreaking *Jurassic Park* (1993).

Left: The Photoshop image for Lee Binding's spectacular Series Seven poster comprised over a thousand layers with a total file size of 4.9 gigabytes.

Below (inset): The Marshal's badge from *A Town Called Mercy*.

Below: Binding's film-style poster image for *A Town Called Mercy*. The episode was partly recorded in Almería, Spain – the traditional location for Spaghetti Westerns – in March.

Bottom left: Character Options released an Ironside Dalek, complete with tea tray, as part of its *Victory of the Daleks* collectors' set in July.

Russell T Davies evoked 1972 disaster movie *The Poseidon Adventure* in *Voyage of the Damned*, and from 2012 to 2013 the cinematic elements of every story in Season Seven were emphasised in a memorable campaign commissioned by Caroline Skinner. The broadcast of each episode was accompanied by a poster in a style and format instantly familiar from cinema foyers. They were created by graphic designer Lee Binding.

"As Series Seven was being split across two years, Caro told me she wanted each episode to be treated as an event in order to maximise the publicity," he says. "So it was decided to advertise each one in a 'Movie of the Week' style."

The first of Binding's posters resembled a 'teaser', without the title or credit block familiar from the subsequent designs. "The Series Seven generic poster showed the Doctor carrying Amy in *Asylum of the Daleks*. The central image was suggested by Caro, who told me 'This moment in the episode is utterly iconic.' I almost had stage fright when I sat down to do it, because it was such a big task. It was a long three-month process to get every element right, and I think everyone was very happy with the way it all came together."

Steven Moffat rejects the notion that he and the show's other writers now look to the big screen for inspiration. "You couldn't get two more television-centric people than myself and Russell," he says, pointing out that he gave up the chance to continue scripting Steven Spielberg's *The Adventures of Tintin* (2011) to become the showrunner of *Doctor Who*. "Working in films isn't a great aspiration. It *is* an aspiration, however, to grab some of those production values and some of those techniques and bring them to the small screen. It's important to consider that we're competing with big movies that turn up on television.

"So we're not trying to abandon all the other values of good writing, good acting and so on. We're trying to do all that and at the same time make it look exotic and beautiful."

Above left: The prop novel that featured in *The Angels Take Manhattan*. Justin Richards went on to ghost-write a prequel adventure on Malone's behalf: *The Angel's Kiss* was released as an e-book on 4 October, before BBC AudioGo issued a talking book version read by Alex Kingston.

Above: This publicity photo of Rob David (Sam Garner) being pursued by a Weeping Angel was taken on location at Royal Fort House, Bristol University on 7 April.

Left: The set for Dr Simeon's laboratory at the Great Intelligence Institute was designed by Michael Pickwoad. *The Snowmen* was the first Christmas special completed at Roath Lock.

Below: Dr Simeon's calling card, announcing the return of a *Doctor Who* villain last encountered in *The Web of Fear*.

Bottom left: The prop for Rory and Amy's gravestone from *The Angels Take Manhattan*. Recording for this sequence took place at Llanelli District Cemetery in Camarthenshire on 19 April.

GI
THE GREAT INTELLIGENCE INSTITUTE
2 BLOOMSBURY LANE + LONDON N31

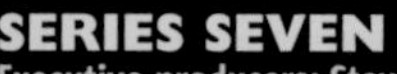

SERIES SEVEN
Executive producers: Steven Moffat, Caroline Skinner
Producer: Marcus Wilson

Asylum of the Daleks
15 March – 1 May
written by Steven Moffat
directed by Nick Hurran
1 September

Dinosaurs on a Spaceship
17 February – 23 March
written by Chris Chibnall
directed by Saul Metzstein
8 September

A Town Called Mercy
7 March – 21 March
written by Toby Whithouse
directed by Saul Metzstein
15 September

The Power of Three
30 April – 28 June, 29-30 July
written by Chris Chibnall
directed by Douglas Mackinnon
22 September

The Angels Take Manhattan
23 March – 30 April, 28 June
written by Steven Moffat
directed by Nick Hurran
29 September

The Snowmen
6 August – 22 September, 18 October, 9, 21, 23, 28 November
written by Steven Moffat
directed by Saul Metzstein
25 December

2013

The question 'Doctor Who?' would resonate throughout the second half of Series Seven.

The Bells of Saint John brought the Doctor and the modern-day Clara together, culminating in the unexpected return of the Great Intelligence. The mystery surrounding 'the impossible girl' deepened. Could this be the same Clara who had died in *The Snowmen* and, as the Doctor would later realise, in *Asylum of the Daleks*?

The Rings of Akhaten took the Doctor and his new companion into deep space before *Cold War* trapped them inside a submarine with a vengeful Ice Warrior. The creatures' comeback was the first story to show one of the Martians out of its armoured shell.

Hide took the Doctor and Clara back to 1974, and a ghost story with a time-travelling resolution. *Journey to the Centre of the TARDIS* went further beyond the control room than we'd ever been before as the malfunctioning ship was threatened by hard-hearted salvage operators.

The Doctor and Clara worked alongside Vastra, Jenny and Strax in *The Crimson Horror*. Mark Gatiss' arch script evoked the black humour of *Carry On Screaming!* before revealing the chilling truth about Mrs Gillyflower (Diana Rigg) and her blind daughter Ada (Rachael Stirling).

Nightmare in Silver featured a new generation of virtually indestructible Cybermen, now accompanied by Cybermites, before the season came to an end with Steven Moffat's *The Name of the Doctor*. On the planet Trenzalore the Great Intelligence forces River Song to use the unheard name to open the Doctor's tomb. Inside the dying TARDIS is the last resting place for all the Time Lord's incarnations and the solution to the mystery surrounding Clara. The revelation of a previously unseen Doctor (John Hurt) led directly into the events of the 50th anniversary special, broadcast in 3D on 23 November.

This milestone was also commemorated by *An Adventure in Space and Time*, a sensitive and meticulous dramatisation of the programme's early years starring David Bradley as William Hartnell. Its commissioning suggested that *Doctor Who* was now regarded not only as a commercial hit, but as an important part of the BBC's cultural heritage.

Matt Smith had announced his intention to leave the series on 1 June. The Christmas special would introduce a new Doctor, and a new chapter in the saga of television's best-loved character.

POLICE PUBLIC CALL BOX
POLICE TELEPHONE
FREE
FOR USE OF
PUBLIC
ADVICE & ASSISTANCE
OBTAINABLE IMMEDIATELY
OFFICERS & CARS
RESPOND TO ALL CALLS
PULL TO OPEN
ST JOHN AMBULANCE
POLICE PUBLIC CALL BOX

BIGGER ON THE INSIDE

Above right: Referring to his notebook, the Doctor tells Clara what she missed in *The Bells of Saint John*.

Above: In *The Bells of Saint John* we first see the Doctor living as a monk in Cumbria in 1207. He paints this portrait of Clara, which the Abbott describes as "The woman twice dead, and her final message… If he truly is mad, then this is his madness."

Left: Amelia Williams' *Summer Falls*, the book read by Artie in *The Bells of Saint John*. The story was subsequently ghost-written by James Goss and released as an e-book on 4 April.

Given *Doctor Who*'s untidy development over the last 50 years it should come as no surprise that debate surrounds the precise meaning of the acronym TARDIS. Since the TV Movie, however, it has remained exactly as Susan described it in the first episode: "I made up the name TARDIS from the initials," she tells Ian and Barbara, "Time And Relative Dimension In Space."

Ian and Barbara learned more about the TARDIS when it exerted a hypnotic influence over everyone on board during *Inside the Spaceship*. In the second half of this two-part story the Doctor deduced that the ship had been trying to warn its passengers that the 'fast return' switch on its central console had jammed, setting it on a course for oblivion at the birth of the Solar System.

The TARDIS' warnings may have been oblique and inarticulate, but they were a strong suggestion that some form of sentient life lay beneath control panels that even the Doctor seemed unable to operate.

"You speak as if she were alive," says Mike Yates in *Planet of the Spiders* Part Three. "Yes," says the Third Doctor. "Yes I do, don't I?"

The TARDIS certainly seems to know just what the newly regenerated Fifth Doctor wants in *Castrovalva*, conjuring up a set of cricketing clothes and a wheelchair which briefly dispel his confusion.

More recent episodes have offered further clues about the TARDIS' status and origin. In *Boom Town* the Ninth Doctor tells Rose "The ship's alive," and in *The Impossible Planet* the Tenth explains that it wouldn't be possible to construct another one: "They were grown, not built."

In *The Doctor's Wife* his frequent references to the "old girl" prove only partially correct when it briefly assumes corporeal form as the inquisitive young Idris. The ensuing conversation seems to

Far left: Clara (Jenna Coleman) and her father Dave (Michael Dixon) visit Ellie's grave in *The Rings of Akhaten*.

Left: The book originally owned by Clara's mother Ellie. Clara took it with her on her first journey in the TARDIS.

Below: Actor Spencer Wilding wears the full Skaldak mask and make-up as he prepares to record a scene in *Cold War*.

Below left: Imprisoned aboard a Soviet submarine in 1983, Skaldak leaves his empty armour as a decoy.

Bottom left: One of the 3D card diorama playsets released by Character Options to tie in with the second half of Series Seven. This scenario, showing the submarine from *Cold War*, is pictured with an action figure of the armoured Skaldak.

Top: The damaged console in *Journey to the Centre of the TARDIS*. Michael Pickwoad had redesigned this set for *The Snowmen*, reflecting his ambition to make the TARDIS once more resemble a machine.

Top right: Elements from the book that Clara discovers in *Journey to the Centre of the TARDIS*.

Above: The Doctor tries to protect the circuits of the architectural reconfiguration system in *Journey to the Centre of the TARDIS*.

Right: One of the circuits which the Doctor describes as part of the TARDIS' "basic genetic material."

Far right: *Spearhead from Space* became the series' first archive Blu-ray release on 15 July. The packaging featured this artwork by Lee Binding.

confirm that the partnership between the Eleventh Doctor and his TARDIS entails rather more than devotion to a mere machine. "You stole me. And I stole you," says Idris.

"I borrowed you," corrects the Doctor.

"Borrowing implies the intention to return the thing that was taken," she says. "What makes you think I would ever give you back?"

This possessive attitude resurfaces in *Hide*, when the TARDIS refuses to let Clara inside. The Voice Visual Interface flickers into life to explain why, and it looks just like Clara. "I'm programmed to select the image of a person you esteem," it says. "Of several billion such images in my databanks, this one best meets the criterion."

"Oh," says Clara. "You *are* a cow. I *knew* it."

The TARDIS' excuse for not wanting to enter the pocket universe where the Doctor is trapped uses some telling terminology. "The entropy would drain the energy from my heart," says the Interface. "In four seconds, I'd be stranded. In ten, I'd be dead."

Perhaps the most accurate description of the ship is the one given by the Third Doctor in Part One of *Death to the Daleks*. "The TARDIS is a living thing," he tells Sarah. "Thousands of instruments." This may sound like the programme hedging its bets, but it's certainly the case that scriptwriters have paid far more attention to the ship's mechanical functions than any notion it may be a character in its own right. Principally, the TARDIS is the Doctor's most impressive gadget. And being bigger on the inside means it has the space to pack quite a few gadgets of its own.

The Food Machine appeared in the First Doctor episodes *The Dead Planet* and *The Edge of Destruction*. An oven-sized contraption that produced foil-wrapped blocks of sustenance to the accompaniment of flashing lights and a loud beeping noise, the Food Machine supplied 'bacon and eggs' for Ian after Susan turned dials selecting 'J62' and 'L6'.

Rather more sophisticated was the Time Scanner that the Second Doctor used to visualise future events in Episode 4 of *The Moonbase*. While foresight would doubtless have been useful on numerous occasions, its compromising effect on the show's storylines meant that the Time Scanner was never seen again.

The TARDIS defence mechanisms include a melting lock and isomorphic controls programmed to respond only to the Doctor's touch, but these are just some of the features to have been inconsistently applied over the years. One of the ship's most useful tricks is its ability

Top: A ticket to Hedgewick's World of Wonders from *Nightmare in Silver*.

Above: The Doctor, Clara, Angie (Eve De Leon Allen) and Artie (Kassius Carey Johnson) arrive at *Nightmare in Silver*'s dilapidated amusement park.

Left: The Cybermen prepare to upgrade Clara and Ha-Ha (Calvin Dean) in *Nightmare in Silver*.

Below: One of the soldiers' guns from *Nightmare in Silver*. The Cybermen soon develop an immunity to its effects.

to remove itself from the path of danger, but so far the Doctor has only remembered to set the HADS (Hostile Action Displacement System) in *The Krotons*, and 45 years later in *Cold War*. Like the Time Scanner, it's one of those in-built gadgets that makes things a little too convenient.

"The problem with the TARDIS is that it's an invulnerable mobile fortress with infinite space and a supply of everything," says Steven Moffat. "If the Doctor was really upset about everybody on *The Ark in Space* why didn't he just move them all inside?

"We have endless script meetings about how to get round this. One solution is to make sure that once the Doctor gets involved in events he doesn't use the TARDIS to alter the outcome because that can cause damage to the timelines – and if he does

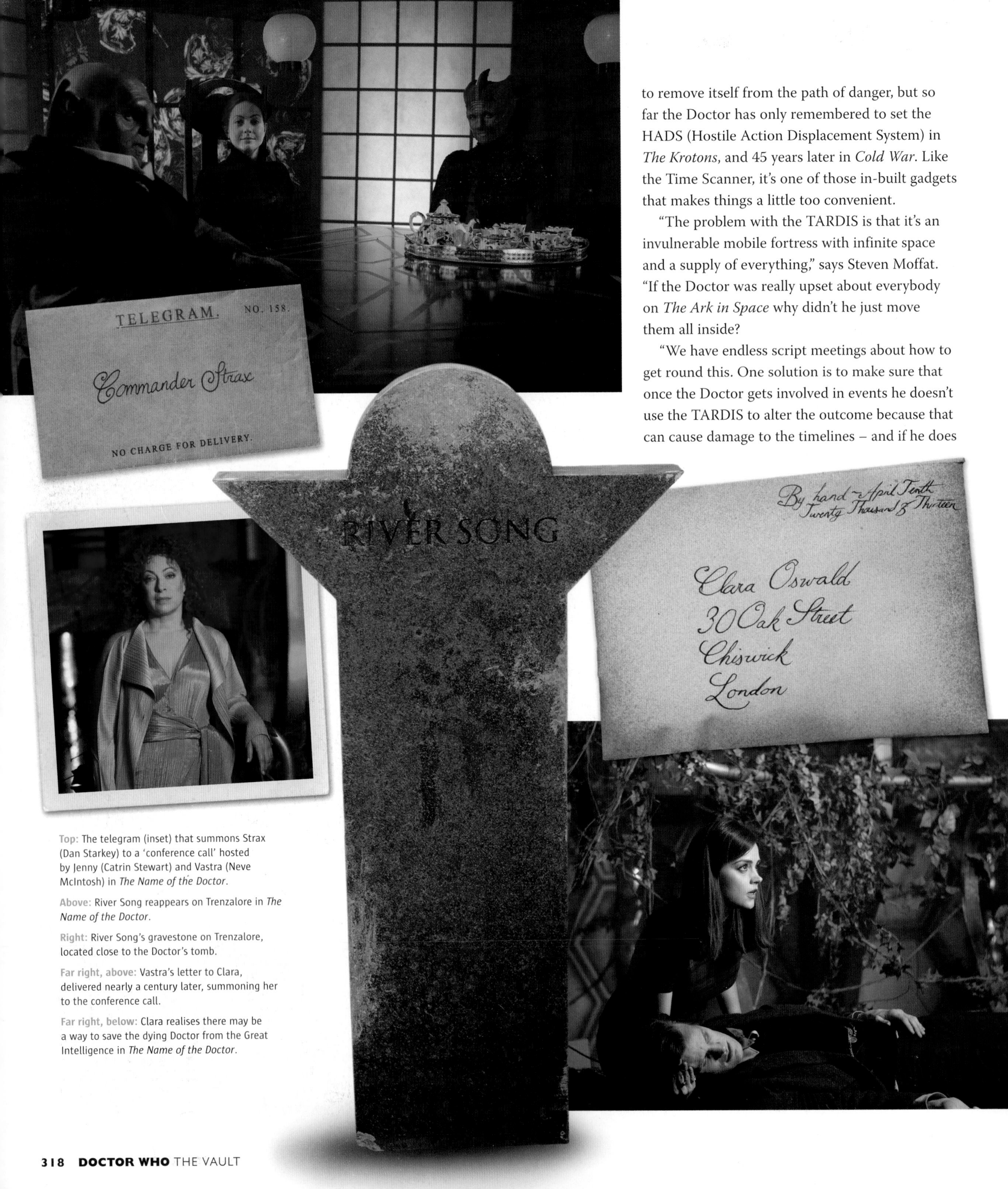

Top: The telegram (inset) that summons Strax (Dan Starkey) to a 'conference call' hosted by Jenny (Catrin Stewart) and Vastra (Neve McIntosh) in *The Name of the Doctor*.

Above: River Song reappears on Trenzalore in *The Name of the Doctor*.

Right: River Song's gravestone on Trenzalore, located close to the Doctor's tomb.

Far right, above: Vastra's letter to Clara, delivered nearly a century later, summoning her to the conference call.

Far right, below: Clara realises there may be a way to save the dying Doctor from the Great Intelligence in *The Name of the Doctor*.

Above and above right: On 26 March the Royal Mail issued these stamps to mark 50 years of *Doctor Who*. Pre-registration numbers were three times higher than for any previous commemorative set.

Right: David Bradley as William Hartnell and Claudia Grant as Carole Ann Ford in Mark Gatiss' drama *An Adventure in Space and Time*. The TARDIS control room and console were faithful recreations of the originals, first seen 50 years before.

too much damage they will eventually fall apart. There can be a character progression in that; as the years go by and the Doctor wanders around time, doing all sorts of work on it, he understands it better. And the more he knows about established history the more he can do with it."

At the end of Series Seven, Moffat's *The Name of the Doctor* showed the Doctor's last resting place. His tomb on the planet Trenzalore was the TARDIS, now colossal following a malfunction in its dimension dams. "All the bigger-on-the-inside starts leaking to the outside," a mournful Doctor tells Clara.

Whether a living creature, or a machine so extraordinarily complex that it can seem sentient, the TARDIS is still the Doctor's companion, even in death.

It was fitting that Time And Relative Dimension In Space took such a prominent role in the 50th anniversary year. The TARDIS remains the series' greatest artefact; the gateway to the Doctor's adventures since he and his granddaughter stole it from a repair shop on Gallifrey, "a very long time ago..."

SERIES SEVEN
(continued)
Executive producers: Steven Moffat, Caroline Skinner
Series producer: Marcus Wilson
(unless otherwise stated)

The Bells of Saint John
24 September, 8-24 October, 9, 21 November, 2, 6, 8-9 December 2012, 18 February
written by Steven Moffat
directed by Colm McCarthy
Producer: Denise Paul
30 March

The Rings of Akhaten
22 October – 30 November 2012
written by Neil Cross
directed by Farren Blackburn
Producer: Denise Paul
6 April

Cold War
13-29 June 2012
written by Mark Gatiss
directed by Douglas Mackinnon
13 April

Hide
22 May – 19 June, 22 September, 27 November 2012
written by Neil Cross
directed by Jamie Payne
20 April

Journey to the Centre of the TARDIS
4-24 September, 18 October, 27-28 November 2012
written by Steve Thompson
directed by Mat King
27 April

The Crimson Horror
2 July – 29 August, 18, 25 October, 16, 26 November 2012
written by Mark Gatiss
directed by Saul Metzstein
4 May

Nightmare in Silver
7 November – 1 December 2012
written by Neil Gaiman
directed by Stephen Woolfenden
Producer: Denise Paul
11 May

The Name of the Doctor
16 November – 1 December 2012, 26 March, 5 April
written by Steven Moffat
directed by Saul Metzstein
Producer: Denise Paul
18 May

TITLE INDEX

Doctor Who television episodes and related mini-episodes, animated stories, radio productions, webcasts, feature film adaptations, stage plays, spin-off series and documentaries. Numerals in **bold** type refer to entries in the episode guide.

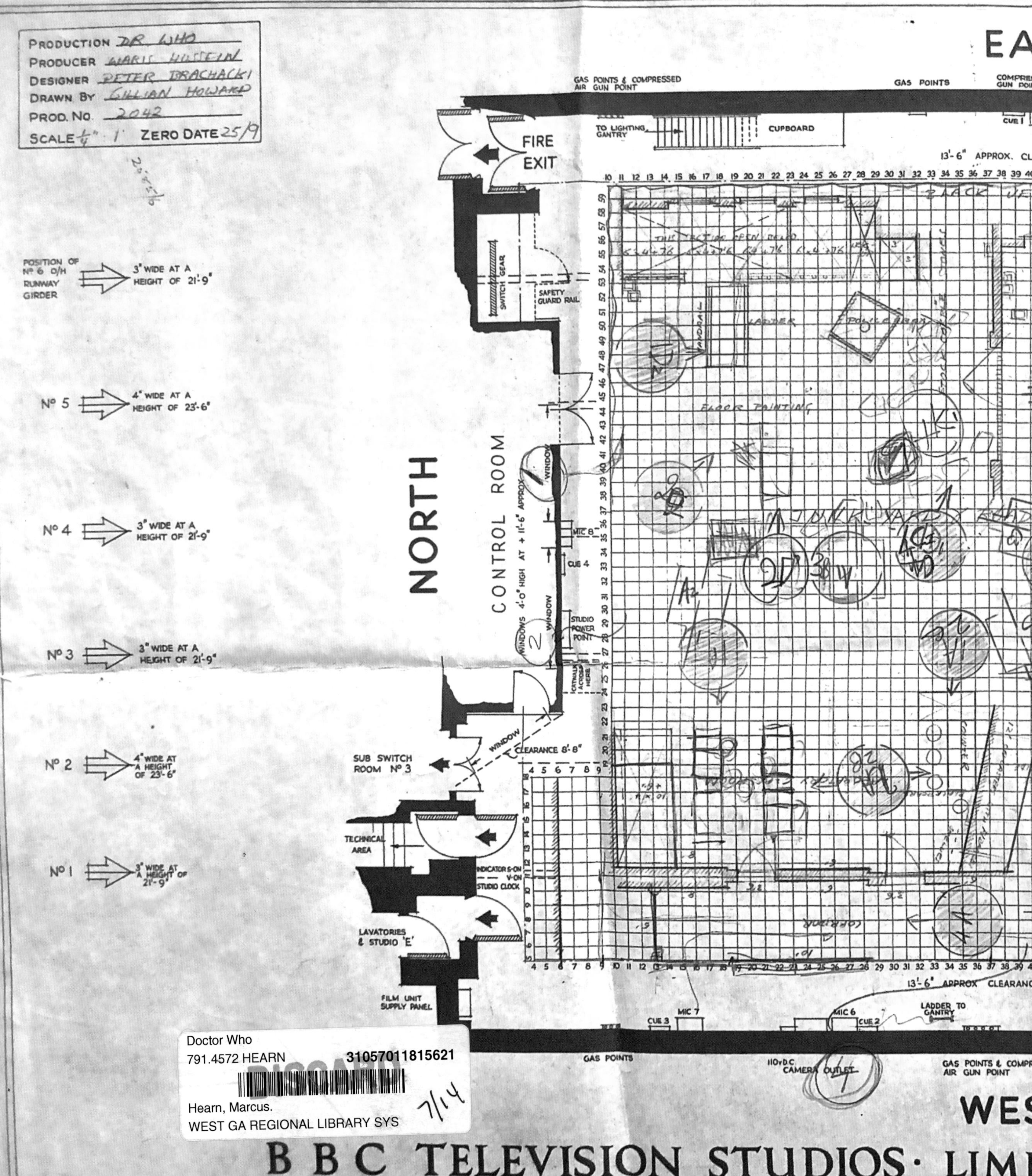
PRODUCTION DR WHO
PRODUCER WARIS HUSSEIN
DESIGNER PETER BRACHACKI
DRAWN BY GILLIAN HOWARD
PROD. NO. 2042
SCALE 1/4" 1' ZERO DATE 25/9
EA
GAS POINTS & COMPRESSED AIR GUN POINT
GAS POINTS
TO LIGHTING GANTRY
CUPBOARD
CUE 1
FIRE EXIT
13'-6" APPROX. CLE
POSITION OF Nº 6 O/H RUNWAY GIRDER
3" WIDE AT A HEIGHT OF 21'-9"
Nº 5
4" WIDE AT A HEIGHT OF 23'-6"
Nº 4
3" WIDE AT A HEIGHT OF 21'-9"
Nº 3
3" WIDE AT A HEIGHT OF 21'-9"
Nº 2
4" WIDE AT A HEIGHT OF 23'-6"
Nº 1
3" WIDE AT A HEIGHT OF 21'-9"
NORTH
CONTROL ROOM
WINDOWS 4'-0" HIGH AT + 11'-6" APPROX
WINDOW
SWITCH GEAR
SAFETY GUARD RAIL
MIC 8
CUE 4
STUDIO POWER POINT
CATWALK ACCESS HERE
WINDOW
CLEARANCE 8'-8"
SUB SWITCH ROOM Nº 3
TECHNICAL AREA
INDICATOR S-ON V-ON STUDIO CLOCK
LAVATORIES & STUDIO 'E'
FILM UNIT SUPPLY PANEL
FLOOR PAINTING
LADDER
13'-6" APPROX CLEARANCE
LADDER TO GANTRY
CUE 3
MIC 7
MIC 6
CUE 2
GAS POINTS
110v D.C. CAMERA OUTLET
GAS POINTS & COMPRE AIR GUN POINT
WES
BBC TELEVISION STUDIOS · LIME